CAN REGULATION WORK?
The Implementation of the 1972 California Coastal Initiative

ENVIRONMENT, DEVELOPMENT, AND PUBLIC POLICY

A series of volumes under the general editorship of
Lawrence Susskind, *Massachusetts Institute of Technology,
Cambridge, Massachusetts*

ENVIRONMENTAL POLICY AND PLANNING

Series Editor:
Lawrence Susskind, *Massachusetts Institute of Technology,
Cambridge, Massachusetts*

THE LAND USE POLICY DEBATE IN THE UNITED STATES
edited by Judith I. de Neufville

CAN REGULATION WORK? The Implementation of the 1972
California Coastal Initiative
Paul A. Sabatier and Daniel A. Mazmanian

PATERNALISM, CONFLICT, AND COPRODUCTION
Learning from Citizen Action and Citizen Participation in
Western Europe
Lawrence Susskind and Michael Elliott

Other subseries:
CITIES AND DEVELOPMENT
Series Editor:
Lloyd Rodwin, *Massachusetts Institute of Technology,
Cambridge, Massachusetts*

PUBLIC POLICY AND SOCIAL SERVICES
Series Editor:
Gary Marx, *Massachusetts Institute of Technology,
Cambridge, Massachusetts*

CAN REGULATION WORK?

The Implementation of the 1972 California Coastal Initiative

Paul A. Sabatier
University of California, Davis
Davis, California

and

Daniel A. Mazmanian
Pomona College
Claremont, California

PLENUM PRESS • NEW YORK AND LONDON

Library of Congress Cataloging in Publication Data

Sabatier, Paul A.

　Can regulation work?

　(Environment, development, and public policy. Environmental policy and planning)
　Bibliography: p.
　Includes index.
　1. Coastal zone management—California. I. Mazmanian, Daniel A.,
II. Title, III. Series.
HT393.C3S2 1983   333.91'717'09794   83-9537
　　ISBN 978-1-4684-1157-7　　ISBN 978-1-4684-1155-3 (eBook)
　　DOI 10.1007/978-1-4684-1155-3

Dedicated to Dolores,
without whose skill and dedication
this book would never have come to fruition.

PREFACE

Although local governments have traditionally exercised almost exclusive regulatory control over land development within their jurisdictions, throughout the 1970s state governments began to amass considerable authority over local land use decisions. Studies of the implementation of these new programs, however, have generally had the drawbacks of not being integrated into theoretical literatures in the social sciences and suffering from substantial methodological problems. On the basis of a review of literatures on policy making by regulatory agencies, the implementation of federal social/distributive programs, and the situational variables affecting agency behavior, in this study we develop a general conceptual framework of the implementation process of regulatory statutes. The framework is then applied to the implementation of one of the more novel and comprehensive state-level land use statutes, the California Coastal Zone Conservation Act of 1972.

In this undertaking we received considerable assistance in the design stage and later in review of preliminary drafts of our report from a National Advisory Committee composed of persons actively involved in the implementation of land use policy, representatives of groups affected by land use regulation, and scholars in the field. For serving on the committee we are indebted to James Carroll, California Council for Environmental and Economic Balance; Paul Culhane, Department of Political Science, University of Houston; Grant Dehart, Office of Coastal Zone Management, Washington, D.C.; Erwin Hargrove, Institute for Public Policy, Vanderbilt University; Helen Ingram, Department of Political Science, University of Arizona; Karl Kurtz, National Conference of State Legislatures, Denver, Colorado; Rob MacDougall, National Conference of State Legislatures, Washington, D.C.; Daniel Mandelker, School of Law, Washington University, St. Louis; H. Milton Patton, Council of State Governments, Lexington, Kentucky; Donald Priest, Urban Land Institute, Washington, D.C.; Lettie Wenner, Department of Political Science, University of Illinois at Chicago Circle; and Lawrence Susskind, Department of Urban Studies and Planning, Massachusetts Institute of Technology.

The extensive interviewing and data collection central to the study would not have been possible without the cooperation of the California Coastal Commissions, their staff, related government agencies, concerned groups and citizens, and the sample of the public surveyed. While it is impossible to mention individually all who cooperated, we would particularly like to acknowledge the invaluable assistance of Bob Brown, Joseph Bodovitz, Melvin Carpenter, Norbert Dall, David Dubbink, Bruce Haston, Donald Hedrick, Jack Lahr, Melvin Lane, Celia von der Muhll, Don Peterson, Joseph Petrillo, Rick Rayburn, Jack Schoop, Stan Scott, Bob Testa, Bill Travis, Lucille Vinyard, and, especially, Peter Douglas and Judy Rosener.

Finally, we are grateful for the dedication and resourcefulness of our consultant on economic impacts, Robert Kneisel; our research assistants, who worked with us at various stages of the study, Dana Alden, Dean Curry, Jack Meek, and Sara Wyant; and our secretaries, Dolores DuMont, Susan Reames, Jane Russell, Beverly Biedenbach, and Penny Arns.

This report is based upon research supported by the National Science Foundation under Grants No. DAR77-14589, DAR77-14589 A01, and DAR77-20077. We gratefully acknowledge the encouragement and assistance of our program manager, Terry Sopher. All opinions, findings, and conclusions expressed in the study are those of the authors and do not necessarily reflect the views of the National Science Foundation.

Paul Sabatier
Daniel Mazmanian

Contents

1

THE PROSPECTS FOR EFFECTIVE IMPLEMENTATION OF REGULATORY POLICY

One of the dominant features of American government in the 20th century has been the emergence of a plethora of regulatory agencies. Beginning with efforts to regulate the prices set by utilities and transportation oligopolies, governments at all levels have gradually become involved in a large variety of programs involving consumer and environmental protection, occupational health and safety, and the protection of civil rights. The scope and number of these programs are matched, however, by criticisms of them: They restrict individual liberty within a market economy. They lead to an inefficient allocation of resources. They raise serious equity problems, especially against lower income classes. And, probably most important of all, they are not even able to achieve their statutory objectives.

This study is addressed primarily to the last criticism. Supporters of any regulatory program will generally acknowledge that it is necessary to restrict someone's economic freedom in order to enhance, for example, public health. Likewise, arguments concerning economic efficiency can generally be countered by pointing to difficulties in measuring all the costs and benefits of a particular program and/or by focusing on equity concerns. Disproportionate costs on the poor can be alleviated through transfer payments. But it is difficult for even the most ardent supporters of a regulatory program to justify the use of the police powers of the state and the administrative and other costs involved if the program is not at least moderately successful in achieving its objectives.

Despite the prevailing pessimism concerning the efficacy of governmental regulation, we would argue that even a regulatory program that requires substantial departure from the *status quo* can achieve its objectives.

Our argument will be examined through a detailed case study of the implementation of one of the nation's most important land use regulatory laws of the past decade, the 1972 California Coastal Zone Conservation Act. In the remainder of this introductory chapter, we shall flesh out the general argument by examining in greater detail the rationale and prevailing pessimism concerning regulatory programs, with specific attention to the substantive area of most relevance to our specific case, state land use regulation. We shall present a more detailed discussion of the six conditions for successful implementation of a regulatory program and suggest the extent to which they were achieved in the case of the California Coastal Commissions during the 1973–1976 period. Finally, we shall provide a brief synopsis of the organization of this book.

I. Governmental Regulation of Economic Behavior, with Special Reference to Land Use Control

Throughout the 20th century, governments in most Western nations have increasingly resorted to governmental control of private economic activities through state ownership and operation of basic industries and social services, through the setting of goals and production/service quotas for private firms, and through direct participation in management of private firms.[1] This has been done in pursuit of public objectives that, for any number of reasons, have not been met by the economic market. In the United States, however, the ownership and management of business enterprises and property has by and large been left in private hands.[2] When problems linked to the operations of the market have arisen, the solution has been to impose government regulations that guide and direct industries or sectors of society but that fall far short of outright management or ownership, and leave individuals and firms still partially dependent on market forces.[3] While this arrangement obviously reflects a philosophical preference on the part of Americans for minimal government intrusion in the workings of the economy, the scope of regulation has nevertheless grown to be quite substantial.

It has been estimated that as of the mid-1970s, for example, the number of federal agencies engaged in regulating some aspects of private activity had grown to 83, to say nothing of the thousands of state and local agencies with regulatory functions.[4] Many of these programs are concerned with regulating prices of monopolies and oligopolies, primarily among utilities and in the transportation industry. Others deal with labor–management relations, financial institutions, consumer protection, air and water pollution control, civil rights, occupational health and safety, land use, and a host of other activities.

The existence—although not necessarily the substance—of most of these programs can be justified on the grounds of some form of market failure. Among the conditions necessary to assure an economically efficient market are (1) that no buyer or seller have a large enough share of the market to influence significantly the market price, (2) that buyers and sellers be sufficiently informed about product prices and qualities to make efficient choices, and (3) that the production costs and consumption activities reflect their true social costs. Monopolies and oligopolies clearly violate the first assumption. Most consumer protection legislation is premised on the inability of consumers to meet the second condition.[5] And environmental protection and land use legislation is usually warranted by violations of the third assumption.[6] Finally, some regulatory legislation is based not upon market imperfections *per se* but rather upon dissatisfaction with violations of civil rights or with the distributional effects of market transactions.

Land use regulation is a good example of the expansion of regulatory control in recent years. Throughout most of the 19th century, strong popular beliefs concerning the inviolability of private property rights and a largely agrarian economy provided little impetus for governmental regulation of the use of land. Increasing urbanization and industrialization made people more interdependent, however, and thus eventually led to a variety of controls that became fairly common in urban areas in the early decades of this century. These included restriction of nuisance industries to specific zones; the establishment of building codes; and, in the 1920s and 1930s, the passage of general zoning ordinances. These regulatory efforts were enacted for a number of reasons, including the control of externalities (e.g., odors from a tannery or noise from commercial zones), consumer protection (e.g., building codes), and the preservation of the socioeconomic character of specific neighborhoods or localities.[7]

Whatever their nature, until quite recently these controls were exercised almost exclusively by local governments because of their proximity and accountability to the regulated population. Most local jurisdictions have, however, been very deferential to market forces, especially in rural areas and on the urban-rural fringe. General plans and zoning ordinances have usually been generous in designating lands for higher intensity use; for example, a great deal of agricultural land on the urban-rural fringe is generally zoned residential and/or commercial. And if zoning restricts development much below market price, ordinances have been quickly changed and/or variances given.[8]

In the last 15 years, however, both the scope and the locus of controls on land use have undergone considerable modification.[9] The scope of control has been expanded in many jurisdictions beyond the earlier preoccupation with public health and neighborhood amenities to include the protection of particularly scenic or ecologically sensitive areas such as wetlands; restrictions on growth because of limited water or sewage capacity, overcrowded schools, or simply a desire to preserve the general character of rural areas and small towns on the urban fringe; the protection or enlargement of public access to valuable recreation areas, particularly along the seashore; and the siting of power plants and other large industrial facilities. This expansion of the scope and rationale of land use control has often been accompanied by a transfer of (some) authority from local governments to state and/or regional agencies, as the latter are presumably less preoccupied with expanding local tax bases, more able to deal with inter-jurisdictional spillover effects, and more capable of marshaling the expertise necessary to deal with complex ecological and socioeconomic issues. The extent of restrictions on private use of land vary enormously among local jurisdictions and among different states, however, with the 1972 California Coastal Zone Conservation Act representing one of the most extensive cases of state reviews over local land development decisions.

While on the one hand the pressures continue to mount on government to regulate private economic behavior, as with the recent surge in land use regula-tion, there appears at the same time to be a growing dissatisfaction with govern-mental regulation among scholars, public officials, and the general public.[10] Criticisms fall into at least four different categories.

First, any regulatory program involves restrictions on individual behavior, which, in turn, arouse resentment. The factory owner confronted with regula-tions concerning product safety, air and water pollution, and/or employee safety is understandably displeased with these intrusions on his freedom, particularly if they involve costs that cannot be fully passed on to consumers without affecting his market position. Moreover, even those agencies that are normally supported by the regulated because of the agency's restrictions on competition—e.g., the Civil Aeronautics Board—restrict the freedom of entrepreneurs who wish to enter the field and of consumers who prefer a different mix of services.[11]

Second, it is often argued that regulatory programs have adverse effects on the efficient utilization of resources. The Civil Aeronautics Board, the Interstate Commerce Commission, and many other agencies that control rates and entry within a given industry have been criticized for creating higher consumer prices within the regulated industry than would be the case under conditions of greater competition.[12] The detailed regulations of the Occupational Health and Safety Administration have, in many instances, almost certainly involved greater ad-ministrative and compliance costs than were warranted by the safety benefits to employees.[13] The very stringent goals, tight deadlines, and uniform nationwide emission standards incorporated in recent federal air and water pollution control legislation very likely involve a lower ratio of benefits to costs than would be the case under a more flexible approach.[14] Agencies as diverse as the Interstate Commerce Commission and the U.S. Environmental Protection Agency have been accused of impeding technological innovations that would improve cost-effectiveness in the regulated industries.[15] In fact, there are very few regulatory programs that have not been criticized by economists (many of them sympathetic to the programs' objectives) for their suboptimal allocation of resources.

Third, many regulatory programs have been criticized on equity grounds, particularly concerning their adverse effects on lower and moderate income classes. Certainly CAB policies that inflate air fares are likely to disproportion-ately affect the poor. The same can be said for the added consumer costs imposed by consumer protection, environmental protection, and occupational health and safety programs, although many of these programs may also disproportionately benefit lower and moderate income classes.[16]

Not surprisingly, the expansion of state (and local) land use controls has been subjected to the same criticisms. Concern with restrictions on personal freedom have been expressed both in terms of vigorous opposition to the erosion of individual property rights and in complaints that state controls reduce inter-community diversity and thus limit the "mix" of communities available to consumers.[17] Several studies have addressed the issue of economic efficiency in terms of the costs of such controls on housing and the inefficiencies involved when large bureaucracies deal with a subject as particularistic as land development.[18] Land use controls have been criticized on equity grounds, in terms of their effects both on the poor (e.g., via exclusionary zoning) and on their ability to drastically affect property values, the so-called windfall and wipeout problem.[19]

Yet many of these criticisms are fundamentally normative. The extent to which restrictions on entrepreneurial freedom are warranted to protect health or consumer safety is fundamentally a question of competing values. Many benefit-cost analyses are ultimately based upon assumptions concerning the value to be attached to human health or safety. While certainly susceptible to some empirical analysis, this is ultimately a moral judgment. And judgments concerning the proper distribution of costs and benefits of any program depend upon the normative criteria employed.[20] In any democratic system, these value choices are ultimately for elected officials and the general political process to make.

There is, however, a fourth frequently voiced and extremely important criticism of regulatory programs on which we focus in this study—namely, their alleged inability to achieve their statutory objectives. For instance, it is not at all clear whether water pollution control regulations and billions of dollars in sewage treatment construction grants over the past two decades have resulted in any appreciable improvement in water quality.[21] Evidence suggests that federal efforts to encourage private employers to hire minority workers through conditional grants have been a dismal failure.[22] A recent analysis of federal safety regulation concluded that, while the relevant agencies have probably had some beneficial impacts on improving worker and consumer safety, they have also wasted a great deal of resources on trivial problems; moreover, alternative strategies—including changes in workmen's compensation—could probably have been more effective.[23] An alternative to governmental regulation of privately owned utilities—the presence of a publicly owned "yardstick"—has proven to be a more effective stimulus to lower consumer electrical rates.[24] In fact, the conventional wisdom may well be that any governmental effort to significantly alter behavior through regulation is unlikely to have any appreciable success.

The reasons cited for the inability of regulatory programs to achieve statutory objectives have been many and varied.[25] They include inconsistencies in the objectives themselves, inadequate causal theories linking the program to the intended effects, insufficient funds and/or sanctions, reliance on the regulated for crucial information or innovations, bureaucratic resistance, inadequate constituency support, unfavorable court review or legislative oversight, and changes in social and economic conditions that substantially undermine political support for the program. For example, Charles Schultze has argued persuasively that many consumer and environmental protection programs are fundamentally deficient in design: Their reliance on detailed regulations rather than pollutant/injury taxes impede technological innovation and are too inflexible to deal with extremely varied situations, thereby decreasing efficiency, increasing opposition, and ultimately decreasing program effectiveness.[26] Jacoby and Steinbruner contend that the very tight deadlines in the auto emmission standards of the 1970 Clean Air Amendments, the absence of any effective sanctions against the auto companies, and inadequate funding of research into alternative power systems for automobiles have locked the program into tinkering with the internal combustion engine, an inherently high-polluting source.[27] Similarly, Ackerman and Hassler argue that detailed regulations and insistence on smoke-stack-scrubber technology ironically may have worsened the problem of air pollution from coal-fired utility plants.[28] Periodic charges that the Food and Drug Administration has been "captured" by the drug companies have generally focused on the absence of an effective consumer constituency and the agency's reliance on the regulated companies for crucial information.[29] And the early efforts of the Interstate Commerce Commission to lower railroad fares were emasculated by the courts.[30]

In the area of land use policy, while there have been several studies of the effects of local zoning on development patterns, there have been very few systematic analyses of the effectiveness of state land use statutes as regulatory mechanisms.[31] Most case studies briefly discuss the history and content of regulatory laws, examine a few controversial cases, identify some important obstacles to implementation, and discuss in rather imprecise terms some of the impacts.[32] Even the few exceptions to these essentially descriptive overviews have not made a serious effort to develop a reasonably coherent framework of the implementation process linking statutory and political variables, the policy outputs of the implementing agencies, and the actual correspondence of behavioral change with statutory objectives.[33]

Of the many possible ways of analyzing the implementation of regulatory statutes, we have chosen first to develop a conceptual framework derived from the literatures on regulation and policy implementation and then to examine the utility of that framework in explaining the behavior and impacts of a set of interrelated institutions.[34] In the following sections we shall discuss the framework and then explain our choice of the California coastal commissions as the case study.

II. CRITERIA FOR EFFECTIVE IMPLEMENTATION OF A REGULATORY PROGRAM

Implementation is the carrying out of a basic policy decision, usually made in a statute (although also possible through important executive orders or court decisions). Ideally, that decision identifies the problem(s) to be addressed, stipulates the objective(s) to be pursued, implies a causal theory of the manner in which those objectives can be attained, and then "structures" the implementation process to reach its goals. In the case of a statute regulating private economic behavior, the implementation process normally (but not always) runs through a number of stages, beginning with passage of the basic statute, followed by the policy outputs (decisions) of the implementing agencies, the compliance of target groups with those decisions, the actual and perceived impacts of those decisions, and, finally, important revisions (or attempted revisions) in the basic statute.

In our view, the crucial role of implementation analysis is to identify the factors that affect the achievement of statutory objectives throughout this entire process. More specifically, a statute or other major policy decision seeking a substantial departure from the *status quo* will achieve its objectives under the following set of conditions:

1. The enabling legislation or other legal directive mandates policy objectives that are clear and consistent or at least provides substantive criteria for resolving goal conflicts.
2. The enabling legislation incorporates a sound theory identifying the principal factors and causal linkages affecting policy objectives, and gives implementing officials sufficient jurisdiction over target groups and other points of leverage to attain, at least potentially, the desired goals.
3. The enabling legislation structures the implementation process so as to maximize the probability that implementing officials and target groups will perform as desired. This involves assignment to sympathetic agencies with adequate hierarchical integration, supportive decision rules, sufficient financial resources, and adequate access to supporters.
4. The leaders of the implementing agency possess substantial managerial and political skill and are committed to statutory goals.
5. The program is actively supported by organized constituency groups and by a few key legislators (or a chief executive) throughout the implementation process, with the courts being neutral or supportive.

6. The relative priority of statutory objectives is not undermined over time by the emergence of conflicting public policies or by changes in relevant socioeconomic conditions that undermine the statute's causal theory or political support.

The conceptual framework underlying this set of conditions has been presented elsewhere in greater detail[35] and is based upon a theory of public agencies that views them as bureaucracies with multiple goals that are in constant interaction with interest (constituency) groups, other agencies, and legislative (and executive) sovereigns* in their policy subsystem.

Before elaborating on each of these conditions, it should be noted that the "strength" of the conditions necessary to achieve policy objectives is itself a function of several factors, including (1) the difficulty and expense of change required in target group behavior, (2) the predisposition of target groups toward the mandated change, and (3) the diversity in proscribed activities of target groups. In other words, the greater the mandated change, the more opposed the target groups, and the more diverse their proscribed activities, the greater the clarity of objectives, soundness of underlying theory, degree of statutory structuring, skill in implementing officials, support from constituency groups and sovereigns, and stability in socioeconomic conditions necessary if statutory objectives are to be attained. Within this context, the set of six conditions should always be sufficient to achieve policy objectives. Moreover, each condition is probably necessary if the change sought is substantial and requires 5–10 years of effort, although it may be possible to get by with moderately clear (but clearly prioritized) objectives if the other conditions are all met. With more tractable problems, it may be possible to omit one or two of the conditions.

Condition 1: The enabling legislation mandates policy directives that are clear and consistent (or at least provides substantive criteria for resolving goal conflicts). Statutory objectives that are clear and consistent serve as an indispensable aid in program evaluation, as unambiguous directives to implementing officials, and as a resource available to supporters of those objectives both within and external to the implementing agencies.[36] For example, implementing officials confronted with objections to their programs can sympathize with the aggrieved party but nevertheless respond that they are only following the legislature's instructions. Clear objectives can also serve as a resource to actors external to the implementing institutions who perceive discrepancies between agency policies and those objectives—particularly if the statute also provides them with

---

*A "sovereign" is any institution that has the authority to directly affect the legal and/or budgetary resources of an agency. They would include the legislature, chief executive, courts, and, in the case of state and local agencies, their intergovernmental superiors. The term is taken from Anthony Downs, *Inside Bureaucracy* (Boston: Little, Brown, 1967), pp. 44–46.

formal access to the implementation process through, for example, provision for citizen suits. If the statute mandates multiple objectives that are sometimes in conflict, it is important that some formal criteria (e.g., a ranking) or more detailed policy directives be provided for resolving those conflicts.

Condition 2: The legislation incorporates a sound theory identifying the principal factors and causal linkages affecting policy objectives, as well as the changes in target group behavior and other conditions necessary to attain the desired goals. Every statute or other basic policy decision that seeks to attain some desired end-state is based upon an often-implicit set of assumptions relating the achievement of that goal back through a set of factors responsible for the problem and ultimately to the behavior of target groups and other conditions subject to change. Following, for example, is a rather simple model of the factors affecting air quality:

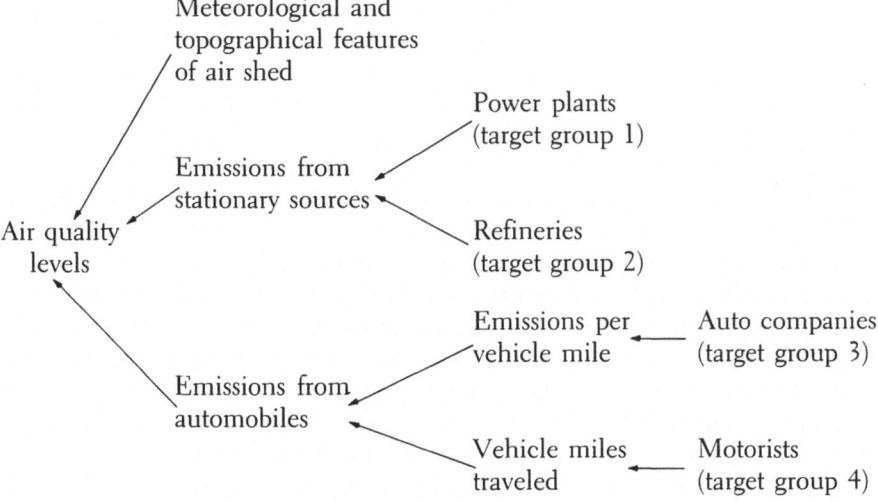

A statute seeking to improve air quality levels must be based upon an adequate understanding of the major causal linkages and must identify changes in the behavior of target groups sufficient to bring about desired air quality levels. For example, early efforts to improve air quality in Los Angeles assumed that the principal sources were power plants, refineries, and backyard incinerators. Even substantial reductions in emissions from these sources proved insufficient to correct the problem because the statute did not address the major source of pollutants, namely, the automobile. And subsequent efforts to reduce automotive emissions to acceptable levels also proved inadequate because they focused entirely on getting the automobile companies to reduce emissions per mile while largely ignoring the enormous increase in vehicle miles traveled by local motorists. [37]

This does not mean that the underlying theory has to identify all of the important causal factors or critical points of intervention. But it does mean that sufficient points of intervention must be identified to achieve statutory objectives. If not, the costs borne by regulated (target) groups will be resented all the more because they have proven to be not only onerous but wasted, and public disillusionment with the program will almost certainly increase.

Condition 3: The statute not only gives implementing agencies sufficient jurisdiction over target groups (and other critical areas of intervention) but also structures the implementation process so as to maximize the probability that target groups will perform as desired. Even the clearest objectives and soundest underlying theory will prove inadequate unless the statute also (a) gives implementing agencies jurisdiction over sufficient target group behavior to at least theoretically permit the attainment of those objectives and (b) structures the implementation process so as to maximize the probability that target groups will perform as desired. The former is implicit in the concept of a sound theory. Knowledge of critical points of intervention is academic without the jurisdiction to intervene.

And the latter involves a variety of mechanisms—too often neglected by behaviorally oriented social scientists—through which a statute can affect the policy outputs of the implementing agencies and ultimately the behavior of target groups. [38] Specifically, for a statute to maximize the probability of target group compliance it should (1) assign implementation to agencies supportive of statutory objectives that will give the new program high priority[39]; (2) provide substantial hierarchical integration within and among implementing agencies by minimizing the number of veto/clearance points and by providing supporters of statutory objectives with inducements and sanctions sufficient to assure acquiescence among those with a potential veto and among target groups[40]; (3) provide adequate financial resources to the implementating agency(agencies) to hire the staff and conduct the technical analyses involved in the development of regulations, the administration of permit systems, and the monitoring of target group compliance[41]; (4) bias the decision rules of the implementing agency(agencies) in favor of adherence to statutory objectives[42]; and (5) provide ample opportunity for interest groups and sovereigns supportive of statutory objectives to intervene in the implementation process, e.g., through liberal rules of standing to agency and judicial proceedings. [43] As we shall see shortly, much of the basis for our expectation concerning the successful implementation of the 1972 California Coastal Initiative resides in the extent to which that legislation consistently structured the implementation process in a manner conducive to the attainment of its goals.

Condition 4: The leaders of the implementing agency(agencies) possess substantial managerial and political skill and are committed to statutory objectives. Any regulatory program seeking to substantially change target group behavior requires implementing officials who are not merely neutral but also sufficiently committed and persistent to develop and enforce new regulations in the face of resistance from target groups and from public officials reluctant to make the mandated changes. A statute can take several steps to increase the probability that implementing officials will be committed to its objectives; these include assigning implementation to a new (or at least generally supportive) agency and stipulating that top implementing officials be selected from social sectors that generally support the legislation's objectives. Moreover, legislative (and executive) supporters often play a major role in appointing top officials. On the other hand, the use of statutory levers is often severely constrained in practice, and appointments may be made by executive (and legislative) officials who are not committed to implementation of the basic policy decision. There is also considerable evidence that the overall commitment of agency officials tends to decline over time as strong supporters become burned out and/or disillusioned with bureaucratic routine, to be replaced by officials much more interested in job security and organizational maintenance than in taking risks to attain ambitious statutory objectives.[44] In short, the support of top implementing officials is thus sufficiently important and problematic to constitute being highlighted as a separate condition.

Policy support is, however, essentially useless if not accompanied by political and managerial skill in utilizing available resources. Political skill involves the ability to develop good working relationships with sovereigns in the agency's subsystem, to convince opponents and target groups that they are being treated fairly, to mobilize latent supportive constituencies, to adroitly present the agency's case through the mass media, etc. Managerial skill involves developing adequate controls so that the program is not subject to charges of fiscal mismanagement, to maintaining high morale among agency personnel, and to managing internal dissent in such a way that dissidents are convinced they have received a fair hearing.[45]

Condition 5: The program is actively supported by organized interest groups and by a few key legislators (or the chief executive) throughout the implementation process, with the courts being neutral or supportive. Any program that seeks substantial change in the *status quo* needs constant and/or periodic infusions of political support if it is to overcome the inertia and delay inherent in seeking cooperation and acquiescence among large groups of people, many of whom perceive their personal interests to be adversely affected. Affecting the amount of potential support in a rather diffuse fashion are the amount of media attention to the problem addressed by a statute and variations over time in public support for statutory objectives. Both are susceptible to an episodic "issue-attention cycle," in which concern with a specific problem is soon replaced with other issues, thereby eroding the political support necessary throughout the long implementation process.[46]

Of more direct concern to implementing agencies, however, is the balance of support for statutory objectives among organized constituency groups and the agencies' legislative, executive, and judicial sovereigns. With respect to the former, support for stringent regulation from environmental and consumer groups generally declines over time, while regulated (target) groups normally have the motivation and the resources to continue to intervene actively in agency proceedings and/or to appeal to agency sovereigns when dissatisfied. As a result, most regulatory agencies eventually recognize that survival in an unbalanced political environment necessitates some accommodation with target groups and thus less departure from the *status quo* than envisaged by the original statutory mandate.[47] A similar change in support often occurs among the agency's legislative and executive sovereigns, although this can often be successfully resisted in the short term by a "fixer," i.e., an important legislator or executive official who controls resources important to crucial actors and who has the desire and the staff resources to closely monitor the implementation process and to intervene on an almost continuous basis.[48]

In addition to such variation in political support over time, federal and state regulatory programs are normally confronted with substantial variation among local jurisdictions in support for statutory objectives and, consequently, in the compliance of local implementing officials and target groups with program directives. While such variation can, in principle, be overcome if the statute provides very substantial incentives for compliance and sufficient financial resources to enable federal/state officials to essentially replace local implementors, in practice the system is seldom structured to that degree, and thus federal/state officials are forced to bargain with recalcitrant local implementors.[49] The result is greater sensitivity to local demands and generally a suboptimal achievement of statutory objectives.

Finally, one must not neglect the judicial sovereigns of the implementing agency(agencies). While most federal and state courts are generally reluctant to overturn agency decisions, they certainly have the potential to essentially emasculate implementation through delay in enforcement proceedings, through repeatedly unfavorable statutory interpretations, and, in extreme cases, by declaring the statute unconstitutional.[50] Given the enormous potential role of the courts, we argue that effective implementation of statutory objectives requires that they be either neutral or supportive.

Condition 6: The relative priority of statutory objectives is not significantly undermined over time by the emergence of conflicting public policies or by changes in relevant socioeconomic conditions that undermine the statute's underlying causal theory or political support. Change is omnipresent in most contemporary societies, in part because most countries are immersed in a dynamic international system over which they have only modest control, in part because policy issues tend to be highly interrelated. Land use control, for example, is linked to energy, to inflation and national monetary policy, to transportation, to tax policy, to public lands, and to numerous other issues. As a result of this continuous change, any particular policy decision is susceptible to an erosion in political support as other issues become relatively more important over time. Obvious examples would be the effect of the Vietnam War and inflation on many Great Society programs and the effect of the energy crisis and inflation on pollution control programs. The viability of stringent regulatory programs is tied particularly to the economic health of the regulated industry and to its salience in the local economy.[51] Support for restrictive land use control programs, for example, can be expected to decline during times of high unemployment in the local construction industry, particularly if construction is a major component of the local economy.

The vulnerability of a regulatory program to such changes in relevant socioeconomic conditions is a function of the extent to which its statute has coherently structured the implementation process, and it enjoys support from key legislators, organized constituency groups, and implementing officials. If the statute strongly biases the outputs of implementing agencies and if it is supported by a "fixer" and an organized interest group, then the program will probably weather temporary fluctuations in the economy and in diffuse political support. On the other hand, weak statutes with little support from sovereigns require implementing officials to be very sensitive to such fluctuations if they are to survive.

The discussion thus far has focused on the conditions affecting the implementation process as a whole. But that process must be viewed in terms of its several stages: (1) the policy outputs (decisions) of the implementing agencies, (2) the compliance of target groups with those decisions, (3) the actual impacts of agency decisions, (4) the perceived impacts of those decisions, and, finally, (5) the political system's evaluation of a statute in terms of major revisions (or attempted revisions) in its content.

If one is concerned only with the extent to which actual impacts conform to statutory objectives, only the first three stages are pertinent. Moreover, it is clear that a statute will achieve its desired impacts if (a) the policy outputs of the implementing agencies are consistent with statutory objectives, (b) the ultimate target groups comply with those objectives, and (c) the statute incorporates a sound theory linking behavioral change in target groups to the achievement of the desired end-state. If one is also interested, however, in following the implementation process to its culmination in revisions (or attempted revisions) in the basic statute, then it is the *perceived*—rather than the actual—impacts that are probably critical. In this regard, there is some evidence that the impacts perceived by members of the policy subsystem will be a function of actual impacts as mediated by the values of the perceiver and that actors who perceive the impacts or the policy outputs of implementing agencies to be undesirable—irrespective of their conformity with statutory objectives—will be more politically active than those who perceive the impacts to be desirable. [52]

This concludes our brief overview of the factors affecting the implementation of regulatory statutes. Specific relations will, however, be examined in greater detail in succeeding chapters. It is now time to examine the reasons for selecting the 1972 California Coastal Initiative as our case study.

III. Selection of the 1972 California Coastal Initiative As a Case Study

The California Coastal Zone Conservation Commissions were established by statutory initiative in November 1972 after repeated failures to persuade the legislature to pass legislation providing for state review of local land use decisions along the coast. The Coastal Initiative established a state coastal commission and six regional commissions with the dual tasks of preparing a comprehensive coastal plan to be presented to the 1976 legislature and of regulating all development in the interim from the seaward boundary of the state's jurisdiction to 1000 yards landward of the mean high tide line. The basic objective of the statute was the protection of the scenic and environmental resources along the coast, as well as public access to the beach. Each regional commission was composed of equal numbers of local elected officials and "public" members appointed by state officials (the governor, the Assembly speaker, and the Senate Rules Committee), while the state commission was divided equally between "public" members and representatives of the regional commissions. Each commission was assisted by a full-time staff of 6–15 professionals, while the commissions' activities were funded by a $5 million guaranteed appropriation. [53]

There are basically two reasons for selecting the California coastal commissions as the empirical focus of this study. On the one hand, the commissions represent a particularly propitious data base for analysis. The presence of six regional commissions provides the opportunity to examine the effects of interregional variation in socioeconomic conditions and public support for statutory objectives on the commissions' policy decisions. The genesis of the commissions in the initiative process means that they were exceedingly open agencies, in terms of both the accessibility of their files and the very strong norms that decisions be made in public. The presence of a commission form of government (rather than a line agency headed by a single administrator) provides a large enough number of decision-makers to permit systematic comparison of voting on applications for development permits. And the requirement that the commissions review all permits approved by local governments provides an opportunity to compare the decisions of state/regional and local institutions on precisely the same permits and thus to measure in relatively precise terms the impact of the coastal commissions on development along the coast.[54]

Nevertheless, the major reason for selecting the commissions for detailed analysis is not methodological but rather theoretical. Given the extremely problematic nature of the implementation process and the prevailing pessimism concerning the ability of regulatory statutes to achieve their objectives, the 1972 California Coastal Initiative represents an excellent opportunity to examine our general hypothesis of the conditions under which a regulatory statute that seeks a substantial departure from the *status quo* can effectively achieve its objectives. For the commissions met all but perhaps one of the six conditions for effective implementation at least moderately well.

First, the Coastal Initiative contained a relatively clear mandate to protect the environmental and scenic resources of the coast and to maximize public access to the wet sand beaches. While these objectives were stated in fairly general terms and no attempt was made to reconcile minor conflicts between, for example, access and the protection of tide pools, the statute contained no requirements that public access and environmental protection be balanced against the need for, e.g., additional housing or energy facilities. It thus had a reasonably clear and consistent general orientation.

Second, the Initiative assumed that its goals could be realized (at least during the 4-year interim period) by giving the commissions absolute jurisdiction over all development within 1000 yards (about .57 miles) off the shore. This development review authority was certainly adequate to protect the scenic resources of the coast and most coastal wetlands, although the absence of authority to acquire and manage land would create some difficulties in actually providing public access to the coast. On the whole, however, the underlying causal theory was fairly sound.

Third, the Initiative generally did an excellent job of structuring the implementation process so as to maximize the probability that the policy decisions of implementing agencies and the behavior of coastal property owners (the principal target group) would be consistent with statutory objectives. Implementation was assigned to a new agency, whose professional staff could be expected to be highly supportive of the act's objectives.[55] As for coastal commissioners, since most state appointees and most local officials from areas that had voted strongly in favor of the Initiative could be expected to support statutory objectives, the state commission and most of the regional commissions could be expected to be dominated by statutory supporters (the exception being one or two regions where local officials were strongly antagonistic to the Initiative).[56] The statute created a very hierarchically integrated system, with the commissions having virtually total control over coastal development; moreover, the monetary fines were substantial enough to discourage blatant noncompliance by target groups. The statute provided a guaranteed appropriation from a trust fund (and therefore did not require annual appropriations from the legislature), which the drafters thought would be consistent with statutory objectives; almost all important developments required a two-thirds affirmative vote of the commission membership; and the statute clearly indicated that regional permit decisions could be appealed to the state commission and that the state commission would have final authority over the coastal plan to be presented to the 1976 legislature. And the statute provided very liberal rules of standing for interested parties to participate in agency decisions and to appeal those decisions to the courts.

Fourth, the leaders of most of the commissions—and particularly the state commission—not only were committed to the achievement of statutory objectives but also possessed substantial managerial and political skills.

Fifth, the commissions were actively supported by environmental groups—particularly the Sierra Club—throughout the implementation process. They also had the active support of a few important legislators, although this was not crucial in the 1973–1975 period because the statute explicitly precluded weakening amendments by the legislature. On the other hand, interest group and legislative support was absolutely essential—and, as it turned out, barely sufficient—during the 1976 session, when the original Initiative was to expire, and thus the entire fate of the coastal program was at issue.

Sixth, and finally, relevant socioeconomic conditions underwent some significant changes in the 1973–1976 period, the most important being a substantial worsening of unemployment in the construction industry. The commissions were, however, generally insulated from the direct political repercussions of these changes by the independence from legislative oversight that the Initiative provided—at least until the 1976 session. Their importance did come to the fore, however, during implementation of the successor law, the California Coastal Zone Act of 1976.

Each of these conditions will, of course, be discussed in considerably greater detail in succeeding chapters. But the general implication is fairly clear: On the whole, the implementation of the 1972 California Coastal Initiative substantially met all but perhaps the last of the six conditions for effective implementation, at least during the 1973–1976 period. Thus, if the commissions were able to substantially achieve their statutory objectives, it would at least tentatively validate our general hypothesis concerning the set of sufficient and generally necessary conditions for the effective implementation of regulatory programs. Moreover, it would suggest that the prevailing pessimism concerning the effectiveness of regulation is somewhat premature—that regulatory legislation can achieve its causal objective if the statute incorporates reasonably clear and con-

sistent objectives and a fairly sound theory, if it coherently structures the implementation process, and if a strong supportive constituency is available to monitor
that process. On the other hand, if the commissions failed to substantially
achieve their objectives, this would suggest that there are some serious deficiencies in our framework or that the prospects for effective regulation are dim
indeed, or both.

    In its basic structure, then, this study is an example of what Lipset, Trow,
and Coleman have termed "deviant case analysis."[57] We make no claims that
implementation of the 1972 California Coastal Initiative is at all typical of the
experience of most regulatory programs. In fact, it is probably atypical. But just
as Lipset and his colleagues used the admittedly deviant case of the International
Typographical Union to illuminate the conditions under which multiple parties
can flourish in normally oligarchic trade unions, so we propose to use the
California Coastal Initiative to examine the conditions under which a regulatory
statute can substantially change target group behavior and eventually achieve its
objectives.[58]

IV. SYNOPSIS OF CASE STUDY

    Selected for detailed scrutiny are four of the seven coastal commissions
created under the 1972 act: the state commission as the pivotal coordinating body
and three of the six regional commissions—the North Coast, the North Central,
and the South Coast. The three were selected to provide variation on (a) the
public vote on the Coastal Initiative and (b) the nature of regional development.
The North Coast region (Del Norte, Humboldt, and Mendocino counties) is
largely rural, heavily dependent upon the timber industry, and plagued with
chronically high unemployment. Whereas 55% of the state electorate supported
the Initiative, only 34% of those in the three counties of the North Coast region
did so; this was the lowest among the six coastal regions. The North Central
region (Sonoma, Marin, and San Francisco counties) is a metropolitan area,
although the coastal zone in Marin and Sonoma counties is quite rural. Sixty
percent of the residents in these three counties supported the Initiative, the
highest percentage from any of the six regions. The South Coast region (Los
Angeles and Orange counties) is highly urbanized and much of the beachfront
is already developed. Substantial pressures exist to extend intensive development
along the coast southward to the remaining open space in Orange County and
northward into the Malibu area. The regional support for the Initiative was a
moderate 53%. Figure 1-1 provides a map of the California coast indicating the
location of the three regions and some significant characteristics.

    Focusing on the regional commissions enables us to compare agencies with
similar statutory bases, but varying markedly in socioeconomic conditions, pub-

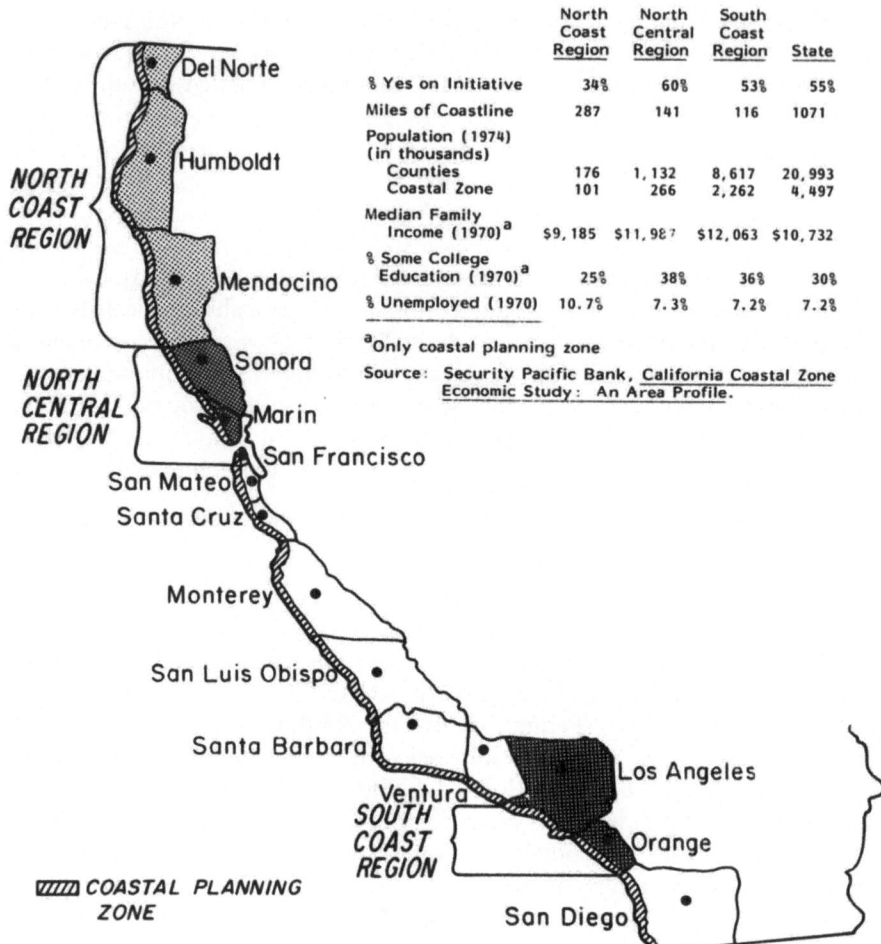

	North Coast Region	North Central Region	South Coast Region	State
% Yes on Initiative	34%	60%	53%	55%
Miles of Coastline	287	141	116	1071
Population (1974) (in thousands)				
Counties	176	1,132	8,617	20,993
Coastal Zone	101	266	2,262	4,497
Median Family Income (1970)[a]	$9,185	$11,987	$12,063	$10,732
% Some College Education (1970)[a]	25%	38%	36%	30%
% Unemployed (1970)	10.7%	7.3%	7.2%	7.2%

[a]Only coastal planning zone

Source: Security Pacific Bank, California Coastal Zone Economic Study: An Area Profile.

COASTAL PLANNING ZONE

FIGURE 1-1.　Location and selected characteristics of the North Coast, North Central, and South Coast regions. [Source: Security Pacific Bank, *California Coastal Zone Economy Study: An Area Profile* (Los Angeles: Security Pacific Bank, 1975).]

lic opinion, commissioner commitment to statutory objectives, and a number of other pertinent factors. Moreover, the interplay between state and regional commissions on planning and permit review provides an opportunity to examine the ability of a relatively "strong" statute to structure decisions in widely different regions and, particularly, in hostile areas such as the North Coast region.

The principal data bases for this study include (1) personal interviews and attitudinal surveys with most commissioners and staff in the three regional commissions and the state commission; (2) analyses of random samples of permit decisions of the four commissions during the 1973–1975 period, including data on the characteristics of the proposed development, participation by outside groups, the staff recommendation and commission decision, and the voting record of commissioners; (3) a detailed analysis of the development of the coastal plan focusing on the development of three of the nine plan elements; (4) an evaluation survey mailed to a random sample of participants in the planning and permit review processes soliciting their evaluation of the commissions' performance and information on their activities; and (5) interviews with interest group leaders, local and state officials, and legislative staff concerning their activities during the 1973–1976 period and the 1976 legislative session. In addition, we attended numerous meetings of the commissions and consulted available reports on the commissions' activities and on a variety of impacts (on, e.g., wetlands, public access, housing costs) that could be related to their decisions.

The focus of our analysis (all but the last two chapters) is on the passage, implementation, and reformulation of the Coastal Initiative during the 1972–1976 period. Chapters 2–4 provide a general overview of the history of the coastal commissions and the factors affecting their decisions during these years. Chapter 2 discusses the formation of the commissions and provides a more detailed analysis of the 1972 California Coastal Zone Conservation Act. Chapter 3 examines the external factors (e.g., socioeconomic conditions, constituency groups, legislative and executive sovereigns, court review, and other agencies) constraining the commissions' behavior, while Chapter 4 focuses on the factors internal to the coastal agency—including staff and commissioner attitudes and state–regional relations—that directly affected policy-making.

The next four chapters then deal with the major decisions of the commissions and their impacts during 1973–1976. Chapter 5 presents some basic data on permit decisions, examines the conformity of those decisions with statutory objectives, and presents the results of a quantitative analysis of a model of the variables affecting the commissioners' permit decisions. Chapter 6 examines the subsequent stages of the implementation process in terms of target group compliance with permit decisions, and the effects of those decisions on the protection of wetlands, public access to the coast, and housing prices within the permit zone. Chapter 7 deals with the other major responsibility of the commissions, the development of a coastal plan outlining future policies and institutions to manage coastal resources. And Chapter 8 examines the 1976 legislature's evaluation of the plan and reformulation of the commissions' mandate.

Chapter 9 then provides a brief overview of the implementation of the 1976 California Coastal Act, in part to update the story through 1980, in part to provide a comparison with the 1973–1976 period.

Finally, Chapter 10 presents our conclusions concerning the prospects for effective regulation and a final discussion of our conceptual framework of the implementation process.

NOTES

1. Andrew Shonfield, *Modern Capitalism: The Changing Balance of Public and Private Power* (London: Oxford University Press, 1965).
2. Major exceptions to this generalization are the Tennessee Valley Authority, the many publicly owned and operated local utilities, and management of wartime economy.
3. James Q. Wilson, ed., *The Politics of Regulation* (New York: Basic Books, 1980); Murry L. Weidenbaum, *Business, Government and the Public*, 2nd ed. (Englewood Cliffs, N.J.: Prentice-Hall, 1981).
4. Charles Schultze, *The Public Use of Private Interest* (Washington, D.C.: Brookings Institution, 1977), p. 7. Our discussion of regulation throughout this book explicitly excludes the criminal justice system.
5. For example, the ordinary consumer cannot be expected to know the relative safety of different types of automotive fuel tank designs. Yet such information would presumably affect the market choices of some consumers. Nina Cornell, Roger Noll, and Barry Weingast, "Safety Regulation," in *Setting National Priorities: The Next Ten Years*, ed. Henry Owen and Charles Schultze (Washington, D.C.: Brookings Institution, 1976), pp. 464–476.
6. For example, the particulate emissions from a coal-burning electrical utility or a steel mill impose costs on adjacent residents. But these costs are not reflected in the price the company charges its customers because the air is treated as a "public good," and it is extremely difficult in a court of law for individual property owners to relate emissions from a specific source to specific costs to them. It is these "externalities" that constitute a principal justification for most pollution control legislation. Paul Barkley and David Seckler, *Economic Growth and Environmental Decay* (New York: Harcourt Brace Jovanovich, 1972), chap. 8.
7. John Delafons, *Land Use Controls in the United States*, 2nd ed. (Cambridge, Mass.: M.I.T. Press, 1969); David Ervin, James Fitch, R. Kenneth Godwin, W. Bruce Shepard, and Herbert Stoevener, *Land Use Control* (Cambridge: Ballinger, 1977).
8. Richard Babcock, *The Zoning Game* (Madison: University of Wisconsin Press, 1969); Daniel Mandelker, *The Zoning Dilemma* (Indianapolis: Bobbs-Merrill, 1971).
9. Nelson Rosenbaum, *Land Use and the Legislatures* (Washington, D.C.: Urban Institute, 1976); Fred Bosselman and David Callies, *The Quiet Revolution in Land Use Control* (Washington, D.C.: Council on Environmental Quality, 1971); Robert Healy, *Land Use and the States* (Baltimore: Johns Hopkins University Press, 1976); Melvin Levin, Jerome Rose, and Joseph Slavet, *New Approaches to State Land Use Policies* (Lexington, Mass.: D. C. Heath, 1974); James Coffin and Michael Arnold, *A Summary of State Land Use Controls: July 1974* (Washington, D.C.: Land Use Planning Reports, 1974); Nelson Rosenbaum, "Growth and Its Discontents: Origins of Local Population Controls," in *The Policy Cycle*, ed. Aaron Wildavsky and Judith May (Beverly Hills: Sage, 1978), pp. 43–61; Office of Coastal Zone Management, U.S. Dept. of Commerce, *State Coastal Zone Management Activities 1975–76* (Washington, D.C.: Government Printing Office, 1976).
10. For the view that government overregulates, see Chris Argris et al., *Regulating Business: The Search for an Optimum* (San Francisco, Institute for Contemporary Studies, 1978). In addition, most issues of the *Public Interest* and *Of Regulation* contain criticisms of specific regulatory agencies. From the opposite vantage point, that regulation should be more stringent; see the plethora of studies sponsored by Ralph Nader. Finally, both houses of Congress have recently completed major studies of regulatory agencies; see, for example, U.S. House of Representatives, Committee on Interstate and Foreign Commerce, *Report on Federal Regulation and Regulatory Reform*, 94th Congress, 2nd session, October 1976.
11. The two examples in this paragraph represent two fundamentally different types of regulation, which James Landis termed "policing" and "managerial," respectively. Policing programs are

those that regulate a specific aspect of a target group's behavior because of its potentially harmful effects on others. For example, a pollution control agency regulates only the emissions from specific point sources; it does not deal with the prices they charge, the wages they pay, or (directly) with their relations with their competitors. In contrast, managerial programs essentially replace market mechanisms by controlling rates, entry, and performance within a specific economic sector. Examples would include the Civil Aeronautics Board, the Interstate Commerce Commission (for railroads and truckers), state public utility commissions, and professional licensing boards. James Landis, *The Administrative Process* (New Haven: Yale University Press, 1938), pp. 16–30. For a general discussion of the freedom argument, see George Stigler, *The Citizen and the State* (Chicago: University of Chicago Press, 1975), chaps. 1–2.

12. George Douglas and James Miller III, *Economic Regulation of Domestic Air Transport: Theory and Policy* (Washington, D.C.: Brookings Institution, 1974); Richard Caves, *Air Transport and Its Regulators* (Cambridge: Harvard University Press, 1962); Ann Friedlander, *The Dilemma of Freight Transport Regulation* (Washington, D.C.: Brookings Institution, 1969).

13. Schultze, *The Public Use of Private Interest*, pp. 55–56.

14. Allen Kneese and Charles Schultze, *Pollution, Prices, and Public Policy* (Washington, D.C.: Brookings Institution, 1975), chaps. 3, 6; National Academy of Sciences, *Air Quality and Automobile Emission Control*, Prepared for the Committee on Public Works, U.S. Senate, 93rd Congress, 2nd session (Washington, D.C.: Government Printing Office, September 1974); National Academy of Sciences, *Air Quality and Stationary Source Emission Control*, Prepared for the Committee on Public Works, U.S. Senate, 94th Congress, 1st session (Washington, D.C.: Government Printing Office, March 1975).

15. William Capron, ed., *Technical Change in Regulated Industries* (Washington, D.C.: Brookings Institution, 1971); Roger Noll, *Reforming Regulation* (Washington, D.C.: Brookings Institution, 1971), pp. 23–27; Kneese and Schultze, *Pollution, Prices, and Public Policy*, pp. 58–64, 82–83.

16. David Harrison, *Who Pays for Clean Air* (Cambridge, Mass.: Ballinger, 1975); S. I. Schwartz, "Distributional Impacts of Automobile Pollution Control Programs," *Journal of Environmental Systems* 5 (1975), pp. 185–202; Martin Krieger, "Six Propositions on the Poor and Pollution," *Policy Sciences* 1 (Fall 1970), pp. 311–324.

17. Benjamin Bobo et al., *No Land Is an Island: Individual Rights and Government Control of Land Use* (San Francisco: Institute for Contemporary Studies, 1975).

18. Dan Richardson, *The Cost of Environmental Protection: Regulating Housing Development in the Coastal Zone* (New Brunswick, N.J.: Rutgers Center for Urban Policy Research, 1976); Robert Kneisel, *Economic Impacts of Land Use Control With Special Reference to the California Coastal Zone Commissions* (Davis, California: Institute of Governmental Affairs, 1978); Vincent Ostrom, Charles Tiebout, and Robert Warren, "The Organization of Government in Metropolitan Areas: A Theoretical Inquiry," *American Political Science Review* 55 (December 1961), p. 837; Robert Hawkins, "Local Land Use Planning and Its Critics," in Bobo et al., *No Land Is an Island*, pp. 101–111.

19. Ervin et al., *Land Use Control*, chaps. 3, 5–7; Bobo et al., *No Land Is an Island*, chaps. 6, 7, 11, 13; Fred Bosselman, David Callies, and John Banta, *The Taking Issue* (Washington, D.C.: Council for Environmental Quality, 1973); John Costonis, Curtis Berger, and Stanley Scott, *Regulation v. Compensation in Land Use Control* (Berkeley, California: Institute of Governmental Studies, 1977).

20. For an excellent discussion of alternative criteria, see Nicholas Rescher, *Distributive Justice* (Indianapolis: Bobbs-Merrill, 1966), chap. 4.

21. Kneese and Schultze, *Pollution, Prices, and Public Policy*, chaps. 3, 5; Harvey Lieber, *Federalism and Clean Waters* (Lexington: D. C. Heath, 1975), chap. 7.

22. Jeffrey Pressman and Aaron Wildavsky, *Implementation* (Berkeley: University of California Press, 1973).

23. Cornell, Noll, and Weingast, "Safety Regulation," pp. 457–504; ·Schultze, *Public Use of Private Interest*, pp. 55–57.

24. Thomas Pelosci, "Organizational Correlates of Utility Rates" (Paper presented at the 1978 Annual Meeting of the Midwest Political Science Association, Chicago, April 20–22, 1978); Stigler, *The Citizen and the State*, chap. 5. The "yardstick" notion in utility regulation is, however, the subject of considerable dispute because of the cost advantages of publicly owned electricity generators through, e.g., below-market-rate interest costs and property taxation advantages.

25. For general reviews of this topic, see Noll, *Reforming Regulation*; Roger Noll, "Government Administrative Behavior and Technological Innovation," Social Science Working Paper No. 62, California Institute of Technology, October 1974; Paul Sabatier, "Regulatory Policy-Making: Toward a Framework of Analysis," *Natural Resources Journal* 17 (July 1977), pp. 415–460; Wilson, "The Politics of Regulation," pp. 135–168; and House Committee on Foreign and Interstate Commerce, *Federal Regulation and Regulatory Reform*.

26. Schultze, *Public Use of Private Interest*; Kneese and Schultze, *Pollution, Prices, and Public Policy*. For reservations concerning this argument, see Giandomenico Majone, "Choice Among Policy Instruments for Pollution Control," *Policy Analysis* 2 (Fall 1976), pp. 589–613; and Lettie McSpadden Wenner, "Pollution Control: Implementation Alternatives," *Policy Analysis* 4 (Winter 1978), pp. 47–65.

27. Henry Jacoby, John Steinbruner, and others, *Clearing the Air* (Cambridge: Ballinger, 1973), chaps. 3, 4.

28. Bruce Ackerman and William Hassler, *Clean Air/Dirty Air or How the Clean Air Act Became a Multibillion-Dollar Bail-Out for High-Sulfur Coal Producers and What Should be Done About It* (New Haven: Yale University Press, 1981).

29. Mark Nadel, *The Politics of Consumer Protection* (Indianapolis: Bobbs-Merrill, 1971), pp. 66–80; James Turner, *The Chemical Feast* (New York: Grossman, 1970); House Commerce Committee, *Federal Regulation and Regulatory Reform*, chap. 8. For the same problem in regulation of the oil industry, see U.S. Senate, Committee on Interior and Insular Affairs, *Hearings on the Energy Information Act*, S. 2782, 93rd Congress, 2nd session, February 1974.

30. Peter Woll, *American Bureaucracy* (New York: W. W. Norton, 1963), pp. 36–40.

31. Ervin *et al.*, *Land Use Control*, chap. 5; Kneisel, *Economic Impacts of Land Use Controls*, *passim*.

32. Healy, *State Land Use Controls*, chaps. 3–5; Luther Carter, *The Florida Experience* (Baltimore: Johns Hopkins University Press, 1974); Phyllis Myers, *Slow Start in Paradise* (Washington, D.C.: Conservation Foundation, 1974); Phyllis Myers, *So Goes Vermont* (Washington, D.C.: Conservation Foundation, 1974); Phyllis Myers, *Zoning Hawaii* (Washington, D.C.: Conservation Foundation, 1976); Ted Lauf, "Shoreland Regulation in Wisconsin," *Coastal Zone Management Journal* 2 (1975), pp. 47–58; Frank Popper, *The Politics of Land-Use Reform* (Madison, Wis.: University of Wisconsin Press, 1981); Daniel Mandelker, *Environmental and Land Controls Legislation* (Indianapolis: Bobbs-Merrill, 1976), chaps. 6–8; Joseph Heikoff, *Coastal Resources Management* (Ann Arbor, Mich: Ann Arbor Science, 1977).

    These criticisms are less true for Healy and for Popper. Healy does make some explicit comparisons concerning, for example, the distribution of authority among state, regional, and local governments in Vermont, Florida, and California. But he does not really relate this to differences in policy outputs, e.g., permit decisions. In general, however, the book is not designed as a rigorous comparative analysis but rather to alert readers to important issues and to alternative strategies—a task that it accomplishes quite well. Popper relates innovations in state and use programs to the critique by Theodore Lowi of "interest group liberalism" [*The End of Liberalism* (New York: W. W. Norton, 1969)].

33. The study that comes closest to meeting our ideal of implementation analysis is Richard Liroff and Gordon Davis, *Protecting Open Space: Land Use Control in the Adirondack Park*

(Cambridge: Ballinger, 1981). For other fairly comprehensive analyses, see Gerald Swanson, "Coastal Zone Management from an Administrative Perspective: A Case Study of the San Francisco Bay Conservation and Development Commission," *Coastal Zone Management Journal* 2 (1975), pp. 81–102; Edmond Costantini and Kenneth Hanf, *The Environmental Impulse and Its Competitors: Attitudes, Interests, and Institutions at Lake Tahoe* (Davis, Calif.: Institute of Governmental Affairs, 1973); and Ken Lowry, "Evaluating State Land Use Control: Some Perspectives and a Case Study of Hawaii's Land Use Law" (Unpublished paper, University of Hawaii at Manoa, 1978).

34. Alternative research designs include (a) the examination of a large number of statutes/agencies so as to provide statistically significant findings across cases and (b) the more detailed comparative analysis of a few statutes/agencies. Both were rejected because we wished to develop more extensive data bases on crucial variables than would have been possible had more than one case (statute) been examined (given the resources available to us). For discussions of the theoretical case study, see Harry Eckstein, *Division and Cohesion in Democracy: A Study of Norway* (Princeton: Princeton University Press, 1966), Preface; and Seymour Martin Litset, Martin Trow, and James Coleman, *Union Democracy: The Internal Politics of the International Typographical Union* (Garden City, N.Y.: Doubleday, 1956), Preface and chap. 1.

35. For more extensive discussions of slightly different versions of the framework, see Paul Sabatier and Daniel Mazmanian, "The Conditions of Effective Implementation: A Guide to Accomplishing Policy Objectives," *Policy Analysis* 5 (September 1979), pp. 481–504; and Daniel Mazmanian and Paul Sabatier (eds.), *Effective Policy Implementation* (Lexington, Mass.: D. C. Heath, 1980), especially chap. 1. See also Paul Culhane, "Bureaucratic Politics Theory and the Open Systems Metaphor" (Paper presented at the 1978 Annual Meeting of the Southwestern Political Science Association, Houston, April 1978).

36. See Leonard Rutman, ed., *Evaluation Research Methods: A Basic Guide* (Beverly Hills: Sage, 1977), chap. 1; and Theodore Lowi, *The End of Liberalism* (New York: W. W. Norton, 1969). For examples in which vague and/or inconsistent objectives hampered implementation, see Jerome Murphy, "Title I of ESEA," *Harvard Educational Review* 41 (1971), pp. 35–63; and Pressman and Wildavsky, *Implementation*, pp. 25–26, 71–77.

37. For a discussion of Los Angeles, see George Hagevik, *Decision-Making in Air Pollution Control* (New York: Praeger, 1970), chap. 5; and Eli Chernow, "Implementing the Clean Air Act in Los Angeles," *Ecology Law Quarterly* 4 (1975), pp. 537–581. For excellent discussions of the need for sound causal theory, see Pressman and Wildavsky, *Implementation*, Preface and chap. 7; Eugene Bardach, *The Implementation Game* (Cambridge, M.I.T. Press, 1977), pp. 250–252; and Paul Berman, "The Study of Macro- and Micro-Implementation," *Public Policy* 26 (Spring 1978), p. 163.

38. For discussions of implementation that, in our opinion, neglect statutory variables, see Berman, "Micro- and Macro-Implementation Process"; Bardach, *The Implementation Game*; and Donald Van Meter and Carl Van Horn, "The Policy Implementation Process," *Administration and Society* 6 (February 1975), pp. 445–488.

39. Implementing institutions are most likely to be supportive of statutory objectives if they are new agencies developed specifically to administer this program or if they are existing agencies who view this program as consistent with their traditional mandate but who feel a need to expand their program base. This, of course, assumes that organizations have policy biases and that it is very difficult to overcome bureaucratic biases; see Richard Elmore, "Organizational Models of Social Program Implementation," *Public Policy* 26 (Spring 1978), pp. 199–216; and Herbert Kaufman, *The Limits of Organizational Change* (University, Ala.: University of Alabama Press, 1971). For an example of considerable debate over the choice of an implementing agency, see Zigurd Zile, "A Legislative-Political History of the Coastal Zone Management Act of 1972," *Coastal Zone Management Journal* 1 (1974), pp. 235–274.

40. The concept of veto/clearance points is taken from Pressman and Wildavsky, *Implementation*,

chap. 5. A veto point occurs when an actor has the capacity to substantially frustrate and/or to veto the compliance of target groups with statutory objectives. In a regulatory program, one must sum the veto points involved in rule promulgation, rule adjudication, and enforcement. In principle, however, resistance from specific veto points can be overcome if sanctions and/or inducements are great enough; see William Gamson, *Power and Discontent* (Homewood, Ill.: Dorsey, 1968), chap. 3; and Harrell Rodgers and Charles Bullock, *Coercion to Compliance* (Lexington, Mass.: D. C. Heath, 1976), chaps. 2, 4.

41. For an example of an otherwise carefully drafted statute whose inadequate financial provisions caused substantial implementation difficulties, see Richard Weatherly and Michael Lipsky, "Street Level Bureaucrats and Institutional Innovation: Implementing Special Education Reform," *Harvard Educational Review* 47 (May 1977), pp. 180–192.

42. Biasing the decision rules toward stringent regulation can be accomplished in a number of ways, including (a) assigning the burden of proof to applicants, (b) requiring a two-thirds vote on important permits, and (c) assigning final authority to the agency (or agency subunit) most supportive of the statute. On the other hand, the Interstate Compact creating the Tahoe Regional Planning Agency created strong obstacles to change; see Costantini and Hanf, *The Environmental Impulse and Its Competitors.*

43. Whereas members of target groups (the regulated) normally do not have problems with legal standing, the beneficiaries of most consumer and environmental protection legislation often do not have a sufficiently direct and salient interest at stake to obtain legal standing and/or to bear the costs of appealing adverse agency decisions to the courts. Thus, statutes, such as the 1970 Clean Air Amendments, that provide liberal rules of standing for statutory supporters can have a significant effect on implementation. See Joseph Sax, *Defending the Environment* (New York: Knopf, 1970), chaps. 3–4; Karen Orren, "Standing to Sue: Interest Group Conflict in the Federal Courts," *American Political Science Review* 70 (September 1976), pp. 723–741; Bruce Kramer, "Economics, Technology, and the Clean Air Amendments of 1970: The First Six Years," *Ecology Law Quarterly* 6 (1976), pp. 161–230; and Marc Mihaly, "The Clean Air Act and the Concept of Nondegradation: Sierra Club v. Ruckelshaus," *Ecology Law Quarterly* 2 (Fall 1972), pp. 801–836.

44. Marver Bernstein, *Regulating Business by Independent Commission* (Princeton: Princeton University Press, 1955), chap. 3; Anthony Downs, *Inside Bureaucracy* (Boston: Little, Brown, 1967), chaps. 2, 8, 9. For some ambivalent evidence, see Kenneth Meir and John Plumlee, "Regulatory Administration and Organizational Rigidity," *Western Political Quarterly* 31 (March 1978), pp. 80–95.

45. For discussions of leadership and illustrations of its importance, see Hugh Heclo, *A Government of Strangers* (Washington, D.C.: Brookings, 1977), chaps. 5–6; Francis Rourke, *Bureaucracy, Politics, and Public Policy,* 2nd ed. (Boston: Little, Brown, 1976), pp. 94–101; Richard Bolan and Ronald Nuttall, *Urban Planning and Politics* (Lexington, Mass.: D. C. Heath, 1975); Eugene Bardach, *The Skill Factor in Politics* (Berkeley: University of California Press, 1972); Andrew McFarland, *Power and Leadership in Pluralist Systems* (Stanford: Stanford University Press, 1969), chap. 8; Phillip Selznick, *Leadership in Administration* (New York: Harper & Row, 1957); and Victor Vroom and Philip Yetton, *Leadership and Decision-Making* (Pittsburgh: University of Pittsburgh Press, 1973).

46. Anthony Downs, "Up and Down with Ecology—The Issue-Attention Cycle," *Public Interest,* Summer 1972, pp. 38–50; Wilson, "The Politics of Regulation," pp. 135–168.

47. For the general argument, see Paul Sabatier, "Social Movements and Regulatory Agencies: Toward a More Adequate View of 'Clientele Capture,'" *Policy Sciences* 6 (Fall 1975), pp. 301–342; and Bernstein, *Regulating Business by Independent Commission,* chaps. 3–8. Much the same process seems to occur in programs designed to help the poor; see Murphy, "Title I of ESEA," pp. 37–38, 61–62; and Frederick Lazin, "The Failure of Federal Enforcement of Civil Rights Regulations in the Public Housing," *Policy Sciences* 4 (September 1973), pp. 263–274.

48. The concept of a "fixer" is taken from Bardach, *The Implementation Game*, pp. 268–283; an example would be Senator Edmund Muskie with respect to federal pollution control policy. For the argument concerning declining support among legislative sovereigns over time, see Barry Weingast, *A Positive Model of Public Policy Formulation*, Working Paper No. 25 (St. Louis: Center for the Study of American Business, 1978).

   In the long term, however, regulatory objectives mandating a substantial departure from the *status quo* can probably be attained only if police power regulation is supplemented by a variety of side-payments (e.g., subsidies and tax incentives) to the most vociferous or adversely affected members of target groups.

49. For examples of the effects of variation in local support on the implementation of federal programs, see Murphy, "Title I of ESEA," pp. 35–63; Rodgers and Bullock, *Coercion to Compliance*, chaps. 2–4; Paul Berman and Milbrey McLaughlin, "Implementation of Educational Innovation," *The Educational Forum* 40 (March 1976), pp. 345–370; Robert Thomas, "Intergovernmental Coordination in the Implementation of National Air and Water Pollution Policies," in *Public Policy Making in a Federal System*, ed. Charles Jones and Robert Thomas (Beverly Hills: Sage, 1976), pp. 129–148; Lieber, *Federalism and Clean Waters*, chaps. 6–7; Berman, "Macro- and Micro-Implementation," pp. 168–179; Elmore, "Organizational Models of Social Program Implementation," pp. 199–216; Jeffrey Pfeffer and Gerald Salancik, *The External Control of Organizations* (New York: Harper & Row, 1978); and Helen Ingram, "Policy Implementation Through Bargaining. The Case of Federal Grants-in-Aid," *Public Policy* 25 (Fall 1977), pp. 499–526.

50. Louis Jaffe, *Judicial Control of Administrative Action* (Boston: Little, Brown, 1965); Edward White, "Allocating Power Between Agencies and Courts," *Duke Law Journal*, 1974 (April 1974), pp. 195–244.

51. See Jane Jacobs, *The Economy of Cities* (New York: Random House, 1969).

52. This is really a corollary of the general theory of cognitive dissonance; see Roger Brown, *Social Psychology* (New York: Macmillan, 1965), pp. 584–604. Some evidence for negatively skewed participation comes from the legislative oversight literature; see L. C. Mainzer, *Political Bureaucracy* (San Francisco: Scott, Foresman, 1973), pp. 82–83.

53. For a brief introduction to the genesis and content of the 1972 Coastal Initiative, see Peter Douglas, "Coastal Zone Management—A New Approach in California," *Coastal Zone Management Journal* 1 (Fall 1973), pp. 1–25. This topic will be discussed in considerably greater detail in chapter 2.

54. For an excellent discussion of the need for a counterfactual (i.e., a reasonable basis of comparison), see A Myrick Freeman, *Environmental Management as a Regulatory Process* (Washington, D.C.: Resources for the Future Discussion Paper D-4, January 1977).

55. The reasoning here is fairly straightforward: First, the creation of a new agency provided the opportunity for a large infusion of new personnel. Second, the commissions were formed only after a very hard-fought initiative campaign (see chapter 2); supporters would thus be highly motivated to fill those new staff positions. Finally, the formula for appointing regional and, especially, state commissioners meant that most would probably be supporters of the Initiative and thus that the criteria for appointing commission staff would probably be likewise weighted toward supporters.

56. Studies of the voting behavior of members of the Tahoe Regional Planning agency had indicated that state-appointed officials in California were much less likely than their locally elected colleagues to approve potentially harmful developments (see Costantini and Hanf, *The Environmental Impulse and Its Competitors*, pp. 55–58). As we shall see in chapter 4, this same pattern was repeated on the regional coastal commissions.

57. Lipset, Trow, and Coleman, *Union Democracy*, Preface and chap. 1, especially pp. 12–13.

58. This does not presume that the commissions did *in fact* substantially achieve their statutory objectives. But just as Lipset *et al.* had a strong *prima facie* basis at the commencement of their

study for believing that the ITU was a basically democratic union, so we have strong grounds for believing that the decisions of the coastal commissions on controversial permit cases were substantially different from those of local governments and generally consistent with the act's objectives. [See Paul Sabatier, "State Review of Local Land Use Decisions: The California Coastal Commissions," *Coastal Zone Management Journal* 3 (1977), pp. 255–290.] The actual extent to which applicants complied with those decisions and the extent to which the commissions achieved their objectives will be discussed in chapters 6 and 9.

THE CALIFORNIA COASTAL COMMISSIONS
Genesis and a Preliminary Implementation Assessment

> *Doom broods over wild California: It will be paved from end to end, and parking*
> *spaces painted all over it. This is a part of the California feeling—this certainty of*
> *irreparable loss, of having gotten here almost too late—and somehow the tension*
> *sharpens your awareness of what remains.*
> Peter Beagle and Michael Bry,
> *The California Feeling,* p. 13

The California coast stretches approximately 1100 miles from the Mexican border to Oregon.[1] A visitor driving the southernmost 300 miles of the coastal highway in 1971 would have been struck, on the one hand, by the beauty of an almost unbroken string of excellent swimming and surfing beaches and, on the other, by the rapidly expanding urban development along most of the coastline. Driving north from Tijuana, one soon reaches San Diego with its large naval base and exclusive residential areas along the coast. The northern part of the county contains picturesque flower farms and some excellent beaches, although much of it is under the jurisdiction of the Camp Pendleton Marine Base. Portions of southern Orange County are still undeveloped, although much of the coast is dominated by wealthy residential enclaves and crowded public beaches. Los Angeles County is heavily urbanized but varies in character from the industrial ports of Long Beach and Los Angeles to moderate-income residential areas in Venice and Long Beach to massive condominiums at Marina del Rey to the very exclusive residential areas of Palos Verdes and Malibu. The drive north to Santa Barbara passes through several moderate-sized cities and military bases, the agriculturally rich Oxnard Plain, and several state beaches—some of them dotted with oil wells.

North of Santa Barbara, the coast becomes much more rural. One passes some popular surfing beaches before the highway turns inland, leaving the coast to the ranches around Point Conception and to Vandenburg Air Force Base. The coast is rejoined at Pismo Beach and again at Morrow Bay on the outskirts of San Luis Obispo. North of here one travels for miles through ranches that slope gradually to the sea before the coastal highway rises to the spectacular bluffs of

FIGURE 2-1A.   The Southern California coast. Photograph: Access restrictions, Orange County.

FIGURE 2-1A. (*Continued*)

the Big Sur. After passing through Monterey, the drive around Monterey Bay is notable chiefly for the Moss Landing power plant surrounded by artichoke fields. After the university town of Santa Cruz, there is another 100-mile stretch of ranches and coastal bluffs before reaching San Francisco. North of the Golden Gate, one travels for almost 300 miles through ranches, magnificent coastal bluffs, and occasional small villages and second-home subdivisions before again turning inland, leaving the coast to the Douglas fir and the redwoods. The ocean is rejoined at the towns of Eureka and Arcata on Humboldt Bay and then followed for another 150 miles through redwoods and coastal plain, interrupted only by the small town of Crescent City, until one finally reaches the Oregon border (see accompanying pictures and map in Figures 2-1A and 2-1B).

Because the California coast in 1971 was a magnificent aesthetic, recreational, and commercial resource by any standard, it was subject to enormous demands for often conflicting uses. Should wetlands be left in their natural state to serve as wildlife habitat or should they be turned into boat harbors, ports, and residential areas? Should the coastal bluffs and beaches of Malibu in northern Los Angeles County become the exclusive preserve of the very wealthy or should this area be available as a recreational resource to the remaining 7 million people in the county? Should the magnificent coast of Northern California (exclusive of San Francisco) retain its wild character with majestic vistas of mountain meeting ocean or should it be dotted periodically with second-home subdivisions? Should power plants be placed along the coast where water is plentiful and adverse impacts on air quality are minimal—even though the effects on water habitat and on aesthetic resources are often considerable—or should they be placed on inland waterways, where competition for air and water resources is considerably

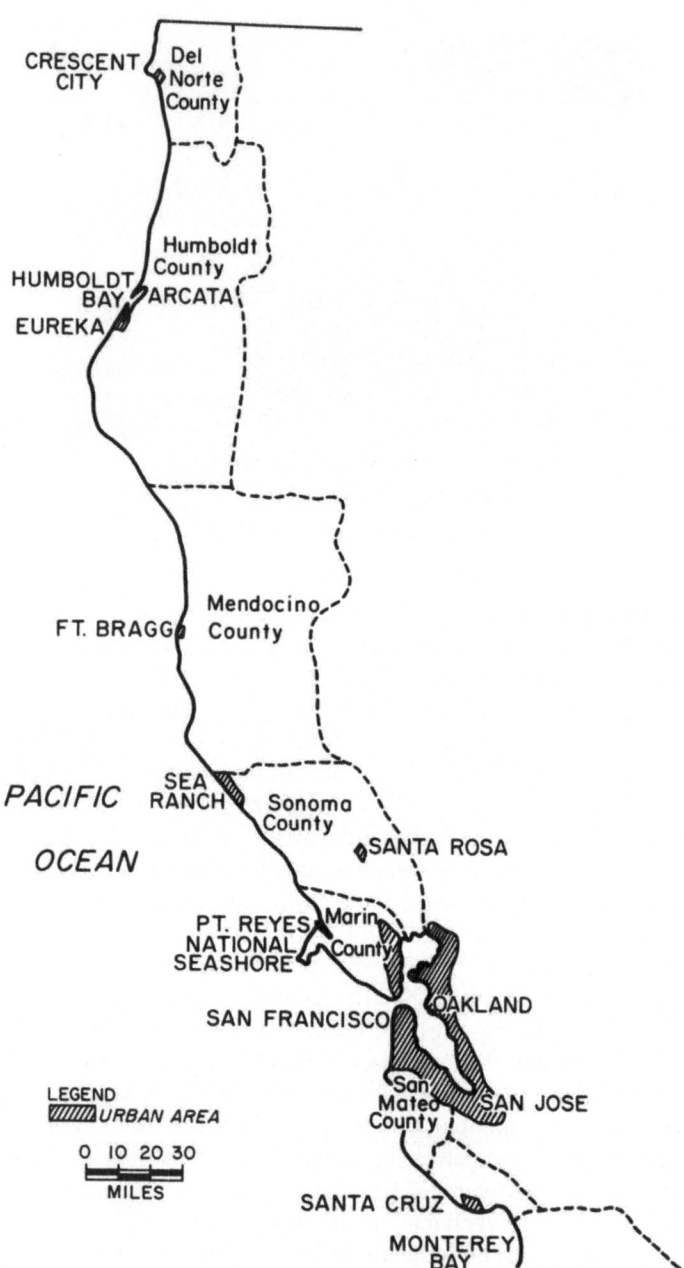

FIGURE 2-1B.   The Northern California coast. Photographs (top): Mendocino Coast, just north of
Sea Ranch; (bottom): Marin Coast, between Pt. Reyes and San Francisco.

FIGURE 2-1B.   (Continued)

greater? Should the extensive oil and gas reserves on the continental shelf off Southern California be developed, even at the increased risk of oil spills and pumping platforms visible from crowded beaches?

It was to resolve conflicts such as these that the people of California established the coastal commissions in November 1972. The first part of this chapter discusses some of the major perceived problems with coastal land use patterns in the early 1970s, as well as some alternative institutional arrangements available to critics. The next section examines the efforts of environmental groups and other critics to obtain the passage of legislation transferring some authority over coastal land use from local to state institutions, first unsuccessfully through the legislature and ultimately through the initiative process. The final section examines in some detail the substance of the 1972 Coastal Initiative and other factors that have led us to expect that it would be rather effectively implemented.

I. Perceived Coastal Land Use Problems and Possible Solutions, c. 1971–1972

Conflicts over the proper utilization and allocation of coastal resources were rooted ultimately in the large and rapidly growing population along the coast. In 1970 California's 15 coastal counties had a population of 12.8 million people, or approximately 63% of the state's total of 20 million. Within these counties, the population that was ultimately to be included in the planning jurisdiction of the coastal commissions (from the coast to roughly 5 miles inland) was 4.2 million— or greater than that of 34 states. About 82% of this population was concentrated in the five southernmost counties (San Diego, Orange, Los Angeles, Ventura, and Santa Barbara). Densities in the coastal (planning) zone of these southern counties averaged 16,896 persons per mile of coastline, which stood in stark contrast to the 286 persons per mile of coast in the five counties to the north of San Francisco (Marin, Sonoma, Mendocino, Humboldt, and Del Norte).[2] Furthermore, the population of the 15 coastal counties increased 26% in the 1960–1970 period, while that of the coastal planning area grew at an even greater rate, 31%. Population increases were highest in the five southernmost counties, while San Francisco and the three northernmost counties in the North Coast region showed very slight declines during this period.[3]

The large and growing population in the coastal counties created demands for more housing and employment opportunities. Population increases resulted in increased demands for energy, which meant pressures for additional power plants, refineries, and offshore drilling. People also needed transportation, which, in California, generally meant highways. They needed recreational opportunities, of which the coast was one of the most important. Finally, population growth—much of it young married couples with children—meant greater demands for schools, fire and police protection, and a variety of other governmental services and the tax revenues to pay for them.

These changes could be seen most dramatically in places like Orange County, once an agricultural plain running between Los Angeles and San Diego, and the last remaining substantial amount of open space (other than Camp Pendleton) along the entire Southern California coastline. In the years between 1960 and 1970, Orange County's population doubled, from 719,500 to 1,432,900 people. In the process, vast amounts of land were converted from citrus groves and ranches to residences and factories and shopping centers, until by the mid-1970s only portions of the once-vast Irvine Ranch remained as open space along the coast. In fact, in 1970 slightly over 110,000 people lived within

1000 yards of the coast, or an average of 2900 people per mile within this very limited area of the county.[4] As might be expected, the residents of what would become the (1000-yard) permit zone under the coastal commissions had income and educational levels slightly above the county average and considerably above the average for the state as a whole.[5] This growing population within the county had some significant impacts on coastal resources, including the construction of a coal-fired (and eventually a nuclear) power plant at San Onofre and considerable construction within and adjacent to the county's wetlands, including a major controversy involving Upper Newport Bay.[6]

The demands on coastal resources created by this growing population in Orange and the other 14 coastal counties created considerable unease among environmental groups, planners, civic associations such as the League of Women Voters, legislators, and many other people, both among political elites and among the general public.[7] There were, however, limits to what could be done. Population growth in coastal counties and the resultant pressures to develop along the coast could not easily be reversed. Housing starts, commercial developments, and the accompanying infrastructure might be restricted in the immediate coastal area and infilling might be emphasized to preserve some open space, but it was unlikely that the basic pattern of intensive urban development could fundamentally be altered, at least not in the southern portion of the state and the urbanized areas of the north. Much of the coastal area—particularly in San Diego, Orange, Los Angeles, and San Francisco counties—was already very intensively developed, and public purchase of the remaining open space was completely beyond the financial capability of the state. Nevertheless, there were a number of specific substantive and institutional problems over which there was widespread dissatisfaction and that could feasibly be addressed. It was these issues that eventually led to the formation of the coastal commissions.

First, probably the most widespread concern among the general public was a general unease that the beauty of the coast was being eroded by the rate and type of development. An August 1972 poll of the state's electorate revealed that 96% of respondents were concerned with what was happening to land in California. When asked the nature of their concerns, over 60% mentioned such things as "overdevelopment," "too much commercial and/or residential development," "preservation of natural environment or open spaces," etc. With regard to the coast, the same poll indicated that by far the most widespread perceived benefit of stopping or slowing coastal development was the preservation of the natural beauty of the coast—volunteered by 44% of respondents.[8] More concretely, several of the people who became active in the Proposition 20 (Coastal Initiative) campaign mentioned a large Holiday Inn constructed on the shores of Monterey Bay in the early 1970s and the Sea Ranch, an enormous second-home subdivision stretching for 10 miles along the rugged Sonoma County coast with a projected buildout of 5200 residences. Other issues included a large residential development in Santa Barbara County, which was defeated in November 1970 by the local electorate, and the deleterious effects of petroleum development and power plants on the scenic resources of the coast.[9]

A second major concern was public access to the coast. By California law, all land seaward of the mean high tide line is in the public domain, subject to disposition by the legislature.[10] As residential and commercial development along the coast increased, however, the public's ability to reach the wet sand beaches was impaired. In the Los Angeles area, this was graphically illustrated by Malibu—an extremely exclusive 27-mile string of houses between the coastal highway and the ocean—in which both lateral and vertical access to the wet sand had been severely restricted by residents. In the north, essentially the same issue was illustrated by the Sea Ranch, which had greatly restricted public access to 10 miles of coast and had been the direct catalyst for the formation of COAAST—a local environmental group that subsequently waged a 3-year campaign to require public access easements in coastal subdivisions and eventually was one of the principal components of the California Coastal Alliance.[11]

Another aspect of public access involved the shortage of public beaches and campgrounds. Although an appreciable 47% of the coast in the early 1970s was publicly owned, most of the federal land was set aside for military and other nonrecreational purposes.[12] This left about 25% of the mainland coast, or 263 miles, legally available for public access.[13] This rather impressive figure masked, however, several important problems. First, the demand for coastal recreation was increasingly outstripping the supply. Thus, in the 1965–1975 period, while new coastal park land acquisitions increased about 18%, visitations to state-operated coastal sites increased over 90% (from 17 million to 33 million visitor days annually). This resulted in campgrounds at most facilities being filled to capacity over 80 nights each summer, with many campgrounds fully reserved months in advance.[14] Second, there was a basic mismatch between the area of largest population and demand for facilities in the south and the concentration of coastal parks in the north. One therefore finds that in 1973 in the sparsely populated North Coast region the state owned 230 thousand acres of oceanfront property, and in the moderately populated Central and South Central regions this rose to 247 thousand and 241 thousand acres, respectively. Yet in the North Central region, which included the San Francisco Bay area, the figure dropped to 176 thousand acres. In the rapidly growing San Diego area—the second most populated coastal region—the state owned only 91 thousand acres. And as for the most populous South Coast region, the state owned a mere 110 thousand acres.[15] Third, public beaches in Los Angeles and Orange counties usually did not have sufficient on-site parking to handle the million or so users on summer weekends. While users had previously parked in the adjacent neighborhoods, this was becoming increasingly difficult because of the gradual conversion of many neighborhoods to high-density commercial and apartment complexes.[16] Not surprisingly, therefore, in the August 1972 public opinion poll mentioned previously, the second most important perceived benefit of restricting coastal development was the possibility of obtaining additional public recreation areas and campgrounds (mentioned by 12% of respondents).

A third source of dissatisfaction with the prevailing patterns of coastal land use during this period was its effect on wildlife habitats, particularly wetlands. While this was apparently a minor issue in the eyes of the general public, it was an important motivating factor for many of the environmental activists who provided much of the resources in the Proposition 20 campaign.[17] Like the other issues, this one involved a number of specific concerns, ranging from oil spills to thermal pollution from power plants to the cumulative erosion of wetland habitat. A 1971 report by an official of the California Department of Fish and Game estimated a 67% reduction in coastal wetland habitat statewide and over 75% in southern California since the turn of the century because of the construction of commercial and pleasure craft harbors and general commercial and residential development. Seventy percent of the remaining wetlands are in San Francisco Bay.[18] Specific controversies during this period included large oil spills in the Santa Barbara Channel and San Francisco Bay, a large residential development proposed for Upper Newport Bay in Orange County, and the gradual erosion of the mudflats and other habitat in Humboldt Bay between Eureka and Arcata, one of the most important waterfowl habitats along the Pacific flyway.[19]

In addition to these three substantive issues, which fueled much of the dissatisfaction with coastal land use "policy" in the early 1970s, there were two institutional problems that were widely perceived (at least among political elites) to be at the root of much of the problem. The first was the generally *laissez faire* and prodevelopment bias of most local governments arising from their dependence upon property and sales taxes for most of their revenue, as well as the prominent role of the construction industry and building trades unions in financing local political campaigns. This view was graphically expressed by Janet Adams, leader of the Coastal Alliance[20]: "Conservationists involved in the fight to save the coast came to regard local government as the enemy and an obstacle to reason. So long as coastal cities, towns, and counties were forced to rely for their financial base upon the property tax dollar, the California coast was fair game for unrestrained and irreversible commercial development." Although California had passed enabling legislation regarding zoning, the development of general plans, provisions for open space easements, etc., these statutes all left tremendous discretion in the hands of local governments, which, in turn, were very reluctant to use them to assure public access to state-owned tidelands (wet beaches), to protect water quality and wildlife habitat, or to prevent the natural tendency of the market to allocate valuable coastal property to intensive uses impairing the natural beauty of the coast.[21] Whatever the specific grounds for dissatisfaction, the alleged deficiencies of local governments were apparently widely shared among the general public. An August 1972 poll of the state's electorate revealed that only 30% of respondents felt that planning and development of coastal areas should be controlled by cities and counties, while 11% favored a regional approach and 51% felt that principal authority should reside in state government.[22]

The other perceived deficiency of existing governmental institutions was the lack of coordination among the multiplicity of local, state, and federal agencies that affected the coast. Although estimates of the precise number of institutions varied, several writers during this period cited the approximately 15 county, 45 city, 42 state, and 70 federal agencies with jurisdiction over some aspect of coastal resources and called for a more coordinated approach that would deal in an integrated fashion with the coast as a special resource.[23] Part of the problem was single-purpose public works agencies, such as highway departments and wastewater treatment districts, which tended to build facilities to meet projected increases in demand with little consideration of the effects of such demand on coastal resources.[24] There was also considerable concern with a whole variety of spillover effects where the actions of one jurisdiction affected the citizens of another. For example, development deemed desirable by one local government because of its contribution to the local tax base often restricted access to a beach used by citizens of adjacent jurisdictions.

In short, the perceived deficiencies in land use practices—i.e., the Miami Beachization of the coast, inadequate public access, and destruction of wetlands—could be related to deficiencies in which the private market, local governments, and existing state agencies made decisions concerning the utilization of limited coastal resources.

Fortunately, critics of the existing practices could point to a number of alternative institutional arrangements. The late 1960s and early 1970s was a period of rapid rise in environmental concern and, with it, the passage of numerous state statutes protecting wetlands and other critical habitat areas.[25] On a more general level, this period witnessed what has been termed "a quiet revolution in land use control," as numerous states passed statutes increasing the scope of governmental regulation of land use and transferring authority for such regulation from local to regional and/or state institutions.[26] This trend also received considerable support in Congress, where several bills encouraging a more active state role in land use planning and particularly in coastal zone management passed at least one house during the early 1970s.[27]

Closer to home, a number of regional and/or state land use institutions had recently been established in California that provided possible models for critics seeking a greater state voice in coastal resource decisions.[28] By far the most important of these was the San Francisco Bay Conservation and Development Commission (BCDC), established by the legislature in 1965 on an interim basis after extensive grass roots opposition to the gradual erosion of the San Francisco Bay shoreline through dredging and filling. The 1969 legislature approved

BCDC's plan for the bay and made permanent its permit authority over all developments within 100 feet of the shoreline. The agency was widely regarded as a success, and many of its features—including the mixture of local government representatives and state-appointed officials on the governing board, its permit authority over developments in the near-shore area, and its creation for a limited period sufficient to prepare a general plan with continuation dependent upon legislative approval—were incorporated into the 1972 Coastal Initiative.[29] Other precursors of the coastal commissions included the system of state and regional water quality control boards established by the legislature in 1970 and the Tahoe Regional Planning Agency (TRPA), established as a bistate compact (between California and Nevada, with congressional approval) in 1969. The former provided the drafters of the Coastal Initiative with a reasonably successful model of state and regional boards, while a few years' experience with the TRPA suggested that little change in land use patterns would result if effective control remained in the hands of local elected officials.[30]

Armed with these alternative models, critics of coastal land use practices turned first to the legislature and eventually to the initiative process for a solution.

II. The Struggle for Coastal Regulation: Frustration in the Legislature 1970–1972

Beginning in 1964 with the Governor's Conference on California and the World Ocean, concern by state officials with the use of coastal resources began to surface.[31] The result was a series of advisory committees, reports, and legislative hearings over the next 8 years, culminating in publication of the Comprehensive Ocean Area Plan (COAP) in May 1972. Though COAP made some attempt to develop policy guidelines, these were exceedingly vague and the 1967 enabling legislation contained no regulatory authority. COAP did, however, result in the gathering of a large amount of data on coastal resources that was subsequently utilized by the commissions established by the 1972 Coastal Initiative.

The first serious efforts to establish state/regional review of local land use decisions occurred during the 1970 session of the legislature.[32] Four bills were introduced, all of them modeled to some extent after BCDC. Three of them (AB 730, Sieroty, Dem-Los Angeles; AB 640, Milias, Rep-Santa Clara County; and SB 371, Nejedly, Rep-Contra Costa County) would have created a state coastal commission and several regional commissions with large (30–40) memberships more or less equally divided between local officials, on the one hand, and state officials and "public" members, on the other. The state commission would have been primarily responsible for preparing a coastal plan (to be eventually submitted to the legislature for approval) and for interim permit review over local government's land use decisions. The commissions' jurisdiction generally extended approximately 5 miles inland for planning purposes and 1 mile from the shore for permit review, although several of the bills excluded BCDC and most urbanized areas. In the Republican-controlled legislature, however, all three of these conservationist-oriented bills were killed in the first policy committee. In contrast, AB 2131 sponsored by Assemblyman Pete Wilson (Rep-San Diego) left much more authority in the hands of local governments and thus had a better chance of passage. It would have created a state coastal commission charged with preparing planning guidelines for local governments, but principal authority would have remained in local hands. Moreover, planning and permit jurisdiction was limited to 1000 yards from the mean high tide line, with urban areas and petroleum reserves excluded from the commissions' authority. Wilson's bill was approved by the Assembly but killed in the Senate Government Operations Committee on the last day of the session.

Defeat of comprehensive coastal legislation during the 1970 session was generally attributed to the lack of support from Governor Reagan and to the absence of any concerted effort on the part of environmental groups, whose attention during this session was focused primarily on water pollution control and environmental impact statement legislation. Nevertheless, one coastal-related bill was approved. Culminating a 3-year effort launched by Bill Kortum of COAAST after county approval of the Sea Ranch, the legislature approved AB 493 by Assemblyman John Dunlap (Dem-Napa County) requiring coastal subdivisions to provide "reasonable" access to the public.

While proponents of active state oversight of coastal land use decisions had not been notably successful during the 1970 session, a number of events during the winter of 1970–1971 gave them considerable grounds for optimism. First, the November 1970 elections saw Democrats gain control of both the Assembly and the Senate. Thus, conservationists could now expect not only greater diffuse support within the legislature but also the active support of the new leadership, Assembly Speaker Bob Moretti (Dem-Los Angeles) and Senate Pro-Tem James Mills (Dem-San Diego).[33] Second, the absence of any coordinated strategy among conservationists during the 1970 session was apparently remedied with the formation of the Coastal Alliance by an impressive group of people, including Assemblymen Sieroty and Dunlap, James Pardau (the chief consultant to the Assembly Natural Resources Committee), E. Lewis Reid (formerly minority counsel to the U.S. Senate Interior Committee), Janet Adams (an advertising and campaign executive who had been active in the formation of BCDC), and Bill Kortum. Although Kortum was named the first president, the guiding force behind the organization was Adams, who was named the executive secretary. The Alliance was formed as an umbrella organization for coastal conservation efforts and as a vehicle for organizations and individuals who did not wish to work through the two major statewide environmental organizations, the Sierra

Club and the Planning and Conservation League (PCL). By March 1971 the Alliance included 34 organizations among its members, including (at least informally) both the Sierra Club and the PCL.[34] Third, all of the conservation organizations agreed that coastal legislation to be carried by Sieroty would be their first priority during the 1971 session of the legislature.

The 1971 session witnessed a plethora of coastal bills, all of which provided less state review than the conservationist bills of the previous session.[35] Assemblyman Wilson introduced legislation similar to the bill previously approved by the Assembly except that it now contained a majority of local officials on the state commission. It soon became clear, however, that his bill would not be favorably received in a Democratic legislature, and he eventually supported the Coastal Alliance bill carried by Sieroty (AB 1471). That legislation—which was cosponsored by Dunlap and by Speaker Moretti, together with a companion bill carried by Senator Alquist (Dem-San Jose)—was very similar to the 1972 Initiative eventually approved by the voters. It created a state coastal commission composed of representatives of the six regional commissions and six public members appointed by state officials. The regional commissions had 12–14 members, equally divided between local elected officials and public members appointed by the governor, the Assembly speaker, and the Senate Rules Committee. The commissions had two principal functions: first, to prepare a comprehensive coastal plan to be presented to the legislature and, second, to regulate all development in the interim. While the commissions' planning jurisdiction remained essentially a 5-mile stretch inland from the coast, the permit zone was reduced (in comparison with the Sieroty bill of the previous session) from 1 mile to 1000 yards from the mean high tide line. Not surprisingly, it was vigorously opposed by realtors, land developers, industrial organizations, most labor unions, utilities, and most local governments. After amendment on the floor to provide an additional local official on each regional commission, it was approved by the Assembly in September.

The Senate again, however, proved to be a graveyard for coastal legislation. Given the more conservative nature of the upper house, Senator Alquist—who was carrying the Senate counterpart to the Sieroty bill—decided to combine his bill with one carried by Senator Grunsky (Rep-Monterey County), which relied more heavily on local governments in the planning process and restricted interim permit review to projects directly affecting public beaches, view corridors to the ocean, and water quality. The combined bill—developed in conjunction with the Planning and Conservation League and some members of the Senate Natural Resources Committee—would have restricted the planning jurisdiction to 1000 yards, excluded urban areas, and added six local elected officials to the state commission (which now had 18 members: 6 "public," 6 regional representatives, and 6 locals). While this compromise obtained the support of the PCL and local government associations, it was vigorously opposed by the Coastal Alliance, the

Sierra Club, the realtors, and the Chamber of Commerce. On virtually the same day that the Sieroty bill was passing the Assembly, the Alquist-Grunsky bill—as well as two other bills supported by the construction industry—was defeated by the Senate Natural Resources Committee. This still left the Sieroty bill (AB 1471), which, it will be recalled, now had a majority of local officials on the regional commissions. In an effort to win approval, it was further amended to exclude several harbors and the cities of San Francisco and Los Angeles; in addition, policy statements throughout the bill now sought to balance conservation and development objectives. Nevertheless, on November 16 the bill died in the Senate Natural Resources Committee on a 4–4 vote—primarily because of opposition from developers, title companies, realtors, and the Reagan administration.[36]

Supporters of coastal legislation were by now quite discouraged and had informally concluded that a significant state role would probably require circumventing the legislature via the initiative process. In consultation with legislative supporters, however, they decided to give the legislature one more chance. Thus, a bill very similar to the original Sieroty bill of the 1971 session was jointly introduced in 1972 by Sieroty and by Senator Grunsky with a total of 57 cosponsors (42 assemblymen, 15 senators). It provided for six regional commissions (with members equally divided between state appointees and local elected officials) and a state commission with planning authority over a 5-mile zone and interim permit review over all development in the 1000-yard area inland from the mean high tide line. The only significant alteration was in the funding process; rather than a state appropriation, funds were now to come from a $5 million allocation from the Bagley Conservation Fund. But 1972 was simply a repeat of 1971. The Sieroty bill easily passed the Assembly, but even a very watered-down version was defeated in the Senate Natural Resources Committee on a 4–4 vote—with lobbyists from the building trades unions and the Chamber of Commerce taking credit for its demise.[37]

III. The Struggle for Coastal Regulation: The 1972 Coastal Initiative

Fortunately for conservationists and other supporters of coastal protection, Article IV of the California Constitution established the initiative process and thereby provided an alternative to the multiple vetoes of the legislature. This required, however, (1) that a bill be drafted, (2) that the necessary signatures be obtained to place it on the November ballot as a statutory initiative, and (3) that the initiative proposition obtain the approval of a majority of the electorate in November.

The first task was easily accomplished. The 1972 Sieroty-Grunsky bill had been drafted with the purpose of using it as an initiative if the legislature failed to substantially approve it. All concessions made during the 1972 session were simply deleted and the original version used as the initiative version. The second task was considerably more difficult. Because supporters had agreed to give the legislature one last chance, they had only about a month's time after the May 15 Senate committee vote to obtain the 325,000 signatures necessary to qualify it for the November 1972 ballot. Nevertheless, the 2½-year campaign to pass coastal legislation had resulted in fairly widespread public awareness of the issue and a rather substantial number of 100 or so organizations loosely grouped under the coordinating umbrella of the Coastal Alliance—not to mention the thousands of Sierra Club members in local chapters throughout the state—that were potentially available to circulate petitions. While the Coastal Alliance as an organization was responsible for printing the petitions, a wide variety of local and state organizations—many of them loosely affiliated with the Alliance—actually gathered the signatures. In Los Angeles, San Diego, and Monterey counties, it was primarily the Sierra Club chapters. In San Jose and in Orange County, members of the League of Women Voters were extremely active during the petition drive. In Eureka, Berkeley, and Santa Cruz, local organizations were primarily responsible for gathering signatures. As a result of this flurry of volunteer activity, on June 19 the secretary of state announced that 418,000 valid signatures had been obtained (from 47 of the state's 58 counties) and thus that Proposition 20, the Coastal Initiative, had qualified for the November ballot.[38]

This relative ease in qualifying the Coastal Initiative for the ballot masked, however, considerable uncertainty concerning its fate at the hands of the electorate. On the one hand, statewide public opinion polls conducted prior to the actual start of the campaign indicated a rather substantial plurality in favor of Proposition 20. For example, an August 1972 poll conducted by Whitaker and Baxter (the opposition's campaign firm) found 59% in favor of Proposition 20, 13% opposed, and 28% undecided.[39] A Field poll conducted the same month obtained similar results: 62% in favor, 14% opposed, and 25% undecided.[40]

On the other hand, almost 40% of respondents in the Whitaker-Baxter poll had no idea of the content of the Initiative and 55% were unaware that the Coastal Initiative would appear on the November ballot. Moreover, two recent environmental initiatives that had begun with similar reservoirs of widespread but rather uninformed public support had been defeated at the polls by rather substantial margins after heavy media campaigns by the opposition. This pattern was aptly summarized by Mervin Field:

> An initiative ballot proposition as it is first presented to voters in summary form appears to fill a need or correct a situation in which a large segment of the public is in sympathy. Initially, while not many people fully grasp all the details and ramifica-

tions, their instinctive reaction is generally favorable. . . . Typically, the public only becomes fully aware of the opposition to the measure relatively late in the campaign, sometimes only a few weeks before Election Day. . . . And more times than not the original instinctive support of the idea is replaced by a negative view.[41]

The opposition in the Proposition 20 campaign must have been reasonably confident it could repeat this scenario, as it continued successfully to oppose a watered-down version of the Sieroty-Grunsky bill in the legislature (*supra*, note 37).

In keeping with a strategy that had worked in the past, the opposition—composed largely of land development firms, the building trades unions, oil companies, utilities, and, to a lesser extent, local government associations—hired a very successful campaign management firm, Whitaker and Baxter, to persuade voters to oppose the Initiative. The opposition campaign was heavily centralized, very well financed, and almost entirely media-oriented. Of the $1.2 million raised—much of it in large contributions from the business community—over $900,000 was spent on publicity.[42] Overall, their publicity budget was six times that of the proponents. In contrast, they had only very modest local campaigns (consisting largely of speakers' bureaus, many of them paid professionals) and very little public support from state-elected officials; in fact, virtually the only support came from two state senators (Dennis Carpenter and George Deukmejian), and even Governor Reagan generally stayed out of the fray.[43]

In point of fact, Whitaker and Baxter were confronted with a very difficult task. They had little time to plan a campaign, as all of their energies until July had been devoted to the defeat of Proposition 9, the Clean Environment Initiative. And the detailed August 1972 poll that they commissioned revealed "few vitally strong concerns around which to organize voter resistance."[44] As indicated previously, there was widespread public support for a strong state role in land use planning, at least with respect to the coast. Moreover, there was a general public perception that the Coastal Initiative was a responsible piece of legislation that would not impose a moratorium on coastal development and would not involve disastrous economic consequences if approved.[45] The only points of attack suggested by the poll involved some public concern about the adverse economic consequences of the Initiative and strong support for the protection of property rights.[46] Thus, the major slogan of the opposition campaign became "Conservation Yes, Confiscation No—Vote No on Proposition 20." But the absence of any other viable issues led Whitaker and Baxter to some other themes that subsequently backfired. These included the contention that the coastal commissions' jurisdiction would include property in the Sierra Nevada and the Sacramento-San Joaquin Valley adjacent to rivers that eventually drained into the Pacific Ocean—an argument that was publically shot down by the Republican attorney general, Evelle Younger—and the exceedingly dubious slogan, "The Beach Belongs to You, Don't Lock It Up, Vote No on Prop 20," a theme that was roundly criticized as potentially misleading and even received a mild rebuke from Governor Reagan.[47]

Despite the media blitz and outspending proponents by approximately five to one, the opposition campaign failed to convince voters that Proposition 20 should be rejected. In fact, the Coastal Initiative was approved by a solid 55% majority of the electorate. The crucial question then becomes, why was Proposition 20 approved while two previous environmental ballot propositions (and two of three subsequent ones) had been soundly defeated?[48]

One of the most important reasons was simply the presence of very widespread public support for coastal protection and for an active state role if this were to be accomplished. In fact, there is some evidence from the Whitaker-Baxter poll that even a complete moratorium on coastal development pending approval of a comprehensive coastal plan might have been approved by the voters.[49] By the same token, the Initiative was not perceived as adversely affecting important values, with the possible exception of property rights. And even that issue probably had only limited potential, as the commissions' decisions would directly affect only a very small percentage of the state's property owners. In short, the general campaign slogan of "Save the Beach" had very widespread appeal.

A second, and probably equally important, reason for Proposition 20's approval was the widespread perception that it was a reasonable and responsible effort to address an important public issue. Political elites were certainly aware that virtually the same legislation embodied in the Initiative had twice passed the Assembly by wide margins, only to be defeated in Senate committee, and there was widespread public support for the legitimacy of an initiative in situations where a few legislators were able to veto a measure with such widespread public and legislative support.[50] Moreover, Proposition 20 received unprecedented public support from legislators and other well-known public officials. It was publicly endorsed by both U.S. senators and by 60 state legislators. Senator Pro-Tem James Mills made a widely publicized bike ride from San Francisco to the Mexican border. An Assembly committee conducted well-publicized legislative hearings late in the campaign concerning the dubious advertising practices of Whitaker and Baxter, as well as possible violations of campaign contribution laws. And the list of organizations endorsing Proposition 20 grew to over 100, with campaign brochures prominently displaying the more notable ones (see Figure 2-2).[51]

Third, the proponents conducted a rather skillful campaign, despite substantial internal divisions.[52] They consistently portrayed the Initiative not as an environmental issue but rather as a public access issue (see campaign brochures in Figure 2-2).[53] They had sufficient funds to present a fairly effective media campaign, about equally divided between radio-television and newspapers-mailings-bumper stickers.[54] While they did not possess the hordes of volunteers that have sometimes been portrayed, they had sufficient volunteer support to run

California needs Proposition 20.

California's coastline is our greatest natural resource. It is probably the most beautiful coastline in the entire world. Only 263 miles of our 1072-mile coastline are open to the public. California has done less to preserve its coast than any other state in the nation.

Proposition 20 will save the coast

By requiring a state commission, and six regional commissions, to develop a plan for the entire California coast. This plan will be submitted to the Legislature in three years. During the planning period, developers must obtain a permit for development from the commission. Proposition 20 is patterned after similar legislation now working to safeguard San Francisco Bay and Lake Tahoe.

"DENNIS THE MENACE"

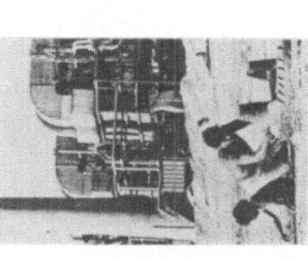

WE WENT TO THE BEACH... BUT IT WAS *GONE !*

Created specially by Hank Ketcham

103

Proposition 20 is supported by:

• Both our United States Senators, Alan Cranston and John Tunney.

• Fifty-seven State Senators and Assemblymen of both parties, including Assembly Speaker Bob Moretti, Senate President Pro Tem James Mills, and coastal lawmakers Donald Grunsky and Alan Sieroty.

• Mayor Pete Wilson of San Diego and many city councils

• American Institute of Architects
American Association of University Women
American Institute of Planners
Associated Sportsmen of California
California Federation of Women's Clubs
California Parent & Teachers Association
Common Cause
League of Women Voters
National Council of Senior Citizens
Women For
United Auto Workers and other unions
California Medical Association

• California Conservation Organizations including California Roadside Council
Federation of Western Outdoor Clubs
Planning and Conservation League
Sierra Club

CALIFORNIA COASTAL ALLIANCE
POST OFFICE BOX 4161
WOODSIDE, CALIF. 94062

SIERRA CLUB
220 BUSH STREET
SAN FRANCISCO, CALIF. 94104

FIGURE 2-2. Selected materials from the "Yes on 20" campaign.

decent local campaigns involving leafleting, phone banks, and/or speakers' bureaus in many areas, especially the crucial population centers of Los Angeles and Orange County.[55]

Fourth, the Coastal Initiative received some favorable media coverage, at least in the press.[56] Early press coverage focused on the content of the Initiative and on lists of supporters and opponents, with lead paragraphs often contrasting "the coalition of over 100 conservation and civic groups" with "land developers, building trade unions, construction firms, and various chambers of commerce." Then in October coverage repeatedly emphasized the slick Madison Avenue style of Whitaker and Baxter's campaign, as well as charges of deceptive advertising by legislators and other proponents. Thus, the campaign was often portrayed as a David versus Goliath struggle of a broadly based but underfunded proponent coalition against a few monied interests that were spending enormous sums, often in a rather unscrupulous fashion. While it is true that a majority of the state's dailies editorially opposed the Initiative, most of the major metropolitan papers endorsed it.

Finally, the Proposition 20 campaign undoubtedly benefited from several fortuitous events. The most important occurred a week before the election when the Federal Communications Commission declared that the unbalanced distribution of television advertising time constituted a violation of its equal-time provision. On this basis, it ordered two Bay Area television stations to provide the proponents with free advertising time; several other stations in major metropolitan areas followed suit. The story was picked up by all the major papers throughout the state, with a typical headline being "TV offers free time to Proposition 20 to counter foes' $500,000 drive." While it is impossible to estimate the impact of the FCC ruling, it certainly reaffirmed the People versus the Special Interests image and provided proponents with some additional free ad time, thereby providing a buffer against last-minute erosion of support.[57]

At any rate, the long struggle to pass coastal legislation in California ended on November 7, 1972, when Proposition 20 was approved by 55% of the electorate. The Coastal Initiative carried 12 of the 15 coastal counties—including normally conservative areas such as Orange and San Diego counties—while losing only in the three northernmost counties later to become the North Coast region (see Table 2-1). It ran about as strongly in inland areas, with particularly strong support in the Bay Area and in the Sacramento Valley (especially the capital). In fact, outside of the north coast, the only geographic region that it did not carry was the San Joaquin Valley. On an individual level, some indication of the types of voters who probably supported Proposition 20 can be gleaned from a statewide Field poll conducted a week before the election. It revealed that by far the best predictor was political ideology, with (self-defined) liberals far more supportive than conservatives. Support for the Coastal Initiative was also somewhat more prevalent among Democrats than Republicans, among people under 30, among college students, and among whites and Orientals rather than blacks and Chicanos. Somewhat surprisingly—given the prominent role of the building

TABLE 2-1
VOTE ON COASTAL INITIATIVE[a]

Region/county	Yes vote Number	Yes vote %	No vote Number	No vote %
I. Coastal counties				
North Coast region	24,176	34%	46,331	66%
Del Norte	912	17%	4,319	83%
Humboldt	15,842	36%	28,373	64%
Mendocino	7,422	35%	13,639	65%
North Central region	289,504	60%	189,611	40%
Sonoma	55,978	55%	45,628	45%
Marin	62,117	62%	37,573	38%
San Francisco	171,409	62%	106,410	38%
Central Coast region	220,845	57%	168,863	43%
San Mateo	136,222	56%	107,718	44%
Santa Cruz	39,776	59%	27,748	41%
Monterey	44,847	58%	33,397	42%
South Central region	171,581	55%	139,207	45%
San Luis Obispo	25,936	54%	22,893	46%
Santa Barbara	72,710	62%	44,727	38%
Ventura	72,935	51%	71,587	49%
South Coast region	1,789,781	53%	1,502,069	47%
Los Angeles	1,469,186	55%	1,185,314	45%
Orange	320,595	51%	316,755	49%
San Diego region				
San Diego	300,988	52%	276,201	48%
Subtotal coastal counties	2,796,875	55%	2,322,282	45%
II. Noncoastal areas				
Southern California[b]	222,089	53%	196,611	47%
Bay Area[c]	725,657	61%	459,394	39%
Sacramento Valley[d]	264,806	59%	182,884	41%
San Joaquin Valley[e]	259,246	46%	300,001	54%
Mountain counties[f]	93,083	51%	88,008	49%
Subtotal noncoastal areas	1,564,881	56%	1,226,898	44%
Total	4,361,756	55%	3,549,180	45%

[a]Source: California Secretary of State, *Statement of Vote*, General Election of November 7, 1972, p. 31.
[b]Riverside, San Bernardino, and Imperial counties.
[c]Santa Clara, Alameda, Contra Costa, and Solano counties.
[d]Sacramento, Yolo, Sutter, Colusa, Yuba, Butte, Glenn, Tehama, and Shasta counties.
[e]San Joaquin, Stanislaus, Mariposa, Merced, Madera, Fresno, Kern, Tulare, King, and San Benito counties.
[f]Amador, Calaveras, Mono, Napa, Plumas, Placer, Siskiyou, Modoc, Lassen, Sierra, Nevada, El Dorado, Alpine, Tuolumne, Inyo, Trinity, and Lake counties.

trades and longshoremen unions in the opposition campaign—labor union membership was uncorrelated with policy preference on this measure.[58]

In concluding this section on the 3-year effort to pass coastal legislation, we need to return to a variable given considerable emphasis in the implementation framework presented in the introductory chapter—namely, the importance of the number and type of veto points in inhibiting change. As we have seen, coastal legislation never made it out of the legislature, where passage required the approval of a policy committee, a fiscal committee, and the general membership of each house, plus the governor's signature, or seven veto points in all. In fact, in all three legislative sessions (1970–1972) coastal bills made it through the multiple vetoes in the Assembly, only to die in Senate policy committee. In contrast, the initiative process has only two veto points, signature-gathering and the vote of the electorate. The importance of the type and number of veto points is illustrated by the fact that the voter-approved Coastal Initiative involved *greater* change than bills killed in the legislature in both 1971 and 1972.

IV. PROSPECTS FOR EFFECTIVE IMPLEMENTATION

Given the enormous effort that went into passage of the 1972 California Coastal Zone Conservation Act, what would an informed observer in the Winter of 1972–1973 have estimated to be the prospects for effective implementation of its objectives?

In the first chapter we posited a set of six generally necessary and sufficient conditions under which a regulatory program seeking a substantial change in target group behavior will achieve its statutory objectives. We also suggested that because the Coastal Initiative looked as though it could meet most of these conditions, it was adopted as our case study. After reviewing the history of the act, we are now ready to indicate what an observor in December 1972 would have estimated to be the probability of meeting each of these conditions.

A. Clear and Consistent Policy Objectives

There is little doubt that the Coastal Initiative contained a strong "tilt" in favor of environmental protection and public access. Its opening section declared:

> The California coastal zone is a distinct and valuable natural resource belonging to all the people and existing as a delicately balanced ecosystem; the permanent protection of the remaining natural and scenic resources of the coastal zone is a paramount concern to present and future residents of the nation . . . ; it is the policy of the state to preserve, protect, and, where possible, to restore the resources of the coastal zone for the enjoyment of the current and succeeding generations.[59]

Nowhere did the act require that this general environment orientation be balanced against the need for housing, energy facilities, expanding tax bases, etc.[60]

More specifically, the statute clearly indicated that the commissions were to place a high priority in both planning and permit review on (a) maintaining and expanding public access to the beach,[61] (b) preserving visual access to the sea from the state highway nearest the cost,[62] (c) preserving natural habitat areas, particularly wetlands,[63] (d) protecting water quality,[64] (e) preserving land in agricultural use,[65] and (f) protecting the scenic resources of the coast.[66] In none of these instances, however, did the statute provide an unambiguous qualitative—let alone quantitative—substantive standard against which the performance of the commissions could later be measured. Moreover, while the basic objectives were generally internally consistent, conflicts nevertheless arose—the most important being the possibility that increasing public access to the tidelands might destroy tide pools and other sensitive habitat areas. In such instances, the statute provided no clear ranking of priorities.

B. Sound Causal Theory

In developing a general strategy to attain the Initiative's substantive objectives, the framers began with three premises: First, the ultimate policy decisions concerning coastal resource allocation should and would eventually be made by the legislature. But the legislature itself would not have the knowledge or the time to prepare a detailed set of policies and to indicate their application, in at least a preliminary fashion, to specific geographic areas. Second, conditions along the 1100-mile coast were so varied and knowledge about them so incomplete that it would be infeasible to write even preliminary policy guidelines without first having considerable experience dealing with specific resource allocation decisions. Third, the preparation of a coastal plan would be meaningless in the absence of authority to review all development proposals during the 2 or 3 years required for plan development. These three premises in turn led to a twofold process: The coastal commissions would be responsible for presenting a comprehensive coastal plan to the 1976 legislature, which would be serving in its traditional reactive capacity. The commissions would also be responsible for administering an interim permit review system that would serve both to educate commission personnel and to preserve planning options. This overall strategy was borrowed directly from the BCDC experience, where it had worked quite well.[67]

While this twofold process represented a feasible strategy for making policy decisions over an enormous area in the context of great uncertainty, it assumed that the principal objectives of providing scenic and physical access and protecting wetlands could be accomplished during the 1973–1976 period through virtually exclusive reliance on the commissions' control over new development within the 1000-yard permit zone. While generally valid at least during the interim period, this reliance on police power regulation entailed some potentially significant limitations: First, it contained no means for the commissions to affect the demand for coastal resources, e.g., through limiting population increases in coastal counties or through altering the effects of the property tax on pressures for land conversion and intensive development. This meant that really restrictive use of the regulatory power would almost certainly arouse intense resentment by coastal property owners.

Second, the reliance on police power regulation gave the commissions authority to meet some objectives better than others. It was must adequate in protecting scenic vistas from the coastal highway to the ocean, as this goal could be met by requiring height and other restrictions on development west of the coastal highway. The police power would also enable the commissions to provide public access to the coast by giving priority to recreational and visitor facilities over other types of development and by requiring access conditions or easements on new developments between the highway and the beach. But the police power would not enable the commissions to provide the management services, e.g., liability insurance and litter collection, necessary before access-ways could actually be opened to the public (although the provision of such services could conceivably be made a condition for the approval of large commercial developments). Moreover, the commissions did not have authority to purchase beachfront property for recreational use. Instead, the police power only enabled the commissions to preserve potential parkland from development during the interim period, with the hope that state or local parks departments would soon acquire the funds to purchase the land. Finally, while the geographical scope of the commission's permit authority was certainly sufficient to protect wetlands from disturbances arising in the immediate area, it did not cover the upstream area beyond tidal action. Thus, it was conceivable that the act's objective could be undermined by extensive siltation or the release of very toxic chemicals upstream.

In sum, the underlying causal theory was an imaginative approach to decision making under uncertainty and the controls over all development in the nearshore area gave the commissions considerable authority to meet statutory objectives during the interim period. But the lack of management and acquisitions capability and, to a lesser extent, the lack of jurisdiction over upstream areas could create problems in actually providing public access to the coast and perhaps in protecting coastal wetlands. [68]

C. Coherent Structuring of the Implementation Process

The best theory for bringing about change will prove ineffective, however, unless the statute structures the implementation process so as to maximize the probability that implementing officials and target groups will behave consistently with statutory objectives. This, in turn, is contingent upon the ability of the enabling legislation to (a) assign implementation to a supportive agency that will give it high priority, (b) provide substantial hierarchical integration within and among implementing agencies by minimizing the number of veto/clearance points and by providing inducements and sanctions sufficient to assure acquiescence upon implementing officials and target groups with a potential veto, (c) provide decision rules supportive of statutory objectives, (d) provide ample opportunity for supportive interest groups and sovereigns to intervene in the implementation process, and (e) provide sufficient financial resources to hire the staff and conduct the technical analyses involved in the development of regulations, the administration of a permit program, and the monitoring of target group compliance. As we shall see, the Coastal Initiative generally did an excellent job of fulfilling most of these requirements.

First, it assigned implementation to a newly created state coastal commission and six regional commissions for which this program would have the highest priority. Moreover, the environmentalist mandate of the commissions, their creation after an intense political struggle, and their status as a bold experiment in land use planning created a situation in which most of the staff positions were likely to be filled by young supporters of the Initiative.[69] This still left, however, the formal policy-making positions in new organizations, i.e., the members of the state and regional commissions. Here the framers of the Initiative drew again from the BCDC experience in deciding to create about a dozen part-time positions on each regional commission (rather than a few full-time positions), which would be divided equally between elected local government officials and appointees of various state officials (the governor, the Assembly speaker, and the Senate Rules Committee). This would provide a diversity of views and political input on each commission, while freeing most of the funds for staff salaries and program administration.[70] On the other hand, the framers were careful to structure the state commission in a slightly different fashion. There would be no direct participation of local officials. Instead, the membership was equally divided between appointees of state officials and representatives from each of the six regional commissions. Assuming that most state appointees would be sympathetic to the Initiative's objectives and that a majority on at least three of the regional commissions would also be supportive, this would likely assure a fairly strong majority of statutory supporters on the all-important state commission—which was given final authority in the preparation of the coastal plan and review authority over regional permit decisions.[71] Nevertheless, the general policy orientation of each of the commissions would be highly contingent upon the appointment process.

Second, the Initiative created a highly integrated decision process in which the commissions would review all permits approved by local governments and state agencies in the permit zone and would have exclusive authority to prepare the coastal plan—subject only to review by the 1976 legislature.[72] Moreover, the civil penalties provided in the legislation were quite substantial ($10,000 plus up to $500 for each day in which a violation persisted), although their effectiveness as a deterrent against noncompliance by target groups would, as usual, be contingent upon the willingness of the agency and the attorney general's office to expand the resources required by any court suit.[73] In short, during the 1975–1976 interim period the commissions were the ultimate arbiters of all development within the permit zone. They were, however, subject to a "sunset clause" and would go out of existence unless their mandate was renewed by the 1976 legislature in the course of its deliberations on the coastal plan.[74]

Third, the Initiative incorporated decision rules that were highly supportive of its objectives. It clearly placed the burden on permit applicants to prove that their proposed developments would "not have any substantial adverse environmental or ecological effects."[75] It required a two-thirds affirmative vote of the membership of the commissions on any permits involving (1) wetlands, (2)· reductions in the size of, or public access to, beaches, (3) reductions in visual access to the sea from the coastal highway, (4) water quality, (5) commercial or sports fisheries, and (6) agricultural land.[76] And it provided that permit decisions by the regional commissions could be appealed to the state commission by "any aggrieved party," thereby facilitating appeals by environmentalists and other potential opponents. Since the statute clearly gave the state commission final authority in the preparation of the coastal plan, this meant that ultimate decisions on both important permit and planning matters would be made by the commission most likely to support statutory objectives.[77]

Fourth, the Initiative encouraged participation by environmentalists and other statutory supporters by requiring that all permit decisions be made after a public hearing, by requiring extensive public hearings at the regional level in the preparation of the coastal plan, and by providing liberal rules of standing to appeal permit decisions to the state commission and to the courts.[78] Moreover, the Initiative explicitly prohibited weakening amendments by the legislature during the 1973–1976 interim period.[79]

Finally, the Initiative provided a guaranteed source of funding for the commissions' activities through a $5 million appropriation from the Bagley Conservation Fund, thereby precluding the need for annual appropriations from the legislature. Combined with anticipated grants from the Federal Coastal Zone Management Act, this should have been sufficient to fund the commissions during the interim period—although, as we shall see, this eventually proved to be incorrect.[80]

On the whole, then, the Initiative structured the implementation process so that the permit and planning decisions of the commission and the behavior of target groups would probably be consistent with those objectives.[81] There were, however, two areas of significant uncertainty: The general policy orientations of the commissions would be highly contingent upon the appointment process, and the entire fate of the commissions and of the coastal plan was placed in the hands of the 1976 legislature.

It should be noted, moreover, that the permit review system incorporated an interesting solution to the traditional dilemma of regulatory agencies: How does one regulate a diverse set of conditions (or behaviors) while avoiding both the Scylla of detailed regulations oblivious to variation in specific situations and the Charybdis of providing administrators with so much discretion that statutory objectives are ignored in an effort to placate important actors in specific cases?[82] The Coastal Conservation Act attempted to resolve this conundrum by forcing administrators to deal with each permit application on a case-by-case basis— thereby giving the due attention to the specifics of the situation—while at the same time encouraging them to use the permit process to develop general planning policies. In addition, the general bias of the appointment process (especially at the state level) and decision rules virtually assured that most important permit and planning decisions would be roughly consistent with statutory objective.

D. Commitment and Skill of Top Implementing Officials

The Coastal Initiative made little provision for the commissions' leadership except to state that "the [state] commission and each regional commission shall each elect a chairman and appoint an executive director, who shall be exempt from civil service."[83] Although the Initiative elsewhere provided that the state commission would have ultimate authority over plan development and, via the appeals process, over most important permit decisions, the appointment scheme provided the possibility that the regional chairmen and executive directors could largely ignore the state commission. This was particularly likely in regions like the North Coast and, to a lesser extent, San Diego, where a majority of the regional commission would probably be opposed to the Initiative's objectives.[84] This topic will be addressed in Chapter 4 when we discuss the policy orientation and the internal relations of each of the commissions. Suffice it to note for the moment that, at the first state commission meeting, Melvin Lane and Joseph Bodovitz were selected as chairman and executive director, respectively, of that commission. As both were presumably strong supporters of the Initiative and had previously occupied similar posts with BCDC—where their leadership skills were widely admired—this suggested that at least the crucial state commission would be in supportive and skillful hands and that they might be able to overcome the decentralizing tendencies of the appointment scheme.[85]

E. Support from Organized Constituency Groups and Key Sovereigns

The long struggle over coastal legislation—and particularly the initiative campaign—provided a nucleus of active supporters in most coastal counties, with the exceptions of Del Norte and Mendocino counties in the North Coast region and, possibly, Ventura and San Luis Obispo counties in the South Central region. Moreover, the Sierra Club decided to maintain its substantial financial support of coastal conservation efforts in the state by hiring, soon after the election, two full-time coastal coordinators to monitor the state commission and the six regional commissions.[86] Thus, there was a reasonable expectation that environmental groups and other supporters of the Coastal Initiative would be able to closely monitor most of the commissions and to intervene actively in their permit and planning decisions. This was particularly crucial in regions, such as the North Coast, which could be expected to be hostile to the objectives of the act and in which an ability to appeal regional decisions to the state commission would be crucial if those objectives were to be realized.

The legislative and referendum campaigns that culminated in passage of the Initiative also left a reservoir of support among legislators, particularly the President Pro-Tem of the Senate (James Mills) and Assembly Speaker Bob Moretti. This would be very important in the appointment process, as Moretti would appoint a third of the "public" members and Mills was chairman of the five-member Senate Rules Committee, which would appoint an additional third. The remaining third were to be appointed by Governor Reagan, an opponent of the Initiative. This split in the policy preferences of key sovereigns further compounded the uncertainty surrounding the crucial appointment process. Beyond appointments, however, legislative and executive sovereigns were expected to have little effect on the commissions' performance until the crucial 1976 session of the legislature, as the Initiative essentially insulated them from formal influence by precluding weakening amendments and by providing a guaranteed appropriation.

F. Changing Socioeconomic Conditions and Political Support for the Initiative over Time

As a general rule, one must assume that the intense political support needed to pass innovative legislation will not remain indefinitely. This is particularly true for issues such as consumer and environmental protection that have very diffuse beneficiaries but very concentrated and intensely concerned cost-bearers.[87] From the perspective of 1972 it may have looked as though coastal protection would continue as salient into the foreseeable future. Nevertheless, continued population growth in the coastal counties could be expected both to increase demands for public access to coastal recreation areas and to increase demands for housing and other forms of development along the coast. The other crucial socioeconomic variable affecting net support for the coastal commissions would be the health of the state's economy and particularly that of the construction industry in coastal counties. The state's overall unemployment rate in 1972 was a rather high 7.2%, but this had apparently not had a debilitating effect on support for the Coastal Initiative in the 1972 election. Unemployment rates are highly volatile, however, and thus were presumably a source of considerable uncertainty over the 4-year interim period until the crucial 1976 legislative session.

Table 2-2 summarizes this discussion of the prospects for effective implementation of the 1972 Coastal Initiative at the time of its passage in the winter of 1972–1973. It suggests that the prospects for attaining the legislation's objectives during the 1973–1976 interim period were reasonably good. The relatively clear objectives (Condition 1), coherently structured implementation system (Condition 3), and probable presence of a supportive constituency (Condition 5) suggested that the permit and planning decisions of the commissions would largely be consistent with statutory objectives, although this was subject to the uncertain

TABLE 2-2

ASSESSMENT OF THE PROBABILITY OF EFFECTIVE IMPLEMENTATION OF THE 1972 CALIFORNIA COASTAL INITIATIVE AT TIME OF ADOPTION

Conditions conducive to effective implementation of statutory objectives	Application to 1972 Coastal Initiative	
	Overall assessment	Reasons/discussion
1. Statute contains clear and consistent policy directives	Moderate	Statute had a very consistent tilt toward the protection of natural resources and public access to the coast. No mention of any "balancing" with economic development. Nevertheless, objectives were qualitative and not always terribly precise.
2. Statute incorporates sound causal theory identifying sufficient factors and target groups to attain statutory objectives	Moderate	Commissions' authority to regulate all development within the permit zone sufficient to protect scenic access to the coast. Not entirely adequate, however, to provide physical access or to protect all wetlands.
3. Statute not only provides jurisdiction over target groups but also structures implementation to maximize probability of compliance from implementing officials and target groups	Moderate/high	
a. Assignment to a sympathetic agency	Moderate/high	Assigned implementation solely to the newly created coastal commissions for whom this program was their sole responsibility.

(*continued*)

effects of the appointment process on the degree of skill and support for statutory objectives among agency officials (Conditions 3 and 4). On the other hand, the somewhat problematic theory incorporated in the statute with respect to wetland protection and public access (Condition 2) meant that the agency's permit decisions might sometimes not have their intended impacts. And the uncertainties inherent in changing socioeconomic conditions (Condition 6)—together with the previous performance of the legislature on coastal bills—meant that the fate of the coastal plan at the hands of the 1976 legislature was an open question.

Given this background on the struggle for coastal legislation in California and the expectations concerning implementation of the 1972 Coastal Initiative, it is now time to examine the implementation process as it actually unfolded during the 1973–1976 period.

TABLE 2-2   (*Continued*)

Conditions conducive to effective implementation of statutory objectives	Application to 1972 Coastal Initiative	
	Overall assessment	Reasons/discussion
		Genesis in initiative process meant that professional staff would probably be sympathetic to statutory objectives. Even split on regional commissions between state appointees and local elected officials made commissions' policy orientations uncertain, although state commissions would probably be supportive.
b. Hierarchically integrated implementing agencies with few veto points and adequate incentives for compliance	Moderate/high	In the course of interim permit review, commissions had essentially complete control over all coastal development, although enforcement required cooperation of attorney general and courts. Stiff fines for violation of act. However, the entire process of planning and permit review was ultimately subject to amendment and revocation by the 1976 legislature.
c. Supportive decision rules	High	Burden of proof on permit appli-

(*continued*)

TABLE 2-2   (*Continued*)

Conditions conducive to effective implementation of statutory objectives	Application to 1972 Coastal Initiative	
	Overall assessment	Reasons/discussion
		cants. Approval of virtually any potentially harmful developments required a 2/3 vote of commission membership. State commission had final say on coastal plan and on most important permit decisions.
d. Financial resources	Moderate/high	Guaranteed appropriation in the initiative should have been sufficient, although subsequently proved to be inadequate because of inflation and unanticipated costs.
e. Formal access by supporters	High	Strong requirements for public hearings in permit review and planning. Liberal rules of standing for appealing permit decisions to state commission and to courts. Initiative specifically precluded weakening amendments by the legislature prior to 1976, although everything then subject to review by the 1976 legislature.
4. Commitment and skill of top implementing officials	Moderate/ uncertain	Commitment and skill of commission chairmen and executive directors uncertain at this time because so contingent upon appointment process. Early appointment of Lane and Bodovitz suggested at least state commission would be in sympathetic and skillful hands.

(*continued*)

NOTES

1. More precisely, California has 1072 miles of mainland coastline, excluding San Francisco Bay, and about 300 miles of coastline on the off-shore Channel Islands [Peter Douglas, "Coastal Zone Management—A New Approach in California," *Coastal Zone Management Journal* 1 (Fall 1973), pp. 1–2]. The following description of the coast is based upon the authors' own

TABLE 2-2 (*Continued*)

Conditions conducive to effective implementation of statutory objectives	Application to 1972 Coastal Initiative	
	Overall assessment	Reasons/discussion
5. Continuing support from constituency groups and sovereigns	Moderate/high	Referendum campaign required the mobilization of supportive constituencies in coastal counties, which should have been available to subsequently monitor implementation. Sierra Club hired two full-time staff expressly to monitor commissions. In legislature, both the Assembly Speaker and the Senate Pro-Tem supported Initiative and Assembly oversight subcommittee chaired by Initiative's principal sponsor; nevertheless, no powerful "fixer" comparable to Muskie. Governor Reagan opposed Initiative. Except for appointment process, commissions generally insulated from legislature until the crucial 1976 session.
6. Changing socioeconomic conditions and political support for the Initiative over time.	Uncertain	Diffuse political support subject to the "issue-attention cycle." Political support—especially in the 1976 legislature—likely to be highly contingent upon the general health of the state's economy and particularly the construction industry, which were highly uncertain in December 1972.

travels, as well as on Gilbert Bailey and Paul Thayer, *California's Disappearing Coast* (Berkeley: Institute of Governmental Studies, 1971), pp. 8–12.

2. These data are taken from Security Pacific Bank, *California Coastal Zone Economic Study: An Area Profile* (Los Angeles: Security Pacific Bank, 1975), Section 3.

3. Security Pacific, *Area Profile*, pp. 3–11, 3–13.

4. Ibid, pp. 3–12, 3–13.

5. Following are the median education (age 25 and older) and median family income for the permit area in Orange County, the county as a whole, and the state:

	Permit area	County	State
Median education (years)	13.2	12.6	12.4
Median family income	$12,976	$12,245	$10,732

Source: Security Pacific Bank, *California Coastal Zone Economic Study: Statistical Appendix* (Los Angeles: Security Pacific Bank, 1975).

6. For an environmentalist's perspective on the transformation of the county, see David Curry, "Irvine: The Case for a New Kind of Planning," *Cry California* 6 (1970).

7. This discussion of perceived problems is based upon three principal sources. The first is a public opinion poll conducted in August 1972 by Opinion Research of California for Whitaker and Baxter, the campaign management firm in charge of the opposition campaign to Proposition 20. The poll involved a stratified sample of 2496 voting-age adults, although some of the questions used herein were addressed to a subsample of 1247 respondents. Results of the poll (hereafter cited as Whitaker-Baxter poll) have not previously been published. We wish to sincerely thank the firm for making them available to us.

   A second important source of the data in this section and in the subsequent section on the Proposition 20 campaign was a series of interviews conducted during the summer of 1977 by Dana Alden, a research associate in this study, with the following individuals active in the Proposition 20 campaign: (1) Clem Whitaker, Whitaker and Baxter; (2) Will Davidson, Whitaker and Baxter; (3) Janet Adams, executive director, Coastal Alliance; also an interview with Sabatier on December 2, 1975; (4) Vern Yadon, Monterey County coordinator, Pro-20 Campaign; (5) Peter Scott, Santa Cruz County coordinator, Pro-20; (6) Roger Hedgecock, San Diego County coordinator, Pro-20; (7) Sally Spurgeion, Orange County coordinator, Pro-20; (8) George Wagner, Angeles Chapter, Sierra Club; (9) Mary Ferguson, Angeles Chapter, Sierra Club; (10) Linda Wase and Judy Anderson, assistants to Bob Jeans, who was the manager of the Pro-20 campaign in Los Angeles and who unfortunately passed away in the interim; (11) Ted Roberts, Bay Chapter, Sierra Club and Alameda County coordinator; (12) Bill Kortum, Sonoma County coordinator, Pro-20; (13) Selma Rubin, Santa Barbara County coordinator, Pro-20; (14) Bert Muhly, Santa Cruz County, Pro-20; (15) John Jonck, San Francisco coordinator, Pro-20; (16) Phyllis Faber, Marin County coordinator, Pro-20; and (17) Wesley Chesbro, Humboldt County coordinator, Pro-20.

   The third important source involves published articles by two of the principal architects of the Coastal Initiative: Peter Douglas, the chief legislative staff person on coastal legislation and one of the two principal draftsmen of the Initiative; and Janet Adams, executive director of the Coastal Alliance, one of the two major organizations in the Pro-20 campaign. The final major source is the journalistic monograph by Bailey and Thayer, *California's Disappearing Coast*, which was published in 1971 and presumably reflected—and perhaps affected—the attitudes of political elites sympathetic to Proposition 20. We have also relied on the two major published accounts of the struggle for coastal legislation, Fred Doolittle, *Land Use Planning and Regulation on the California Coast* (Davis, Calif.: Institute of Governmental Affairs, 1972); and Stanley Scott, *Governing California's Coast* (Berkeley: Institute of Governmental Studies, 1975), pp. 319–352.

8. The precise text of the question (Whitaker-Baker poll, question 19) is as follows: "Everything has its good and bad points. Can you tell me what you think would be particularly *good* about stopping or slowing development of California's coastal land?" Of the 1247 respondents, 81% volunteered a codable response to this open-ended question. Following are the most frequently-mentioned:

| Preserve natural beauty of coastline | 44% |
| More public recreation areas/parks | 12% |

Less pollution	9%
Less crowded/fewer people	8%
Less commercial and other development	7%
Will give time for better planning	6%
Preserve marine and other wildlife	3%

Surprisingly, no one apparently mentioned public access, although there is widespread agreement that this was the most important issue in the Proposition 20 campaign a couple of months later. Respondents were then asked to indicate the *bad* effects of stopping or slowing coastal development. Only 43% provided a codable response, the most frequently mentioned of which are listed below:

Adverse effects on economy/jobs/taxes	17%
Restrict public use/access	5%
Vacation/recreation facilities needed	3%
Progress needed	3%

Surprisingly, only 2% of respondents mentioned the adverse effects on property rights, although other items in the survey that dealt specifically with this issue indicated that it was a major concern to people.

9. Bailey and Thayer, *California's Disappearing Coast*, pp. 3–4, 18–26, 35–42. Interviews with (1) Bill Kortum, Sonoma County coordinator for Proposition 20, (2) Janet Adams, Coastal Alliance, (3) Clem Whitaker, Whitaker and Baxter, (4) Peter Scott, Santa Cruz County coordinator, and (5) Vern Yadon, Monterey County coordinator. In a private conversation with one of the authors, Charles Warren, an assemblyman in 1972 and chairman of the crucial Assembly Energy and Land Use Committee in 1976, indicated that he became concerned over coastal conservation upon hearing reports of plans for nuclear power plants every 10 miles along the coast by the 1990s.

10. In a very important case, the California Supreme Court reaffirmed this point of law in a 1971 decision, *Marks* v. *Whitney* 6 Cal. 3d 251, 98 Cal. Rep. 790. For discussions of this case, see N. Gregory Taylor, "Patented Tideland: A Naked Fee?" *Journal of the State Bar of California* 47 (September–October 1972), pp. 421–425; and Mark Eikel and W. Scott Williams, "The Public Trust Doctrine and the California Coastline," *The Urban Lawyer* 6 (Summer 1974), pp. 519–571.

11. Of the 17 people involved in the Proposition 20 campaign who were interviewed (see note 7), both of the respondents in Whitaker and Baxter and 8 of the 10 Pro-20 coordinators with whom campaign issues were discussed mentioned public access as one of the two or three most important issues—and usually *the* most important issue—in the campaign. With respect to public access to the wet beaches, it is important to note that much of this access from the coastal highway involved traversing private property, often without the owner's consent. From a political standpoint, however, the *expectations* of the public that they had access to the beach were at least as important as their legal rights. Moreover, the latter were clouded considerably by a 1970 California Supreme Court decision suggesting that property owners who did not take active measures to prevent trespassing for a 5-year period implicitly dedicated a legally enforceable public access easement; see *Gion* v. *City of Santa Cruz*, 2 Cal. 3d (1970); and Jay Shavelson, "Gion v. City of Santa Cruz," *Journal of the State Bar of California* 47 (September–October 1972), pp. 415–419. Whatever the resolution of the legal questions, there is no doubt that access over private property was more effectively impeded when the land was transformed from a ranch into a second-home subdivision—as was done in the case of the Sea Ranch. For a more general discussion of public access to the coast, see Dennis Ducsik, *Shoreline for the Public* (Cambridge: M.I.T. Press, 1974).

12. Security Pacific, *Area Profile*, p. 2-1.

13. California Department of Parks and Recreation, *California Coastline Preservation and Recrea-*

tion Plan, August 1971, pp. 53–75. See also Bailey and Thayer, *California's Disappearing Coast*, pp. 15–16.

14. Letter from Russell Cahill, director, California Department of Parks and Recreation, to Joseph Bodovitz, executive director, California Coastal Commission, December 9, 1977.

15. Security Pacific, *Area Profile*, p. 2-4.

16. Interview with Judy Rosener, member, South Coast Regional Commission, 1975.

17. On the one hand, in the August 1972 Whitaker-Baxter poll only 3% of respondents volunteered wildlife protection as a benefit of restrictions on coastal development (see note 8). On the other hand, wetland protection was mentioned as an important motivating factor by Mary Ferguson, one of the key Sierra Club people in Los Angeles County, and in the Douglas article ("Coastal Zone Management," pp. 9–10). Moreover, wetland protection was the most frequently mentioned objective of Proposition 20 in the eyes of the (elite) respondents to our evaluation survey (see chapter 8); it was mentioned by 80% of respondents, with virtually no difference among the various groups surveyed.

18. John Speth, California Department of Fish and Game (unpublished paper, quoted in Bailey and Thayer, *California's Disappearing Coast*, pp. 14–15).

19. California Department of Fish and Game, *The Natural Resources of Humboldt Bay*, Coastal Wetland Series #6, December 1973.

20. Janet Adams, "Proposition 20—A Citizen's Campaign," *Syracuse Law* Review 24 (Summer 1973), p. 1023. For a similar view, see Bailey and Thayer, *California's Disappearing Coast*, pp. 43–44.

21. For discussions of California land use law c. 1970, see Stanford Environmental Law Society, *California Land Use Primer* (Palo Alto: Stanford University, 1972); California Legislature, Assembly Select Committee on Open Space, *California Land Conservation Act of 1965 and Related Open Space Provisions*, November 1973; California Government Code, Chapters 3–4; California Council on Intergovernmental Relations, *Local Government Planning in California*, March 1972; and John Winters, "Environmentally Sensitive Land Use Regulation in California," *San Diego Law Review* 10 (June 1973), pp. 693–756. There were important amendments to land use law in the 1971 session (AB 1301, amending the Subdivision Map Act) and in the 1973 session when city and county zoning ordinances were required to conform to the general plan.

22. Whitaker-Baxter poll, question 22: "An area of particular concern these days is the matter of development and protection policy along the California coastline. In your opinion should the planning and development of coastal areas be controlled locally, regionally, by the county, or by the state?" The preference for state/regional—rather than local—control was confirmed by questions 15, 28, and 33 of the same poll.

23. Douglas, "A New Approach in California," pp. 1–2; Bailey and Thayer, *California's Disappearing Coast*, p. 45; Adams, "Proposition 20," pp. 1022–1023; Mel Mogulof, *Saving the Coast* (Lexington, Mass.: D. C. Heath, 1975), chap. 1.

24. Bailey and Thayer, *California's Disappearing Coast*, pp. 22–34. For a more scholarly analysis of this process, see John Hollis and James McEvoy, "Demographic Effects of Water Development," in *Environmental Quality and Water Development*, ed. Charles Goldman, James McEvoy, and Peter Richerson (San Francisco: W. H. Freeman, 1973), pp. 216–232.

25. Nelson Rosenbaum, *Land Use and the Legislatures* (Washington, D.C.: Urban Institute, 1976).

26. In addition to Rosenbaum, see Fred Bosselman and David Callies, *The Quiet Revolution in Land Use Control* (Washington, D.C.: Council on Environmental Quality, 1971); and Robert Healy, *Land Use and the States* (Baltimore: Johns Hopkins University Press, 1976).

27. Zigurds Zile, "A Legislative-Political History of the Coastal Zone Management Act of 1972," *Coastal Zone Management Journal* 1 (1974), pp. 235–274; Noreen Lyday, *The Law of the Land: National Land Use Legislation 1970–75* (Washington, D.C.: Urban Institute, 1976).

28. All of these were mentioned by Douglas ("A New Approach in California," pp. 10–12), the legislative assistant to Assemblyman Sieroty and one of the principal drafters of the Coastal Initiative. See also Bailey and Thayer, *California's Disappearing Coast*, pp. 48–50.

29. Harry Jackson and Alvin Baum, "Regional Planning: The Coastal Zone Initiative Analyzed in Light of the BCDC Experience," *Journal of the State Bar of California* 47 (September–October 1972), pp. 426–431; Robert Ditton, John Seymour, and Gerald Swanson, *Coastal Resources Management* (Lexington, Mass.: D. C. Heath, 1977), chap. 9; Rice Odell, *The Saving of San Francisco Bay* (Washington, D.C.: Conservation Foundation, 1972); Joseph Bodovitz, "The San Francisco Bay Conservation and Development Commission," *The Crowded Coast: The Development and Management of the Coastal Zone of California* (Los Angeles: USC Center for Urban Affairs, 1971), pp. 98–111. In addition, a number of key BCDC officials subsequently took similar positions with the state coastal commission, including Bodovitz (executive director), Schoop (chief planner), and Mel Lane (chairman).

30. Douglas, "New Approach in California," pp. 11–12; Doolittle, *Land Use Planning and the California Coast*, p. 58. For an analysis of the early history of the TRPA, see Edmund Costantini and Kenneth Hanf, *The Environmental Impulse and Its Competitors* (Davis, Calif.: Institute of Governmental Affairs, 1973).

31. Doolittle, *Land Use Planning and the California Coast*, pp. 1–12; Scott, *Governing California's Coast*, pp. 314–325.

32. Doolittle, *Land Use Planning and the California Coast*, pp. 43–50; Scott, *Governing California's Coast*, pp. 325–331; Bailey and Thayer, *California's Disappearing Coast*, pp. 53–58; Adams, "Proposition 20," pp. 1023–1024.

33. For a discussion of partisan differences on environmental issues in the California legislature, see Michael McCloskey and John Zierold, "The California Legislature's Response to the Environmental Threat," *Pacific Law Journal* 2 (1971), pp. 574–602.

34. This discussion of the formation of the Coastal Alliance is based primarily upon Scott, *Governing California's Coast*, pp. 331–335; Adams, "Proposition 20," pp. 1024–1026; and an interview by one of the authors with Janet Adams on December 2, 1975.

35. On the 1971 session, see Doolittle, *Land Use Planning and the California Coast*, pp. 51–70; Scott, *Governing California's Coast*, pp. 336–345; Bailey and Thayer, *California's Disappearing Coast*, pp. 58–75; and Adams, "Proposition 20," pp. 1029–1032.

36. Doolittle, *Land Use Planning and the California Coast*, pp. 69–70; Scott, *Governing California's Coast*, p. 345. While the governor was not a vigorous public opponent of the Sieroty bill, his preference for local control and general opposition to regional government were certainly well known. He also offered COAP as an alternative to the Coastal Alliance bills.

37. There was one other coastal bill, SB 860 by Senator Carpenter (Rep-Orange County), which would have created a 15-member state board (9 of whom would be local officials) to issue advisory planning guidelines to local governments. It passed the Senate but was killed in Assembly committees, largely because of conservationist opposition (Doolittle, *Land Use Planning and the California Coast*, pp. 70–73; Scott, *Governing California's Coast*, pp. 345–351; Adams, "Proposition 20," p. 1032). The companion bills (AB 200, SB 100) were introduced separately in their respective houses. When SB 100 was killed in policy committee in May, the initiative drive was launched. Nevertheless, AB 200 passed the Assembly, 60–11, on July 6 with only minor amendments. Even when admended to provide a majority of local officials on the regional commissions, a major change in the regional commission boundaries, a policy balance between conservation and development, and several changes in the permit review procedure, it still was never approved by the Senate Natural Resources Committee (AB 200, as of July 20, 1972). Given that Proposition 20 had already qualified as an initiative, this is an indication of the opposition's confidence and/or strategic blunder.

38. This account is taken from a 53-page chapter in Dana Alden and Paul Sabatier, *Environmental*

Propositions in California, an unpublished analysis of six environmental propositions in California. Principal sources for the synopsis include the 14 interviews with supporters mentioned previously in note 7; a variety of public documents; and Janet Adams's account ("Proposition 20," pp. 1032–1036), which gives the mistaken impression that the Coastal Alliance was a centralized organization coordinating the entire campaign. The petition phase was under serious time constraints because Section 3508 of the California Election Code (West Supplement 1973) requires that the necessary signatures be collected 131 days prior to the election or, in this case, approximately June 25.

39. Whitaker and Baxter poll. Moreover, another question, which did not mention Proposition 20 by name but rather dealt with its content, found that 58% favored "the concept of legislation which would create a new state coastal zone commission" while 31% opposed and 11% were undecided. Support was the same in coastal and noncoastal counties, was moderately correlated with income and education, but was essentially uncorrelated with beach use, partisan affiliation, and union membership. Support for the creation of a state coastal commission correlated strongly with support for Proposition 20. (For example, of those who supported the creation of such a commission, 84% indicated they would vote yes on Proposition 20, while 16% said they would vote against it.)

40. The poll was conducted August 10–14, 1972, and involved 561 respondents on this question. Field Research Corporation, *California Poll 7204* (Berkeley: State Data Program, 1973).

41. Field Research Corporation, "Trend of Public Opinion Has Swung Against Proposition 9," Press release #753, June 3, 1972. Mervin Field is the California equivalent of George Gallup or Louis Harris. The two previous environmental referenda were Proposition 18 in November 1970, which would have permitted cities to divert highway trust funds to mass transit, and Proposition 9 (the Clean Environment initiative) in June 1972, dealing with air pollution control, pesticides, and nuclear power plants. With respect to Proposition 18, a poll conducted the first week in December (Field poll 7006) found 68% of the public in favor, with only 20% opposed. In fact, there was still a 52% to 20% margin of support during the last week of October (Field poll 7007). But the measure was eventually defeated by the voters, 46% to 54%. Much the same thing happened on Proposition 9. An April 1972 Field poll (No. 7202) revealed a 64% to 22% majority in favor. However, a very skillful media campaign by Whitaker and Baxter reversed this to a 35% to 47% plurality in opposition by late May (Field poll 7203), and the measure was eventually defeated at the polls, 35% in favor and 65% opposed. More detailed analyses of these two propositions are found in Alden and Sabatier, *Environmental Initiatives in California*, chaps. 2–3. For discussions of Proposition 9, see Carl Luttrin and Allen Settle, "The Public and Ecology: The Role of Initiatives in California's Environmental Politics," *Western Political Quarterly* 28 (June 1975), pp. 352–371; and S. P. Sethi, *Advocacy Advertising and Large Corporations* (Lexington, Mass.: D. C. Heath, 1977), chap. 5.

42. Whitaker and Baxter had a very impressive record (188 victories and 12 defeats) prior to the Proposition 20 campaign and had, in fact, just come from a very successful campaign to defeat the Clean Environment Initiative (Proposition 9, June 1972). Examination of campaign expenditure reports filed with the secretary of state reveals that the vast majority of the funds raised by Whitaker and Baxter came in the early weeks of the campaign and that only 1% of all monies received came in contributions of $25 or less. Following are some of the major donors to the campaign:

Contributor	Amount
Land developer/Construction/Finance	
Irvine Company	$50,000

(continued)

Contributor	Amount	
Deane and Dean, Inc.	50,000	
Bechtel Corporation	30,000	
General Electric	30,000	
Southern Pacific Land Co.	25,000	
Hioro Corporation	10,000	
Crocker National Corp.	11,250	
Subtotal		$206,250
Utilities		
Pacific Gas and Electric	$30,000	
Southern California Edison	27,633	
Subtotal		$ 57,633
Petroleum		
Texaco	$15,000	
Standard Oil of California	35,000	
Occidental Petroleum	10,000	
Mobil Oil	15,000	
Gulf Oil	15,000	
Subtotal		$ 90,000
Total		$353,883

Other organizations opposing the Initiative included (1) a number of labor unions, including the building trades, teamsters, and longshoremen, (2) the California Real Estate Association, (3) the California Manufacturers Association, and (4) the California Chamber of Commerce. See Scott, *Governing California's Coast*, p. 360; and California Journal, *Ballot Propositions*, November 1972, p. 25.

43. A much more detailed analysis of this and other aspects of the campaign can be found in Alden and Sabatier, *Environmental Initiatives in California*, chap. 5. See also Adams, "Proposition 20," pp. 1036–1042; Luttrin and Settle, "The Public and Ecology," pp. 365–370; and Scott, *Governing California Coast*, pp. 353–363.

44. This was the conclusion of Opinion Research Corporation (the polling firm) in its report to Whitaker and Baxter. In an interview with Dana Alden, Clem Whitaker (president of Whitaker and Baxter) blamed the rather poor campaign run by his firm on insufficient time to do the research needed and to reach important opinion-makers, as the firm really didn't begin its background preparation on Proposition 20 until August 1972.

45. Of the 1247 respondents in the Whitaker-Baxter poll, 64% found it "hard to believe" or "not possible" that "the initiative will impose a complete moratorium on all coastal development." Similarly, when asked if "the coastal initiative is backed by the same extremists who tried to impose Proposition 9 on the people last June," 28% tended to agree, 40% tended to disagree, and 32% didn't know. Finally, fully 65% of respondents disagreed with the statement "The Coastal initiative restrictions are so severe that it would be disastrous if they are passed into law." Nevertheless, in its printed materials the opposition campaign continued to stress that Proposition 20 would impose a moratorium on development with disastrous economic consequences; see their statement in California Secretary of State, *Propositions and Proposed Laws, Together with Arguments*, General Election, November 7, 1972, pp. 51–55.

46. Recall from footnote 8 that, whereas 81% of respondents in the Whitaker-Baxter poll could think

of some benefits of restricting coastal development, only 43% could volunteer any costs—of which the most frequently mentioned (by 17%) were economic. While respondents did not on their own associate restrictions on coastal development with restrictions on the property rights of coastal residents, a number of other questions in the poll suggested sympathy for their plight: 66% felt it would be "unfair to prohibit landowners from improving their property." Fifty-three percent were "in sympathy with coastal property owners who feel they might be denied free use of their own property." Fully 70% of respondents felt that "a rancher, restricted from building on his own property, should be paid for the diminished value of his property." And 76% of respondents felt that "if someone owns a lot on the coast and they want to build a house on it, they should be able to do so if they comply with locally approved building codes."

47. According to one of the senior partners in Whitaker and Baxter, the firm's decision to use the "Don't Lock Up the Beach" slogan was based on information in the opinion survey suggesting some concern with the delaying effects of Proposition 20 on the development of coastal recreational facilities (see note 8), as well as their own analysis that this would in fact occur. When the poll revealed that this concern was more widespread in Southern California and that voters in that area were more confused about Proposition 20's purpose, the slogan was targeted specifically for that region.

48. This is the topic addressed in considerable detail in Alden and Sabatier, *Environmental Initiatives in California*. In addition to the Gas Tax Diversion Referendum of November 1970 and the Clean Environment Initiative of June 1972 (see note 32), environmental ballot propositions that were rejected by the electorate included the Wild Rivers Initiative of 1974 and the Nuclear Power Initiative of June 1976. The only proposition (besides the Coastal Initiative) that passed was a Gas Tax Diversion Proposition in 1976, which was essentially unopposed by "the highway lobby" because the diversion had already taken place at the federal level.

49. In what was probably the most surprising finding of the Whitaker-Baxter poll, 65% of respondents agreed that "it is a good idea to stop all coastal development for five years or more until a comprehensive master plan can be prepared for development of our coastline."

50. In the Whitaker-Baxter poll, 50% of respondents did *not* agree that "land management problems are so complex that it is foolhardy to bypass our elected state representatives and enact land use laws by initiative," while 40% agreed with the statement and 10% were undecided.

51. Among the organizations supporting Proposition 20 were the following: (1) Coastal Alliance, (2) Sierra Club, (3) Planning Conservation League, (4) League of Women Voters, (5) American Association of University Women, (6) Associated Students, University of California, (7) Federation of Western Outdoor Clubs, (8) Congress of California PTA, (9) National Council of Senior Citizens, and (10) American Institute of Planners. See Scott, *Governing California's Coast*, p. 360; and California Journal, *Ballot Propositions: 1972 General Election*, p. 25.

52. Interviews with people active in the Pro-20 campaign (see note 7) indicate that it was a very decentralized operation, with considerable friction between Janet Adams and the Sierra Club. In most coastal counties, *ad hoc* coalitions were put together. While they usually relied on the Coastal Alliance or the Sierra Club for written materials, each organization was largely autonomous. And in Los Angeles County, local coordinators rejected the suggestions of Ms. Adams and hired a professional campaign manager, Bob Jeans, who put together an excellent campaign, relying heavily upon the local Sierra Club chapter. Despite the friction, it is doubtful that the campaign could have succeeded without (1) the financial and volunteer resources of the Sierra Club and (2) the ability of the Coastal Alliance to serve as an umbrella for a variety of groups and individuals, many of whom would have been extremely reluctant to work in a campaign dominated by the Sierra Club.

53. The focus on public access, rather than on environmental protection, was repeatedly stressed by Ms. Adams and characterized even the publicity materials developed and distributed by the Sierra Club.

54. Analysis of campaign expenditure reports filed with the secretary of state reveal that the Pro-20 campaign was able to raise approximately $270,000, with $63,000 coming from the Coastal Alliance and $117,000 from the Sierra Club. Of the total, approximately $137,000 was spent on publicity, about equally divided between radio-television and newspapers-mailings-bumper stickers.

55. Our interviews with county coordinators revealed tremendous variation in the amount and type of local activity. In Santa Cruz the campaign consisted largely of locally funded and locally produced newspaper ads and radio spots featuring endorsements from Senator Grunsky and other local opinion leaders. In Humboldt County there was heavy reliance upon student volunteers from the Arcata Ecology Center. Activities included a bike ride, several debates at local radio and television stations, and probably the most intensive precinct canvassing in the entire state (in conjunction with the McGovern campaign). In Orange County there was extensive distribution of literature at shopping centers and beaches, numerous locally financed and produced radio spots and newspaper ads, and a very extensive phone bank. Most of these same activities took place in Los Angeles County, where the local campaign also sent out 60,000 handouts to targeted precincts the night before the election.

56. This is based upon analysis of the files of a professional clipping service. For a more extensive account, see Alden and Sabatier, *Environmental Initiatives in California*.

57. Based upon interviews, analysis of the clipping service files, and Adams, "Proposition 20," pp. 1039–1041.

58. These interpretations should be viewed with some caution, however, as the poll indicated that a very sizable 37% of respondents were still undecided—compared with 37% who indicated they would vote yes and 25% who would vote no (Field poll 7206, consisting of 1467 telephone interviews conducted October 30 through November 1). As this was a telephone poll, very little information was solicited, which accounts for the absence of normally included variables such as income and education. Unfortunately, Whitaker and Baxter did not commission any polls during the course of the campaign.

For example, 52% of liberals indicated they would vote yes, compared with 29% of conservatives (and 36% of moderates).

59. California Public Resources Code, Section 27001 (1973). The basic plan objectives found in Section 27302 were as follows: (a) the maintenance, restoration, and enhancement of the overall quality of the coastal zone environment, including but not limited to, its amenities and aesthetic values; (b) the continued existence of optimum populations of all species of living organisms; (c) the orderly, balanced utilization and preservation, consistent with sound conservation principles, of all living and nonliving coastal zone resources; (d) avoidance of irreversible and irretrievable commitments of coastal zone resources. Similarly, Section 27402 made the granting of a permit contingent upon finding that it would be consistent with Section 27302 quoted above and that "the development will not have any substantial adverse environmental or ecological effect."

Our analysis of the objectives of the Coastal Initiative is based primarily upon our own reading of the statute. It has also benefited considerably, however, from a workshop held in September 1977 in which our preliminary analysis was discussed with (1) Peter Douglas, (2) Joseph Petrillo, staff counsel to the state commission, (3) Don Peterson, a county supervisor and coastal commissioner in the North Coast region, (4) Judy Rosener, a commissioner from the South Coast region, and (5) Celia von der Muhll, one of the two Sierra Club coastal coordinators. Several people from the business–labor community were also invited to attend, but none were able to do so.

60. The entire permit review process—and the emphasis on permit conditions—was a tacit acknowledgement of the need for residential and other development along the coast. Nevertheless, the only place where the Coastal Initiative explicitly recognized such needs was in Section 27304 (c)

(6), which indicated that the Coastal Plan should contain "a public facilities element for the general location, scale, and provision in the least environmentally destructive manner of public services and facilities in the coastal zone. This element shall include a power plant siting study."

61. Section 27304 (c) (4) requires that the plan contain "a public access element for maximum visual and physical use and enjoyment of the coastal zone by the public." Section 27401 (c) requires a two-thirds vote on any development reducing or imposing restrictions upon public access to beaches or tidelands, while Section 27403 (a) requires that permits be subject to reasonable conditions to ensure that "access to publicly owned or used beaches . . . is increased to the maximum extent possible by appropriate dedication."

62. Section 27401 (d) requires a two-thirds vote on "any development which would substantially interfere with or detract from the line of sight toward the sea from the state highway nearest the coast." While it may be difficult to interpret this (and other) procedural requirements for two-thirds votes as substantive standards, they certainly indicate that the act placed a high priority on this value.

63. Section 27401 (a, d) required a two-thirds vote on any permit involving "dredging, filling, or otherwise altering any bay, estuary, salt marsh, river mouth, slough, or lagoon" or adverse effects upon commercial or sports fisheries. See also sections 27302 (a, b) and 27402 quoted above.

64. Section 27401 (e) required a two-thirds vote on developments adversely affecting water quality.

65. Section 27401 (e) similarly required a two-thirds vote on "a development which would adversely affect . . . agricultural uses of land existing on the effective date of this division."

66. Sections 27001 and 27302 (a) talk about protecting the "scenic resources" and "aesthetic values" of the coast. In addition, Section 27403 (d) required that permits be conditioned to ensure that "alterations to existing land forms and vegetation, and construction of structures, shall cause minimum adverse effects to scenic resources."

67. Peter Douglas and Joseph Petrillo, "California's Coast: The Struggle Today—A Plan for Tomorrow," *Florida State University Law Review* 4 (April 1976), p. 188; also Swanson, "Coastal Zone Management," pp. 81–102.

68. Although Joseph Bodovitz indicated that lack of control over upstream watersheds did not prove to be an important problem in the 1973–1976 period, the potential problem was certainly recognized in the Coastal Plan, which recommended that the commission's geographic jurisdiction be expanded to cover upstream watersheds (see chap. 8).

69. This argument is generally consistent with Downs, *Inside Bureaucracy*, chaps. 2, 8–9. Those familiar with the history of the California Energy Conservation and Development Commission will note that simply creating a new agency does not guarantee a supportive staff. That case can be distinguished from the coastal commissions, however, by its ambiguous mandate and the fact that its creation was not the result of a high-visibility political campaign but rather the skillful work of a few legislators and the governor.

70. Almost every observer of the BCDC experience has noted that the mix of local elected officials and public members provided a fruitful dialogue among individuals with different backgrounds. See Swanson, "Coastal Zone Management," pp. 90–99; and Odell, *The Saving of San Francisco Bay*, p. 30.

71. As noted previously, the framers of the Initiative were well aware of the differences in voting behavior between state appointees and local elected officials among California members of the Tahoe Regional Planning Agency (Costantini and Hanf, *The Environment Impulse and Its Competitors*, pp. 55–58). Remember from Table 2-1 that the Initiative passed with substantial majorities in the North Central, Central, and South Central regions. Presumably about half of the local officials in these regions would be generally supportive of the Initiative, which, if combined with generally supportive state appointees, would produce a supportive majority that would then elect a supportive representative to the state commission.

72. California Public Resources Code, Division 18, Chapters 4 and 5. As Mogulof (*Saving the*

Coast, p. 4) observed, "Within this permit zone the commissions are the closest thing to a regional comprehensive governing authority that exists in this country."

73. California Public Resources Code, Division 18, 1973, Sections 27500–27501.
74. For a skeptical view of the efficiency of sunset laws, see Robert Belm, "The False Dawn of the Sunset Law", *Public Interest*, Fall 1977, pp. 103–118.
75. California Public Resources Code, Section 27402 (a).
76. Ibid., Section 27401.
77. Section 27423 (a) provided that "an applicant, or any person aggrieved by approval of a permit by the regional commission, may appeal to the state commission." "Aggrieved party" was subsequently interpreted by the commission to mean anyone who had participated at the regional hearing or who had sent in a letter prior to the hearing. For general discussions of legal standing, see Karen Orren, "Standing to Sue: Interest Group Conflict in the Federal Courts," *American Political Science Review* 70 (September 1976), pp. 725–741, and Joseph Sax, *Defending the Environment* (New York: Knopf, 1970), chaps. 3–4.
78. California Public Resources Code, Division 18, 1973, Sections 27423–27424.
79. Ibid., Chapter 8, Section 5.
80. Ibid., Chapter 8, Section 4. See also Douglas, "Coastal Zone Management," pp. 20–21.
81. This can perhaps best be illustrated by contrasting the Coastal Initiative with the enabling statute of another regional land use agency. Compared to the consistent conservationist tilt of Proposition 20, the 1969 Bistate Compact creating the Tahoe Regional Planning Agency provided an ambiguous policy mandate—"to adopt and enforce a regional plan of resource conservation and orderly development"—and then structured the implementation process so as to virtually assure minimal departure from the *status quo ante:* The governing board had a majority of local officials. Any permit application was automatically approved unless denied by a majority of *both* the California and Nevada delegations within 60 days. The TRPA had no authority over casino and public works construction in Nevada. And the agency was heavily dependent for its funding from the same local governments that had vigorously opposed its creation. In short, a brief comparison of these two statutes serves as an excellent example of the extent to which statutes can constrain the implementation process.
82. As Charles Schultze has persuasively argued, traditional regulatory approaches that rely heavily upon legal rules to direct very heterogeneous activities run the risk of gross inefficiencies (Schultze, *The Public Use of Private Interest*, pp. 6–27, 46–64). On the other hand, Theodore Lowi and Kenneth Culp Davis have persuasively criticized the extensive discretion that is usually allowed administrators and have called for systems in which case-by-case adjudication is encouraged/required to develop rules; see Theodore Lowi, *The End of Liberalism* (New York: Norton, 1969), chap. 10; and Kenneth Culp Davis, *Discretionary Justice* (Urbana: University of Illinois Press, 1969).
83. California Public Resources Code, Division 18, 1973, Section 27243.
84. Recall from Table 2-1 that the residents of the North Coast region voted overwhelmingly against Proposition 20. One would expect local officials to mirror this sentiment and that, given one or two state appointees who were also opposed, a hostile majority would result. The situation in the San Diego region was somewhat different. Although Proposition 20 passed, it was only by a modest (52%) majority. Thus, many local representatives on the commission would probably be opposed. Moreover, as the region comprised only a single county, a regional perspective was really impossible (interview with Janet Adams, December 2, 1975).
85. Swanson, "Coastal Zone Management," p. 90.
86. Sabatier interview with Celia von der Muhll, one of the two coordinators, 24 October 1975.
87. Downs, "The Issue-Attention Cycle," pp. 38–50.

3

BEYOND THE REACH OF THE COASTAL ACT

External Factors Affecting the Coastal Commissions

The coastal commissions were provided statutorially guaranteed autonomy and had sweeping formal powers over all development in the 1000-yard interim permit area. Nevertheless, there was a wide range of external forces that would affect their decisions. We now turn to a brief sketch of these forces, beginning with the pressures created by the changing needs and resources of the state's population, public opinion, and media attention. Then examined are the roles played by organized interest groups, the commissions' legal superiors (i.e., the legislature, governor, courts and U.S. Office of Coastal Zone Management), and peer agencies in the state government and their local counterparts.

I. NEEDS AND RESOURCES

Political demographers have long argued that the needs and resources—the size, density, level of wealth and education, and consumption patterns—of a population, in the broadest sense, drive its economic and political systems. They supply the "must" and "can't" ingredients of policy decisions.[1] In the words of Heinz Eulau and Kenneth Prewitt, "much of what policy-makers do they do because the relationship between policy environment and ecological environment, as it emerges through time, orients their conduct and actions."[2] The coastal commissioners would be no exception.

The greatest pressures on the coastal commissions were the needs generated by the rapidly growing urban, highly educated, recreation-oriented, and environmentally aware population of the state. Yet they would not always push the commissions in the same direction. As people became better educated, wealthier, and more mobile, their demand for coastal parks and recreational opportunities would grow and their awareness of environmental issues would likely expand. Both factors could be expected to stimulate support for the goals of regulating and preserving coastal areas as a unique public asset. Yet a growing population would also create demands for living space, jobs, roads, utilities, and port facilities, which sometimes by necessity and sometimes by choice were more and more often being located along the coast. And to the extent that coastal protectionist and access policies were seen as contributing to the cost of available living space and/or to major economic dislocations, pressure would build to return to reliance on marketplace decisions for the allocation of coastal resources.

The trends in three key areas during the life of the commissions would, therefore, make more difficult the protectionist efforts of the commissions. To begin with, growth of the state's population continued unabated, consistently surging upward from the approximate one-quarter million annual increase in the late 1960s and early 1970s to a 350,000 annual increase by 1976 (see Table 3-1,

TABLE 3-1
NEEDS AND RESOURCES

A. Relatively volatile factors
  1. Population growth

	State[a]	15 coastal counties[b]
1960–1969 average	401,900	256,860
1971	205,500	136,700
1972	189,000	79,900
1973	235,000	108,500
1974	257,100	108,700
1975	306,500	—
1976	350,400	—

  2. Increase in market price of homes, statewide and in coastal counties of Southern California,[c] April 1970–October 1975

		Planning area		Inland area	
		Semi-annual increase	Cumu-lative increase	Semi-annual increase	Cumu-lative increase
	Apr. 1970–Oct. 1970	2.0%	2.0%	1.5%	1.5%
	Oct. 1970–Apr. '71	2.6	4.7	1.7	3.2
2½ years prior to	Apr. 1971–Oct. '71	3.2	10.8	1.9	5.2
commissions	Oct. 1971–Apr. '72	3.7	14.9	1.7	7.0
	Apr. 1972–Oct. '72	3.0	18.3	2.0	9.0
Transition	Oct. 1972–Apr. '73	5.4	5.4	3.0	3.0
	Apr. 1973–Oct. '73	6.2	6.2	2.9	2.9
	Oct. 1973–Apr. '74	8.1	14.8	4.5	7.5
2½ years during	Apr. 1974–Oct. '74	8.9	25.0	7.2	15.2
commissions	Oct. 1974–Apr. '75	7.1	33.9	7.0	23.3
	Apr. 1975–Oct. '75	6.6	42.7	8.4	33.7

(*continued*)

TABLE 3-1 (Continued)

3. Unemployment/employment, 1970–1976[d]

	% unemployment, statewide	N persons employed, statewide	N persons employed in contract construction, statewide
1970	7.2%	5,581,102	303,013
1971	8.8	5,516,782	301,027
1972	7.6	6,176,931	319,946
1973	7.0	6,647,521	344,797
1974	7.3	6,864,388	329,634
1975	9.9	6,843,350	303,412
1976	9.2	7,144,442	318,245

B. Relatively stable factors, by state and coastal region

	State	North Coast	North Central	Central	South Central	South Coast	San Diego
1. Median years of education—1970[e]	12.4	12.2	12.5	12.5	12.5	12.4	12.4
2. Per capita income—1973[f]	$5,487	4,423	7,157	5,960	4,579	5,782	5,000
3. Per capita acres oceanfront, parks, and recreation—1973[f]	.05	1.32	.16	.26	.30	.01	.06

[a]Security Pacific Bank, Research Department, "Estimated Population Increases in California," January 20, 1978.
[b]Security Pacific Bank, California Coastal Zone Economic Study: An Area Profile (Research Department: Security Pacific Bank, 1975), pp. 3–12.
[c]Includes Santa Barbara, Ventura, Los Angeles, Orange, and San Diego Counties; Security Pacific Bank, Research Department, "Increases in Market Prices of Homes in the Coastal Counties of Southern California," undated.
[d]Figures on unemployment taken from California Statistical Abstracts, Table C-7 series, "Civilian Labor Force, Employment and Unemployment for Metropolitan Areas and Counties"; figures on employment statewide and in consruction taken from California Statistical Abstracts, series on "Average Monthly Employment Covered by the Unemployment Insurance Code by County and Industry Division."
[e]Computed from median education by county figures, 1970, in Security Pacific Bank, An Area Profile: Statistical Appendix.
[f]Computed from income, parks, and population figures in Security Pacific Bank, An Area Profile, pp. 2–4, 3–13, 5–6.

part A). This appeared to corroborate the dire projection by the California Department of Finance that between 1970 and the year 2000 the coastal counties would have to absorb some 7.7 million new residents, 3 million of which would locate in the 5-mile coastal planning area.[3] Directing this population away from the coast would not be easy.

Second, as the demand for coastal locations grew—for recreation, for commercial development, for residential use—so did their costs. In the resale of houses, for example, the trend was quite marked. As shown in part A of Table 3-1, the market price of housing in the 5-mile planning area for the five southern California coastal counties rose 18.3% in the 42 months prior to the establishment of the coastal commissions, compared to only 9.0% in the inland areas of the counties. In the 30 months from April 1973 to October 1975—the first 2½ years of the coastal commissions—housing prices in the planning area jumped 42.7%, precipitating the charge that through restricting building, red tape, and delays the commissions were pricing the middle class out of coastal homes. Yet, during the same period, prices inland rose 33.7%. In other words, although housing prices were rising faster along the coast than inland, the magnitude of the difference (about 9%) was the same before and after the establishment of the commissions. Nevertheless, it was widely believed that the coastal commissions and related regulatory bodies bore a good part of the responsibility for the housing inflation,[4] and thus, the commissions came under sustained attack by builders and coastal property owners.

Third, even more ominous for the commissions was the issue of unemployment. Historically, the northern counties of the state had higher rates of unemployment than those in the south. For example, the 1970 census shows that while the lower five coastal commission regions' unemployment held at approximately 7.2%, unemployment in the North Coast region was running at 10.7%.[5] This situation changed abruptly during the life of the commissions. Unemployment, statewide, went from 7.0% in 1973 to 7.3% in 1974, to a shocking 9.9% in 1975 and 9.2% in 1976 (see Table 3-1, part A).

The construction industry was especially hard hit by the 1974–1975 recession. As a consequence, business and the building trade unions became even more vocal opponents of the commissions. Figure 3-1A and 3-1B trace the plight of construction workers from 1970 through 1976. It shows, first, that despite the high rate of unemployment in 1974 and 1975, the absolute number of employed persons in all categories continued to rise throughout the 1971–1976 period, with only a slight dip in 1975 (Figure 3-1A). Employment in construction, however, after surging forward from 1971 to 1973, plummeted not only in terms of the percent of those looking for but not finding work, but also in absolute numbers in 1974 and again in 1975 (Figure 3-1B). With a resurgence in building in 1976, a partial recovery occurred in the numbers employed in construction, but not to the high mark for the decade reached in 1973.

Overall, construction employment in coastal counties mirrored the trend statewide. Of significance for the kinds of pressures this would bring on the coastal commissions, the decline in employment varied notably from the coastal regions in the north to those in the south. This can be seen by comparing the employment in construction for the counties in the South Coast region to those in the North Central region in Figure 3-1B. Construction employment in the South Coast counties, which accounted during the period for about 40% of all construction employment statewide and 60% of that in the coastal counties, rose from 1971 to 1973, only to fall precipitously to the lowest point of the decade by 1975. Construction employment in the San Diego region followed the same

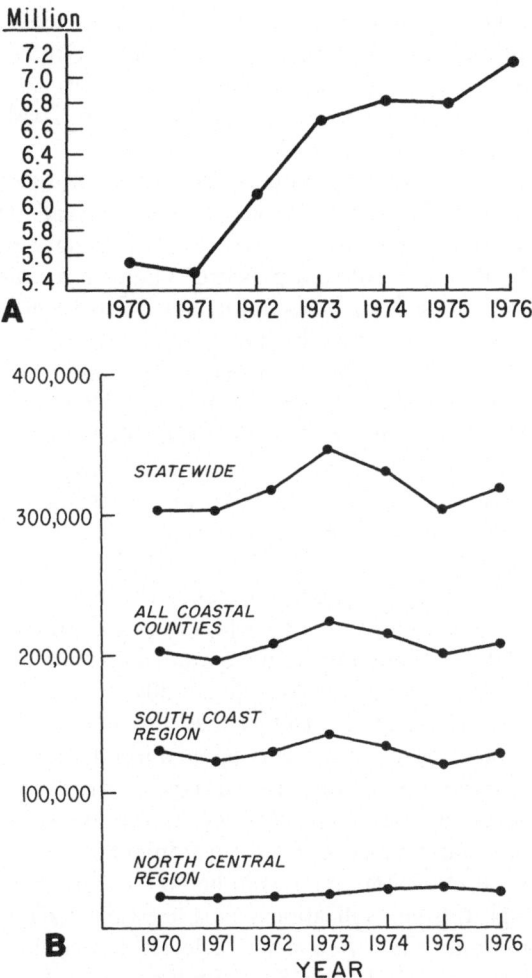

FIGURE 3-1 A. Number of persons employed statewide in all categories, 1970–1976. B. Number of persons employed in contract construction, 1970–1976.

pattern, with an even more dramatic drop-off in the 1973–1975 period. Meanwhile, total employment in construction in the counties of the North Central region continued along a gradual pattern of growth through 1975, thus avoiding the harshness of the 1974–1975 recession, only dropping off slightly thereafter.

Data on construction employment in the counties of the North Coast region are incomplete, but on the basis of figures for Humbolt and Mendocino counties (the two most populous counties in the region) for 1971 and

1974–1976, it appears that construction in the region followed the path of gradual growth through 1975 experienced in the North Central region. In contrast, construction employment in the counties of the Central and South Central regions more closely approached the trend in the South Coast and San Diego regions.

A fourth demographic pressure operating on the commissions throughout its first 4 years was the need for more coastal parks and recreational opportunities, particularly in Southern California. In Table 3-1, part B, parks per capita is listed as relatively stable, that is, a constant pressure. The regional breakdowns reveal the relative wealth in coastal parklands in the North Coast region (1.32 acres per capita) as compared with the rest of the state, particularly the South Coast region (.01 acres per capita). Placing parkland per capita under the relatively stable column is not to say that per capita acres of coastal parkland remained unaltered from 1973 to 1976, only that it remained reasonably constant, particularly in contrast to the more volatile factors such as unemployment. By listing the per capita parkland figures as relatively stable, we are suggesting only that they were minimally affected downward by the additional growth in the state's population and upward by the addition of costal parklands made available for use by the public (as opposed to those purchased or designated for purchase) during the implementation period. Indeed, due to delays in actual purchase and development of new coastal parks, any impact the commissions might have in this area would not become visible until well into the 1980s.

Two additional demographic factors that also remained relatively stable throughout the implementation process need mention, both of which have been found in previous studies to be associated with government activism in general. The level of educational attainment of a community usually serves as a good barometer of a community's receptivity to government activism. The more educated the community, the more it tends to be "public-minded" and trusting of government (though this generalization seems less valid today). Moreover, the more educated a community, the greater its level of environmental concern.[6] However, as shown in part B of Table 3-1, with the exception of the North Coast region, little variation existed in level of education along the coast and, therefore, this factor cannot be expected to account for much of the variation in regional commission behavior.

Just as level of education is usually a good guide to level of support for government activism, financial resources are viewed as setting the bounds of "capability."[7] Real income and property wealth are the common measures of a community's resource level.[8] In the case of the coastal communities, the best available measure of resource capacity is per capita income. California has traditionally been considered a wealthy state, supporting a wide range of governmental activities and services, so that its support for Proposition 20 was not surprising. But variations within the state are noteworthy. For instance, as of

1973, coastal counties as a whole had a higher per capita income than the remainder of the state: $5734 versus $5060. Within the coastal regions a good deal of variation existed too. Per capita income ranged from a low of $4423 for the counties of the North Coast region to a high of $7157 in the North Central region (see Table 3-1, part B). The less wealthy a region, according to customary demographic analysis, the less it would be willing to absorb the costs associated with new public programs, including those likely to be initiated under the Coastal Act. Conversely, the wealthier the community, the more likely it would be to bear added burdens.

Needs and resources as measured in aggregate statistics, of course, do not automatically translate into policy. Yet the high correlations often found among demographic factors, spending, and government activism should alert us to the fact that they do percolate up through the political system. We should therefore expect to find that as population grows and beach recreation is held at a premium, yet availability of coastal property to the general public is diminishing, pressure for public action will grow, e.g., that beach access and provision of recreational opportunities would be of uppermost importance in the South Coast and San Diego regions. These were also, however, the regions where the pressures for development were the strongest. The competing pressures would obviously have to be reconciled by the coastal commissions. In the far less dense and less wealthy counties of the North Coast region, there would probably be much interest in having the state provide greater environmental protection and recreational opportunities and in shouldering the burden of an expanded regulatory role by the state. Yet these tendencies are only suggested in general through a demographic analysis. They would ultimately be affected by the public's attitude toward coastal regulation, the role played by state and local officials, the role of organized interests, and the attitudes of the commissioners and staff recruited to the commissions.

II. Public Opinion and the Mass Media

Anthony Downs has suggested that public support for most issues, and particularly for environmental protection, follows a cyclical pattern.[9] The cycle begins with the preproblem stage, where a "highly undesirable social condition exists but has not captured much public attention." Then comes the alarmed discovery often precipitated by some dramatic event, such as a major oil spill or an accident at a nuclear plant, as well as "euphoric enthusiasm about society's (government's) ability to 'solve the problem' or 'do something effective' within a relatively short period of time." For most major problems, however, solutions are seldom easily attainable, especially when they require basic systemic and social changes that, naturally, are not the kinds of solutions most people had envisioned. Over the course of several years this realization breeds frustration and disillusionment, which in turn lead to a gradual decline in public interest and support for strong government action. Meanwhile, other issues begin to emerge and to capture the public's imagination, becoming the new focus of public debate. Thus, the final stage arrives, where the issue either has been eclipsed or at least now shares the limelight with others, government institutions have been established to deal with it, and the public is no longer as emotionally charged and actively concerned about it. The issue may remain important, but with rare exception it is no longer highly salient.

In many ways, the concern of Californians with coastal protection appears to have moved through the issue-attention cycle between the late 1960s and the middle 1970s. By the late 1960s many Californians became concerned with the rapid deterioration of coastal resources and reduction in beach access. This resulted in several legislative efforts at enacting a coastal protection bill. By 1972 the concern was sufficiently widespread to make viable a mass-based initiative drive, which itself served to galvanize public sentiment, and which resulted in establishment of a government agency—the coastal commissions—to address the problem. But the process was barely under way when other issues entered the public arena—most notably the energy crisis of 1973–1974 and the recession and unemployment of 1974–1975—thereby pushing the entire spectrum of ecology issues, including the coastal commissions, from center stage. The process was accelerated by the vocal opposition to the commissions by labor and business leaders, who opposed the commissions as one of the most visible and powerful environmental regulatory agencies and as a supposedly important contributor to the recession and unemployment in the state, especially in the construction industry. Also, by 1975 it became evident to many that a quick and effective "solution" to the problems of coastal degradation would require more than cosmetic treatment. The coastal commissions through their permit powers might delay or even occasionally curtail the development spiral, but much more would be needed. Changes would be required in the way land use, growth, taxation, and energy consumption decisions were being made not only along the coast but throughout the state. The controversies that ensued were heated, involving complex ecological, economic, and legal issues, and often frustrated and discouraged both participants and the public. It was this chain of events, we believe, that led many to back off from an unswerving conservationist and protectionist stance toward the coast and other natural resources by the mid-1970s and look for a new balance between environmental protection and economic development.

The shift in emphasis is suggested in the public opinion surveys for the period. That environmental concerns had entered the alarmed discovery stage by the early 1970s is shown in the results of a statewide survey conducted in early 1971 by the highly respected California Poll. The survey found more Californians (78.3% of the respondents) ranking air and water pollution as problems on which more government money should be spent than any other issue. Also, 53.4% felt more money should be spent on preserving and protecting wilderness and scenic areas, and 47.6% were willing to spend more on acquiring and maintaining state parks and beaches.[10]

Four years later, however, a January 1975 survey commissioned by the California Council for Environmental and Economic Balance (CCEEB) reflected the public's increasing awareness of the interrelatedness of environmental protection and economic development. The survey revealed, first of all, that a substantial number of Californians remained unshaken in their commitment to environmental protection, at least along the coast. Indeed, fully 38% agreed with the statement that the commissions should impose a building moratorium along the coastline (see part A of Table 3-2). Despite this widespread support for what was clearly an extreme position, the public was at the same time sensitive to economic necessities. Consequently, fully 64% chose "more jobs" when choosing between creating more jobs and protecting the environment; 20% chose "protecting the environment," with 16% either saying neither or not responding (see part B of Table 3-2). Moreover, when asked to identify one statement among five that best approximated their basic perspective, most respondents (43%) selected "We have to strike a balance between the need for environmental controls and the need for jobs," and the second largest group (27%) selected "The environment should be protected where possible, but you have to be flexible where people's jobs are concerned" (part C of Table 3-2).

To know that the public supported environmental and coastal protection was noteworthy but, as is clear by early 1975, insufficient information. A large proportion of the public had adopted the position that protection and jobs were often in conflict and the critical issue had become striking the proper balance between the two. Thus, although environmental and coastal protection issues had not been eclipsed, and the poll results at least suggest that the public may have viewed the coastline as uniquely in need of protection (a point to be returned to in Chapter 8), they now had to share the limelight with the competing issues of jobs and economic growth—particularly during a period of 9% unemployment in the state.

An important facet of the issue-attention cycle is the breadth and kind of attention given an issue by the media. Media reporting requires drama, excitment, and controversy. Indeed, Downs argues that "a problem must be dramatic and exciting to maintain public interest because news is 'consumed' by much of the American public . . . largely as a form of entertainment."[11] Thus, when an issue moves into the alarmed discovery stage, the media pick up on it immediately, often serving as an important catalyst to public attention. In turn, when a new governmental agency is established to address the issue, it is accorded extensive and usually sympathetic coverage, at least by major papers and television. Over

TABLE 3-2
CCEEB SURVEY ON JOBS AND ENVIRONMENT, JANUARY 1975

A. The state and regional coastal commissions should stop entirely any type of construction along California coastline.

	Statewide	L.A./Orange	Rest of So. Cal.	S.F. Bay area	Central Valley
Agree	38%	33%	36%	43%	52%
Neither agree/disagree	25%	23%	33%	24%	23%
Disagree	37%	43%	31%	34%	24%

B. Generally if you had to choose between creating more jobs and protecting the environment, which would you personally choose?

	Statewide	L.A./Orange	Rest of So. Cal.	S.F. Bay area	Central Valley
More jobs	64%	67%	65%	59%	57%
Protecting environment	20%	18%	21%	26%	20%
Neither/no response	16%	15%	14%	15%	23%

C. Which of these statements comes closest to your views?

	Statewide	L.A./Orange	Rest of So. Cal.	S.F. Bay area	Central Valley
We have to protect the environment at all cost, even if it means some people might lose their jobs.	7%	5%	6%	10%	8%
We have to strike a balance between the need for environment controls and the need for jobs.	43%	38%	47%	50%	43%
The environment should be protected where possible, but you have to be flexible where people's jobs are concerned.	27%	30%	24%	22%	26%
The environment is pretty well protected and the main task now is to find more jobs for working people.	11%	13%	14%	6%	15%
In every instance the environment has to take a back seat when people's livelihood is concerned.	8%	10%	7%	8%	5%
No opinion/no response	4%	4%	2%	5%	3%

time, however, the media tends either to lose interest in the subject or to focus on the costs and hardships imposed by the government agency, or both. The interaction between public disillusionment and declining or less supportive media attention can accelerate the downward spiral of the issue-attention cycle.

The experience of the coastal commission fits the pattern only partially. The major metropolitan papers in both Los Angeles and the San Francisco Bay area (e.g., *Los Angeles Times, San Francisco Chronicle*) covered coastal protection issues fairly closely and accurately, with extensive coverage during the campaign for Proposition 20 in 1972 and the early developments under the commissions and then again during the development of the Coastal Plan and the battle over a successor agency in 1975–1976. During the latter period, though, a good deal of coverage was given to the drama of hard-hat protests at planning hearings and opposition arguments. A similar pattern of reporting prevailed in the smaller metropolitan papers such as the *San Rafael Independent Journal* and *Santa Rosa Press Democrat* in the North Central region. Editorial comment was not always supportive at this level, however, as in the case of the *San Rafael Independent Journal*. The sympathies and slant of the small coastal town papers was usually different. The tendency was for far more reporting of the hardships imposed by the commissions on local property owners and developers than of the commissions' efforts to secure access for the general public and the protection of coastal resources. Moreover, a basically negative tone was often reflected in their editorial comment. [12]

On the basis of this admittedly limited and impressionistic data, it appears that the major media coverage followed only in part the attention cycle identified by Downs. Over time, attention to the commissions did decline as an almost inevitable consequence of the media's need to continually uncover new issues and new dramas. Yet support for the commissions by the major papers, especially the *Los Angeles Times*, remained fairly constant throughout the 1973–1976 period. On the other hand, the smaller papers servicing the coastal communities seldom endorsed the commissions editorially, and their coverage tended to emphasize costs on property owners.

The media, of course, do not operate in a vacuum. The governmental agency involved shoulders part of the responsibility for ensuring that the full and, one hopes, impartial story of the agency's activities is placed before the public. Taken to the extreme this could obviously degenerate into a self-serving agency public relations campaign. Yet as an antidote to the complaints by those being regulated aired by the media, a balancing role is certainly warranted. In the case of the coastal commissions, this might have involved establishing continuing liaison with the metropolitan and small town press and television networks and supplying usable stories about the commissions' activities and accomplishments. An example would be material on the commissions in requiring a beach access dedication as an important condition for many of the permits granted along the shoreline and, possibly even more important, how these dedications were in keeping with the objectives of the Coastal Act and maintaining the shoreline for use by the general public. In practice, however, the media typically reported the controversies over access dedications as unconstitutional usurpation by the commissions of private property. But the "usurpation" was not very different from that by any municipal government requiring sidewalk and setback dedications.

There are several reasons why the commissions did not communicate their story as well as they might have to the public. First, the chairman of the state commission, Melvin Lane, and the executive director, Joseph Bodovitz, did carry on occasional conversations with the editors and staff of the *Los Angeles Times*,[13] which they seem to have considered sufficient media contact. But they failed to provide the kind of supportive material on the commissions' accomplishments that could provide an effective defense of the agency. Second, in the initial organization of the commission staff, a decision was made not to employ a formal media liaison, partly due to the confines of a limited budget, but also because it was assumed that both Bodovitz and Jack Schoop (chief planner) as ex-newsmen could attend to the media. Finally, Bodovitz and others were firmly convinced that support by the public, the media, and the active political community would arise if the commissions simply concentrated on doing the best job possible both in permit decisions and in development of the Coastal Plan.[14] This approach did not prove very effective, and thus less of the commission's story was presented by the media than might have been the case. The limited effort and ability of the commissions to forestall the issue-attention cycle from running its natural course would eventually affect public support for the commissions and the 1976 legislature's reception of the commissions' Coastal Plan.

III. Citizen Activists and Interest Groups

A factor not mentioned by Downs in his discussion of the issue-attention cycle is the level of citizen and organized interest group activity in the life of the government agency created to solve the problem. Yet one of the key factors in our six conditions of successful implementation is the presence of an organized and active supportive constituency with the resources to closely monitor a program's implementation, to intervene in agency proceedings, to appeal adverse agency decisions to the courts and the legislature, and to persuade key legislators and agency officials to retain the statutory priorities in the face of declining public attention. To reiterate our hypothesis: The greater the support from interested citizens and active monitoring from organized interest groups, the less likely the life cycle of an agency will follow the issue-attention cycle of the general public; the less likely the initial enthusiastic and zealous agency leaders will be replaced by more career-oriented and accommodating ones; the less likely the agency will become more responsive to the interests being regulated than to those supporting regulation; and the more likely the objectives of a stringent statute, enacted during the alarm and discovery state of an issue cycle, will be realized.[15]

With respect to the coastal commissions, a substantial supportive constituency did exist. It was evident in the campaign for the passage of Proposition 20 (see Chapter 2) and through the implementation of the Coastal Act in both the planning and permit processes. For example, as shown in Table 3-3, opponents to development spoke on nearly one-third of all the permits that went before the North Coast, North Central, and South Coast regional commissions from 1973 through 1975.[16] They participated most frequently in the North Central regional hearings (44.0% of the time) and least frequently in the South Coast (30.7%). Two factors would seem to account for the difference. First is the relatively large number of environmental protectionist groups in the San Francisco Bay area. Second, the larger number of permits handled by the South Coast made monitoring of each by opponents to development more difficult, thus forcing them to be more selective in targeting their opposition.

Naturally, not all permit opponents (or proponents, for that matter) were alike. An especially noteworthy distinction is between participants who spoke at hearings as individuals and typically on an *ad hoc* basis and those representing organized interests.[17] Just how active the organized interests were on permits is shown in Table 3-4, which identifies participation by organized groups across three regions.

The table shows that the formally constituted groups in support of coastal protection—Sierra Club, Audubon Society, PACE (People, Access, Coastal Environment)—entered directly into 9.8% of the nearly 1800 permits heard before the three commissions from 1973 through 1975. It also shows that in terms of the percent of permits participated in, the regional variation was substantial; it ranged from a high of 24.5% in the North Coast and 24.9% in the

TABLE 3-3
PERMITS WITH OPPOSITION BY REGION AND YEAR, 1973–1975

	Permits with opposition[a]							
	North Coast		North Central		South Coast		Three-region total	
Year	%	N	%	N	%	N	%	N
1973	18.5	(17)	33.8	(25)	27.7	(135)	27.1	(177)
1974	54.9	(39)	48.6	(36)	31.5	(164)	35.9	(239)
1975	34.9	(38)	62.1	(18)	34.3	(116)	36.1	(172)
Total	34.9	(94)	44.0	(79)	30.7	(415)	32.7	(588)

[a]Opposition includes both individuals and group representatives who spoke at the permit hearing.

TABLE 3-4
PARTICIPATION BY KIND OF GROUP IN PERMIT REVIEW

Percentage of permits

Group	North Coast region (N = 273)			North Central region (N = 177)			South Coast region (N = 1348)			Three-region total (N = 1798)		
	Supporting	Opposing	Total participation	Supporting	Opposing	Total participation	Supporting	Opposing	Total participation	Supporting	Opposing	Total participation
Statewide environment or planning groups[a]	1.5%	23.4%	24.5%	0.6%	24.9%	24.9%	1.6%	3.3%	4.8%	1.4%	8.5%	9.8%
Homeowners or local environmental group	0.4%	6.2%	6.6%	2.3%	18.1%	19.8%	1.9%	9.7%	11.1%	1.7%	10.0%	11.2%
Business/unions	1.5%	1.1%	2.6%	4.0%	0.5%	4.5%	9.4%	0.1%	9.5%	8.7%	0.3%	8.0%
Recreation group	0.0%	0.0%	0.0%	0.0%	0.0%	0.0%	1.3%	0.0%	1.3%	0.9%	0.0%	0.9%
Government agency	2.6%	3.7%	6.2%	2.3%	2.3%	4.0%	11.7%	0.5%	12.2%	9.4%	1.2%	10.5%

[a]Includes both the overall coordinator and the local chapter of the Sierra Club.

North Central to only 4.8% in the South Coast. In terms of the absolute number of permits this represented, however, the variation is less dramatic; 67 permits in the North Coast, 44 in the North Central, and 65 in the South Coast. These latter figures suggest that despite the greater number of permits handled by the South Coast, the statewide environmental groups spread their resources and energies across all three regions fairly evenly, selecting a roughly equivalent number of permits to challenge in each. Supplementing the efforts of these statewide organizations were at least three or four local homeowners and environmental groups (e.g., the Venice Town Council, the Orange County Environmental Coalition) in each region. In the South Coast region, as an example, these groups participated in over twice the number of permits that involved the statewide groups (11.1% versus 4.8%). Finally, it should be noted that although the supportive constituency directed most of its efforts at preventing development in critical areas and what it considered the worst types of development, in approximately 15% of the cases when either statewide environmental or local homeowner and environmental groups spoke out, it was in favor of a permit.

Table 3-4 also shows that despite the disproportionate opposition to Proposition 20 and coastal regulation by business and labor, they were far less active in providing day-by-day support of development than might have been expected, and especially in the North Coast and North Central regions, they were far less active in permits than was the supportive constituency. In the South Coast, business and labor groups were most active, participating in 9.5% of the permits. Yet this was still less than the combined participation by statewide and local opponents to development. Although, when the substantial support for granting permits by local government agencies in the South Coast is added to the equation—a factor that did not exist in the two northern regions—the level of activity of organized interest groups as indicated by their participation in permits is shown to be the most balanced in this region.

While participation on permits would serve to keep the commissioners ever sensitive to the views of individuals and organizations on specific development projects, it was only part of overall monitoring and coordinated action required of an enduring supportive constituency. However, that also existed. In California there existed a long-standing tradition of local coastal conservation groups,[18] the Sierra Club, and the coalition of other environmentally oriented forces that had come together to win passage of Proposition 20. Thus, there was a substantial constituency in place from the outset to monitor implementation of the Coastal Act. Probably the most important step in coordinating the relatively large but loosely knit constituency was the decision by the Sierra Club to employ two full-time coastal coordinators (funded by the Sierra Club Foundation), one for the north and one for the southern half of the state. The coordinators served as valuable sources of information throughout the supportive community on developments in the Coastal Plan, on developing positions on permits at the regions, and often taking the lead on appealing regional decisions to the state commission. They were, of course, aided in their monitoring of the commissions by the Audubon Society, PACE, Friends of the Earth, the Lake Merced Council, local Sierra Club chapters, and many of the constituent groups and individual members of the Coastal Alliance. With the exception of the Sierra Club's coastal coordinators and a $100,000 trust fund at the disposal of the Lake Merced Council (established by a developer in an out-of-court settlement) eventually used to establish PACE, however, the financial resources of the supportive groups were meager. Nevertheless, they were able to call upon a good deal of voluntary legal, scientific, and organizational talent to aid in developing positions on permits, evaluating elements of the Coastal Plan, and testifying at the numerous regional and then state-level hearings on the plan (see Chapter 8), as well as lobbying in favor of the plan and a successor coastal agency before the 1976 legislature (see Chapter 9).[19]

The supportive constituency was not, of course, the only organized interest concerned with the commissions. Many corporations—particularly among utilities, oil companies, and large development firms—monitored permit hearings and testified on relevant sections of the Coastal Plan. Organized interest groups, involving farmers, timber companies, the oil companies, and others, were also active in the planning process. But by far the most active organization representing a fairly wide spectrum of businesses and labor was the California Council of Economic and Environmental Balance (CCEEB). CCEEB was formed shortly after the passage of Proposition 20 to serve as a concerted voice of the moderate business and labor community, which was progrowth but accepted as inevitable some degree of new state environmental regulation. One of CCEEB's main activities was to monitor the coastal commissions. At the permit level, this was done primarily in the South Coast region and the state commission, although CCEEB intervened directly on only the San Onofre nuclear power plant permit controversy.[20]

CCEEB also played a very active role in the planning process, providing research and communication functions for its member organizations, publishing a monthly newsletter, undertaking analysis of each of the elements of the Coastal Plan, and testifying at virtually all the hearings on the plan. CCEEB eventually played a prominent role in the 1976 legislative consideration of the plan (see Chapter 8) and was instrumental in the passage of the important State Urban and Coastal Bond Act of 1976.

IV. AGENCY SOVEREIGNS

The Coastal Initiative was especially designed to insulate the coastal commissions from the kinds of political pressures that usually flowed from the governor and the legislature as a result of their ability to set and change the agency's legal authority, monitor its activities, and determine its level of funding. For example, with the exception of his appointment powers, neither the governor nor his state Resources Agency had formal authority over the commissions. Nevertheless, some degree of cooperation was required, and from the commission's perspective, support from the governor was needed in certain instances.

Somewhat surprisingly, given Governor Ronald Reagan's public antipathy toward the commissions, his aides worked closely with the commissions in ' initially setting up offices and recruiting staff and subsequently in responding to additional needs as they arose (such as providing budget support for legal counsel from the attorney general's office and endorsing the commissions' request for supplemental funding from the legislature).[21] One exception to the cooperative and basically noninterfering relationship with the governor should be noted. After the commission denied the extremely controversial additional units to the San Onofre Nuclear Plant, Chairman Lane was asked to explain the commission's decision to the governor's entire cabinet.[22] As for the Coastal Plan being developed by the commissions, the governor's office showed little interest.

The state legislature, which exercises control over agencies through oversight hearings, the annual budget review process, and statutory revisions, was also denied much of its usual prerogative through the automatic $5 million provided by Proposition 20 from the Bagley Conservation Fund for funding of the commissions, and the two-thirds vote requirement for weakening an initiative statute. Even when sizable supplemental funding requests were made by the commissions (over $1.6 million between 1974 and 1975), the opportunity was not noticeably used by the legislature to influence commission activities. Finally, interviews with commissioners indicated that the legislative appointing authorities—the speaker and the Senate Rules Committee—seldom attempted to influence permit and planning decisions.

The same pattern of noninterference held for the U.S. Office of Coastal Zone Management (CZM), which, through its funding of state coastal programs, could have exercised a fair amount of influence. However, OCZM was itself just getting established in 1973 and viewed the California commission as the prototype aggressive state coastal resource protection agency that it would like to encourage.[23] With little more than *pro forma* review, OCZM made fairly generous special projects awards to the commissions from 1974 on; $720,000 in 1974, $900,000 in 1975, and $1,200,000 in 1976.[24]

The Initiative could not insulate the commissions from the judiciary, however. Thus, opponents (and proponents) of the act immediately turned to the courts. By the time the commissions had completed the Coastal Plan, the attorney general's office had been involved in nearly 240 cases dealing with implementation of the act.[25] By late spring of 1977, 27 of these cases had been pursued through the court of appeals or the state supreme court.

Although the issue of the vested rights of property owners was by far the most litigated issue,[26] the most important issues raised in the courts went to the basic constitutionality of the commission's powers. These were largely resolved in the *CEEED* case, decided by the California appellate courts in November 1974.[27] Among other things, the court of appeals rejected plaintiffs' contentions that the Coastal Initiative (1) violated the due process rights of affected property owners and (2) constituted an unlawful taking of private property without just compensation. On the former, the court ruled that the Initiative essentially called for the development of a provisional plan and thus did not require all of the procedural safeguards of a zoning ordinance. With respect to the "taking" issue, the court (a) reaffirmed its position in the *Candlestick Properties* case that conservationist objectives constituted a legitimate use of the police power and (b) ruled that the temporary nature of the commissions' permit and planning decisions did not constitute a definitive "taking."[28]

The decision in CEEED set the tone of the California appellate courts in dealing with all subsequent challenges to the commissions' powers. In fact, of the 20 state appellate court decisions issued in the 1973–August 1976 period, the commissions were upheld in all but one.[29] In denying the claims of plaintiffs, the court continued to reiterate its position that placing restrictions on land use in pursuit of a permissible state objective, of which environmental protection was clearly one under the *Candlestick Properties* ruling, was neither unconstitutional nor a matter of "taking." Even where stringent restrictions were imposed, such as denying a current development permit in order to hold the land for possible future public use, the courts held that the restrictions were only an interim measure designed to protect the integrity of the coastal planning process.[30]

Sympathy from the appellate courts did not preclude the necessity of litigating the several hundred cases challenging the powers and decisions of the commissions. Yet the initiative made no provision for in-house legal counsel, nor did the commissions' limited budget allow for any. Even with in-house counsel, suits against state agencies must be argued in court by the attorney general's office. This placed the attorney general in a pivotal position vis-à-vis the commissions. Not to defend them or, even more damaging, to defend them inadequately could easily result in the judicial dismantling of the commissions' powers. Fortunately for the commissions, throughout the 1973–1976 period the attorney general's office provided excellent counsel.[31] Nevertheless, in our litigation-prone society this serves as a reminder that an attorney general's office can be an important veto point over an implementing agency.

The extremely liberal interpretation of the commissions' powers by the appellate courts could not ensure the sympathy of the lower courts, the very level where most cases were resolved. After all, for a private party to take a case to the appellate level usually requires an outlay of $50,000 or more.[32] While we have no evidence of widespread and flagrant violation of the spirit of CEEED by lower courts, one exception is worth noting. It occurred in Mendocino County. Of the two superior court judges in the county who would normally hear suits involving the coastal commissions, one disqualified himself from all such cases, and the second, believing that Proposition 20 was unconstitutional, refused to enforce the act.

Overall, we conclude that the courts played a very supportive role in the implementation of the Coastal Act.

V. Peer Agencies and Their Local Counterparts

Although the governor, OCZM, the legislature, and the courts gave the commissions a rather free hand, the commissions could not operate in an affirmative way without becoming a part of the web of federal, state, and local agencies with planning and resource development responsibilities that affected the coastline.[33] Although the commissions had the authority to review the proposals of other agencies and to veto development proposals that would have an adverse effect on the coastal environment, they did not have the authority or resources to implement their own ideas. For example, while the commissions could veto a proposed private development because of the property's potential value for public recreation, they could not purchase the land and develop a recreation facility. That required the cooperation of the state Parks and Recreation Department and the Public Works Board.

Furthermore, it proved difficult to halt a private project that had been in the pipeline prior to the enactment of the Coastal Act and that had already won the active support of other governmental agencies. This was well illustrated in the controversy over the proposed expansion of the San Onofre nuclear power plant. The plant is jointly operated by Southern California Edison Company and San Diego Gas and Electric Company and is located on an open stretch of the coast north of San Diego, just below San Clemente. The proposed expansion involved building two 1140-megawatt reactors alongside the existing plant. Approval for the project already had been granted by the entire gamut of federal and state agencies: Atomic Energy Commission (AEC), Environmental Protection Agency, Army Corps of Engineers, Coast Guard, California Public Utilities Commission (PUC), State Water Resources Board, State Lands Commissions, and Regional Water Quality Board. When the proposal came before the San Diego regional coastal commission in 1973, the permit was approved on a 9–1 vote. This, however, did not quiet the opposition, which was concerned with the possibility of radiation leaks or a nuclear accident, with the long-term disposal problem of nuclear waste material, and with the substantial reduction of pristine beach area. On these grounds, the permit was appealed to the state commission.

For the antinuclear forces, this was not the most propitious time to be raising their challenge. One plant was already in place at San Onofre, and stopping the expansion would have reduced only marginally the potential dangers cited. Moreover, the United States was in the midst of its first energy "crisis." Yet in California, in 1973, San Onofre was the target of opportunity for the antinuclear movement.[34]

The coastal commissions did not have the authority to judge the merits of the debate over nuclear safety, but only whether the plant would have an adverse affect on the coastal environment and whether it would be an unnecessary, unwise, and irreversible commitment of coastal resources.[35] Constructing the plants would be an irreversible commitment of some coastal resources—that was obvious. Plans called for destroying up to one-half mile of sandstone bluffs and canyons, removing this land from other possible uses, and requiring an enormous amount of sea water for cooling the reactors, with a potentially adverse affect on the near-shore marine life. But was it a necessary commitment of resources? Despite a concerted campaign orchestrated by the utilities and state and federal officials, the state commission in December of 1973 denied the permit by a vote of 6 in favor, 5 against, with 8 votes (⅔ of the members) required for approval.

The state staff argued at the time that the denial was not of the plants *per se*, but only of their location. It proposed that the applicants locate the plants one-half mile inland and that a new kind of cooling system and reactor be used that would require less water to reduce the adverse environmental impacts. The utilities refused to move inland, alleging that this would cause unnecessary and costly delays of up to 6 years. The rebuke also was not appreciated by the AEC, PUC, business community of the state, and several important state legislators. As Chairman Lane had correctly predicted prior to the vote, the commissions could not act aloof of practical political considerations on matters of such importance. As he had foreseen, in denying the permit for San Onofre the commissions immediately became targeted by business and resource development agencies as a major culprit in the growing energy crisis.[36]

Letters and calls from citizens, organized interests, and public officials began coming into the commission, many of which forcefully called for some kind of accommodation with the applicants that would allow development of the nuclear plants. Private negotiations were thereupon begun between the state staff and the applicants (to the exclusion of the project opponents). At the same time, Southern California Edison appealed the commission ruling in the San Diego superior court, contending in part that the commission's denial was because of a concern over nuclear safety, over which it had no jurisdiction.[37] On the basis of this factor and under the advice of the attorney general's office, the commission decided to take the unprecedented step of calling for a revote on the application, with the clear understanding that nuclear safety could not be part of its considerations.

Within a month, the staff returned to the commissions with a recommendation to approve the application with what were obviously less stringent conditions than it had earlier suggested. The staff recommendation no longer called for moving the plants to an inland site. Instead, the original permit was to be modified to reduce destruction of coastal bluffs to .2 mile, to provide greater access to the state beach during the construction of the plants, and to require an independent study of the possible affects on marine life. With only slight modification of these conditions by the commissioners, the revote on the applications at the commission meeting of February 20, 1974, went 10 in favor, 2 against.

This was a clear sign that the commission had bowed to the practical and political realities of the need to balance coastal protection and preservation with energy growth and development, particularly on projects actively supported by other agencies. Also of importance, the commission decided not to impose any conditions that would require the applicants to return for reapproval from any of the several federal and state agencies with responsibilities for nuclear plant siting, e.g., requiring the utilities to extend the sea water intake and outflow conduits to a depth of 40 meters, as had earlier been recommended by the staff and considered by the commission.

Another example of the commission's delicate and sometimes very difficult dealings with peer agencies was its confrontation with the Aliso Water Management Agency. Aliso was actually a conglomeration of water districts and local governments in southern Orange County, created in 1972 to construct and operate an areawide sewage treatment facility. The facility would bring sewage discharge in the area up to state standards as well as accommodate the rapidly expanding population in the southern part of the county, which at the time had reached a little more than 50,000 residents. The plant itself would not be located in the coastal permit zone, but the outfall pipe would traverse it. To deny the Aliso permit would clearly accomplish environmentalists' objective of restricting further development in the southern part of Orange County, or at least placing a moratorium on it until completion of the Coastal Plan. But under pressure from Aliso and its tight deadlines schedule (a schedule imposed by federal funding sources), the South Coast Regional Commission granted the permit. In doing so, it added two important conditions: The size of the outfall pipe would have to be reduced from 54 to 48 inches, limiting the projected growth in the area by the year 2000 from 220,000 persons to 174,000, and the water quality standards for discharge would have to be more stringent than those required by the State Water Resources Control Board (SWRCB).[38] These conditions not only created substantial difficulties for Aliso in retaining its funding support from SWRCB and the federal government but infuriated state water pollution control officials—an affront that was to return to haunt the commission in the 1976 legislature.

Although these cases illustrate that there was sometimes a good deal of acrimony between the commissions and peer agencies, in other instances a far more cooperative and productive relationship evolved. For example, in the

South Coast region several local governments readily incorporated the building code guidelines that began to develop piecemeal at the regional commission. Newport Beach, for instance, incorporated the commission's requirement for two on-site parking spaces for each unit, and Laguna Beach reduced its density and lot coverage provisions for all coastal sites. Redondo Beach went so far as to revamp its entire building code to include the commission's reduced height limitations, sign setbacks, and so forth. [39]

In Marin County, in the North Central region, interagency relations were often quite harmonious. Indeed, it was in Marin that the commissions and the county developed a collaborative permit procedure that coordinated and streamlined the permit process for large development permit applicants. The North Central staff also worked closely with the Bay Area Regional Water Quality Control Board to persuade Marin County to improve enforcement of its septic tank regulations.

Another example of a very close working relationship was between the North Coast region's staff and the Department of Fish and Game. [40] In this case, the coastal commissions' regulatory authority was used to enhance the powers of Fish and Game, which on its own had little authority to halt development and had had practically no influence on local governments in the region. At commission hearings, Fish and Game was extremely successful in blocking developments that it opposed. In return, Fish and Game provided the commissions with expertise on wildlife matters and assisted them in monitoring permit violations, particularly after Fish and Game assigned a staff person to a half-time position as liaison with the commissions.

On the whole, however, the implementation of Proposition 20 was not an auspicious occasion for interagency cooperation. First of all, Proposition 20 was essentially a slap in the face of (most) existing agencies; it was passed because of the belief by many that existing agencies, especially at the local level, had done an inadequate job of protecting coastal resources. Second, Proposition 20 made the commissions very independent of existing agencies, with practically no provisions to require or even encourage coordinated action. Third, given the uncertain future of the commissions beyond 1976, many agencies simply assumed that the commissions would not be around after 1976 and thus represented merely a temporary irritant.

VI. Summary

During the implementation of the Coastal Act from 1973 thru 1976, several of the external factors that had an important bearing on the commissions changed, thereby bringing about pressures for relaxation of the statute's conservationist policies. The energy crisis of 1973–1974 and the nationwide recession resulted in a shift in public concern and media attention from environmental protection in the early 1970s to unemployment and resource development. As shown by public opinion surveys, a new balance was being sought, one that at least temporarily put jobs and economic development above environmental protection. The dire position of the construction industry caused by a slowdown in the economy in 1974 and 1975 placed particularly severe pressures on the commissions to be more lenient toward development. The commissions also often found it extremely difficult to effectively use their veto authority over other agencies to bring about the kinds of basic changes in growth and development patterns required if coastal resources were to be protected.

Despite these factors, the commissions' formal superiors either stayed fairly neutral (the governor and the legislature) or continued supportive (the courts and OCZM) throughout the 1973–1976 period. And even in the face of waning public attention and the inevitable frustration involved in the day-to-day battles over permits and the Coastal Plan, the supportive constituency showed a remarkable tenacity and ability to monitor implementation of the act.

It was within the context of this mix of external pressures, which on balance would make successful implementation more difficult, that the staff and commissioners operated.

NOTES

1. Oliver Williams, *Suburban Differences in Metropolitan Policies* (Philadelphia: University of Pennsylvania Press, 1965), p. 78.
2. Heinz Eulau and Kenneth Prewitt, *Labyrinths of Democracy: Adaptations, Linkages, Representation, and Policies in Urban Politics* (Indianapolis: Bobbs-Merrill, 1973), p. 506.
3. Office of Coastal Zone Management and California Coastal Commission, "State of California Management Program and Revised Environmental Impact Statement" (U.S. Department of Commerce, 1977), p. 112.
4. See, for example, Robert C. Ellickson, "Why Housing Prices Went Through the Roof," *Los Angeles Times*, July 24, 1978; Editorial, "A Clean Environment Is Costly," *Environment and the Economy* (July 1978), p. 3; and Gene Bylinsky, "Pollution Control," *Fortune* (July 1971).
5. The actual 1970 census figures on unemployment are adjusted upward according to the new method of calculating unemployment introduced by the U.S. Department of Labor in 1974. All unemployment figures reported pre- and post-1974 are computed under the new method.
6. Frederick H. Buttel and William L. Flinn, "The Politics of Environmental Concern," *Environment and Behavior* 16(1) (March, 1978), p. 31.
7. Williams, *Suburban Differences*, p. 78; Thomas R. Dye, *Understanding Public Policy*, 3rd ed. (Englewood Cliffs, N.J.: Prentice-Hall, 1978), pp. 270–78.
8. Williams, *Suburban Differences*, p. 78.
9. Anthony Downs, "Up and Down with Ecology—The Issue Attention Cycle," *Public Interest*, 23 (Summer, 1972).
10. *California Poll*, 7101, State Data Program, Institute of Governmental Studies, University of California, Berkeley, pp. 8, 10.
11. Downs, "Up and Down with Ecology," p. 42.

12. These conclusions are based on an analysis of our media files and interviews with Judy B. Rosener, coastal commissioner, January 10, 1979, and Bob Brown, executive director, North Central region, May 4, 1979.

13. Interviews with Joseph Bodovitz conducted by a panel convened by Stanley Scott, Institute of Governmental Studies, University of California at Berkeley, April and May 1979.

14. Interviews with Judy B. Rosener, coastal commissioner, January 10, 1979, and Jack Schoop, May 21, 1975. Schoop was especially concerned about the failure of the commissions to effectively nurture the media and to counter a number of what he considered widespread misconceptions about the treatment by the commissions and the Coastal Plan of property rights, economic impacts, and agricultural land preservation.

15. The classic statement on the life cycle of regulatory agencies is presented by Marver Bernstein, *Regulating Business by Independent Commissions* (Princeton: Princeton University Press, 1955); for the role of supportive constituencies, see Paul Sabatier, "Social Movements and Regulatory Agencies: Toward a More Adequate—and Less Pessimistic—Theory of 'Clientele Capture,'" *Policy Sciences* 6 (1975), pp. 301–342.

16. The data on participation in permit hearings reported on was collected jointly by Rosener and ourselves. For an in-depth examination of participation in the permit process, see Judy B. Rosener, "Citizen Participation in an Administrative State: Does the Public Hearing Work?" (Unpublished Ph.D. dissertation, The Claremont Graduate School, Claremont, California, 1979).

17. The importance of organized group opposition to a permit is indicated by the finding that when statewide groups did register their opposition, the likelihood of a permit denial increased. Rosener, "Citizen Participation," pp. 147–159.

18. Fred S. Farr, "The Role of Citizen Action Groups in Protecting and Restoring Wetlands in California" (Unpublished paper, Carmel).

19. Composite portrait from interviews with statewide and local supportive constituency activists and others.

20. Interview with Michael Peevy and Peter Fearey of CCEEB, 31 October 1975.

21. Interview with Melvin Lane, chairman, State Coastal Commission, 2 December 1975.

22. Interview with Lane, 2 December 1975.

23. Interview with H. Grant Dehart, Pacific regional manager, Office of Coastal Zone Management, 31 August 1977.

24. Letter from William J. Brah, Pacific regional program assistant, Office of Coastal Zone Management, 1 September 1977.

25. Peter M. Douglas and Joseph E. Petrillo, "California's Coast: The Struggle Today—A Plan Tomorrow," *Florida State University Law Review* 4 (1976), p. 212.

26. See Gerald Bowden, "Legal Battles on the California Coast: A Review of the Rules," *Coastal Zone Management Journal* 2 (1976), p. 287; the seminal case concerning vested rights was *San Diego Regional Commission* v. *See the Sea, Ltd.*, Cal, 3rd 888, 109 *Cal. Rptr.* 377 (1973), which was decided in favor of the property owners.

27. *CEEED* v. *The California Coastal Conservation Commissions*, App., 118 Cal. Rptr. 315, decided by Court of Appeals, Fourth District, Nov. 19, 1974. CEEED is an acronym for "California for an Environment of Excellence, full Employment and strong Economy through planned Development." The group was organized informally in early 1972 and incorporated in August of that year.

28. App., 118 *Cal. Rptr.* 315, p. 328. For discussions of the "taking" issue, see Bowden, "Legal Battles," pp. 280–282, and Fred Bosselman, David Callies, and John Banta, *The Taking Issue* (Washington, D.C.: Government Printing Office, 1973).

29. Minutes of State Coastal Commission meeting, August 31, 1976, pp. 6–7.

30. Bowden, "Legal Battles," p. 288.

31. The attitude of the attorney general's office changed abruptly in 1977–1978, as the attorney general, Evelle Younger, began posturing for his gubernatorial bid on the 1978 Republican ticket.

32. Joseph Sax and Joseph DiMento, "Environmental Citizen Suits," *Ecology Law Quarterly* 4 (Winter 1974), pp. 58–61.

33. Banta, in his study of interagency relations, points to five areas where the activities of the commissions would inevitably overlap with other agencies: (1) setting budget priorities, especially for agency projects already on the drawing board when Proposition 20 passed; (2) providing technical expertise; (3) securing grant approvals for development projects within the coastal zone; (4) monitoring and enforcing environmental quality standards; (5) setting future development policies and plans. John S. Banta, "The Coastal Commissions and State Agencies Conflict and Cooperation," in *Protecting the Golden Shore: Lessons from the California Coastal Commissions*, ed. Robert C. Healy (Washington, D.C.: The Conservation Foundation, 1978), p. 103.

34. Though most of the environmental groups in the state opposed further development of nuclear power under present technology by this time, they were not unanimous. The debate over San Onofre, in fact, caused a schism in the ranks of the Sierra Club between those opposed outright to any further nuclear plants and those who saw nuclear as preferable to reliance on either imported oil- or coal-fired plants, and to the air pollution caused by both. See Melvin B. Mogulof, *Saving the Coast: California's Experiment in Governmental Land Use Control* (Lexington Books, 1975), p. 31.

35. The commission had been advised by the attorney general's office that under the U.S. Atomic Energy Act, the federal government had exclusive authority to regulate and control safety and other issues directly related to development of nuclear power plants. Therefore, the commission could not legally take radiation hazard issues into account in deciding on the San Onofre appeal. Minutes of the state commission meeting, December 5, 1973.

36. Robert C. Healy, "Saving California's Coast: The Coastal Zone Initiative and Its Aftermath," *Coastal Zone Management Journal* (1974), p. 382.

37. Minutes of the state commission meeting, January 9, 1974.

38. Healy, "Saving California's Coast," p. 377.

39. Interviews with Commissioner Rimmond C. Fay, 26 August 1975, and South Coast Commission Executive Director Melvin Carpenter, 10 June 1975.

40. Interview with Gary Monroe, California Department of Fish and Game, 20 July 1978.

4

The Internal Process of Policy Implementation
Policy Preferences and Relationships among Agency Officials

Most studies of policy implementation—especially in social programs such as job training, education, and health—have focused on relationships within and among large bureaucracies in which the dominant theme has often been the subtle distortion of program intent through delay and resistance by officials defending their traditional programs or responding to local resistance.[1] From our perspective, however, many of these problems arose because the statutory framers were unable or unwilling to minimize the number of veto/clearance points involved and to assign implementation to sympathetic agencies that would give the new program high priority. In addition, decision procedures in these bureaucracies were sufficiently complex that it was difficult for outsiders to monitor the process and to effectively present their case.[2]

The framers of the 1972 California Coastal Initiative managed to avoid, or at least to minimize, most of these difficulties (1) by assigning implementation almost exclusively to a new agency; (2) by trying to provide that at least a majority of agency officials would be supportive of the Initiative's objectives; (3) by creating a highly integrated internal review process in which most major decisions on both permit review and plan development would be made by the state commission, whose membership was the most likely of all the commissions to support the Initiative's objectives; (4) by creating a small agency(s) in which coordination among subunits would be manageable; and (5) by creating an exceedingly open policy process with clear decision points in which monitoring and effective intervention by outside groups would be facilitated.

The framers placed almost the entire responsibility for implementation of the Initiative in the hands of a new agency. In so doing, they avoided the problematic task of getting an existing agency to give high priority to the new program and/or of attempting to change the basic program thrust of an ongoing bureaucracy.[3] By creating a new agency, they thereby assured that the new program would be given the highest priority, that there would be no need to integrate it into ongoing bureaucratic routines, and that all positions in the new agency would be open at a time when the supporters of coastal protection were at their greatest strength. The last is very important, particularly in the recruitment of commission staff, as there is some evidence that officials in young regulatory agencies come disproportionately from among strong advocates of regulation, particularly if the agency was established after a controversial and highly visible political campaign.[4]

The framers of the Initiative also took a number of steps to increase the probability that a majority of formal policy-makers—the commissioners—would be supportive of statutory objectives. As will be recalled, the membership of each regional commission was evenly divided between local elected officials and "public" members appointed by the governor, the Senate Rules Committee, and the Assembly speaker. Not only were the "public" members to be appointed by state officials—and thus presumably more likely to reflect state/regional rather than purely local concerns—but there were also some loose statutory requirements that they be trained in the environmental sciences or planning, backgrounds that presumably would enhance their sensitivity to statutory objectives.[5] Moreover, since a majority of voters in 12 of the 15 coastal counties had voted in favor of the Coastal Initiative, there was some expectation that several of the local officials selected for the commissions would mirror those preferences and thus that a majority of the overall membership of several/most regional commissions would be supportive of the act's objectives.[6] Finally, the decision—partially based on the successful Bay Area Conservation and Development Commission experience—to have rather large commissions that would include local elected officials meant that commissioners would have to be part-time officials, i.e., receive only expenses and a modest *per diem*. But this preference for part-time rather than full-time commissioners was also motivated by the concern that full-time commissioners might become preoccupied with perpetuating their positions and thus more likely to make concessions to the powerful antiregulation forces that had been successful in preventing legislative approval of strong coastal legislation.[7] Despite these efforts to increase the probability that a majority of regional commissioners would support statutory objectives, the appointment criteria were still sufficiently vague that the general policy orientation of the regional commissions would be heavily dependent upon the vagaries of the appointment process (as well as the external factors discussed in the previous chapter).

Partially because of this uncertainty, the framers of the Coastal Conservation Act clearly provided that ultimate decisions on the coastal plan and most important permit cases would be made by the state commission.[8] As will be recalled, its membership was to be equally divided between "public" members and representatives from each of the regional commissions. In short, there was no direct representation of local governments. Moreover, since statutory supporters would probably control several/most of the regional commissions and since most local government officials would probably not have the time to serve both on the regional commission and as the regional representative to the state commission, it was highly probable—though by no means certain—that a majority of the members of the state commission would be supporters of the act. This would be crucial in counterbalancing appeals from regional commissions, such as the North Coast, which could be expected to be generally hostile to the act's objectives. Again, however, the appointment criteria for state commissioners were sufficiently open that the general policy orientation of the crucial state commission would also be heavily dependent upon the appointment process. The statute also gave the state commission no role in selecting the chairmen and the executive directors of the regional commissions, thereby somewhat weakening the otherwise strong hierarchical integration within the commissions.[9]

In addition to these efforts to bias the policy preferences of implementing officials and to provide an integrated policy process, the rather modest funding provided by the act ($5 million for the 4-year interim period) precluded the creation of a large bureaucracy and all the problems of coordination that ensue. In fact, the regional commission generally had a staff of 6–10 professionals (plus clerical help), with the state commission staff about twice that size.

Finally, the Coastal Conservation Act sought to avoid the indeterminateness (and even secrecy) characteristic of most bureaucratic decision making by establishing clear decision points within specified time limits on both planning and permit decisions and by requiring that all important decisions be made in public after ample opportunities for public testimony and rather extensive notification requirements.[10] These explicit decision points and provisions for public participation would, it was hoped, encourage monitoring and participation by outside actors.

With these expectations based upon the provisions of the Coastal Initiative and the political situation surrounding the genesis of the commissions in the winter of 1972–1973 as a backdrop, we now turn to the actual selection and policy preferences of, first, commissioners and, then, commission staff on the state commission and the three regional commissions included in this study. We shall also provide a general overview of the relationships among commissioners and staff within each of the four commissions examined, as well as relations between the regional commissions and the state commission.

I. The Selection Process and Policy Preferences of Coastal Commissioners

Recall that one of our six conditions for effective implementation is that agency officials be committed to the achievement of statutory objectives and skillful in using available resources to attain those goals. The framers of the Coastal Initiative went to some lengths to structure such support among implementing officials. Yet their efforts were still subject to the vagaries of the process by which both "public" members and local elected officials were actually appointed to the commissions. In this section we shall first briefly examine the appointment process in each commission and then present some aggregate and comparative data on the policy preferences of different types of commissioners. [11]

Starting with the North Coast Regional Commission, most were small businessmen and many reflected the dominant resource base of the region's economy in timber, ranching, and commercial fisheries. [12] The criteria used by the governor, Senate Rules Committee, and Assembly speaker in selecting the "public" member in this region are somewhat difficult to discern, as most of the commissioners did not actively solicit their appointments and thus were not very knowledgeable about the process. Nevertheless, most of the members clearly met through education and/or experience the statutory requirement for expertise in planning and resource management, although some doubts might be raised about labor leader William McHugh and conservationist Mildred Benioff. Each of the appointing officials sought geographical balance, appointing one member from Humboldt and one from Mendocino county (with Del Norte having such a small population that its lack of public members is understandable). There may also have been some effort to provide political balance, as two of the three appointing officials split their appointments between proponents and opponents of the initiative. Even the "public" members who voted for Proposition 20 were not particularly active in the initiative campaign and would generally be considered moderates. Finally, only two of the "public" members (John Mayfield and Bill Grader) were active in partisan politics.

Among local elected officials, all clearly reflected the concern of the region's electorate and local government bodies with expanding the very precarious economic base of the region and protecting property rights and local control. Moreover, all of the municipal representatives came from coastal towns, while all of the supervisors repesented coastal districts. In general, appointment to the coastal commission did not seem to arouse much competition in this region—perhaps because of the general attitudinal homogeneity of local officials, their reluctance to be associated with this alien and unpopular state institution, and/or the substantial time constraints imposed by their regular jobs and their duties as local officials.

Turning from selection of regional commissioners to election of the region's representative to the state commission, this entire process was marked by periodic controversy in the North Coast Commission. Dwight May, a moderate conservationist, was the initial selection after numerous ballots—even though he did not represent the views of the commission majority—apparently because of the desire of members from Humboldt County to have one of the three leadership positions.[13] But he resigned after a year (in part because of time and travel constraints), to be replaced by Mayor Bernard Vaughn of Ft. Bragg, who had the time and whose views were more in keeping with the commission majority. In somewhat of a surprise, however, Vaughn was replaced after only a year by Don Peterson—apparently because of concerns over geographical equity and the need to have a more effective spokesman for the region who was also more or less acceptable to both the property rights and conservationist factions on the commission.[14]

The North Central Regional Commission differed in a number of respects from its counterpart to the north.[15] First, almost all of the public members supported the Coastal Initiative; moreover, while most would be described as moderate conservationists, several of them—most notably Ellen Jonck and Phyllis Faber—were local leaders in the Proposition 20 campaign. Second, local government appointments were much more controversial in this region, largely because in two of the three counties (San Francisco and Sonoma) local elected officials were generally opposed to Proposition 20 while the local electorate had rather strongly supported it. This produced a variety of responses. The Sonoma County Board of Supervisors appointed an avowed opponent. On the other hand, while a majority of the city councilmen in the same county were likewise opposed, they appointed a moderate supporter of the Initiative who had had extensive planning experience. The San Francisco Board of Supervisors was able to reach the Solomonlike compromise of splitting its two votes (as it is both a city and a county). Finally, there was no conflict in Marin County, where a majority of both supervisors and councilmen supported Proposition 20 and appointed representatives who shared this view. The result of this complex situation was that a slight majority of the local commissioners in this region actually supported the Coastal Initiative. The third and final point of contrast with the North Coast Commission was that the North Central's representative to the state commission, San Francisco Supervisor Bob Mendelsohn, was apparently selected with little controversy, kept his position throughout the 1973–1976 period, and supported Proposition 20.

There were, however, several similarities with the North Coast. Most of the "public" members clearly met the expertise requirement and few of the public members were active in partisan politics. Most of the state appointing officials were careful to provide geographical balance in their appointments in the North Coast, so Senator Peter Behr—one of the state's leading environmentalists—played an important role in this region.

Turning to the South Coast, probably the most important of the regional commissions because of its enormous permit load,[16] the appointment pattern was somewhat of a mixture of that found in the two northern regions. Almost all of the originally appointed "public" members met the statutory requirement for expertise in planning and resource management, although this was not true of the two subsequent appointees by Governor Brown. Moreover, most had not been notably active in the Proposition 20 campaign. Several had, however, been active in partisan politics, and one of Governor Reagan's appointees, Donald Phillips, was a member of the Long Beach City Council. Finally, as with the appointments in the North Coast, there seems to have been some attempt by state officials to balance proponents and opponents of development and to split appointments between the two counties.

Among local officials, all except the Los Angeles City representative came from coastal districts. Unlike those of the North Coast, few of the local appointees had expertise in resource management other than the experience that any local official gets in land use planning. Finally, local governing bodies in the region faced the same dilemma as their colleagues in the North Central: Should they appoint representatives who mirrored their own opposition to the Coastal Initiative or their constituents' support of it? Perhaps because popular support was not as strong as in the North Central while development pressures were considerably greater, slightly more than half of the local bodies in this region wound up appointing opponents of the Initiative and, in a few cases such as Rubley and Nowell, very vocal opponents.[17]

As for the South Coast's representative to the state commission, the appointment originally went to Los Angeles County Supervisor Jim Hayes, one of the most powerful officials in the region and a swing vote acceptable to both developers and conservationists. After almost 2 years, however, he resigned from the state commission because of health problems and severe time constraints. There ensued a ferocious battle over his replacement between the development- and environmentally oriented members. Among the latter, only Rim Fay was willing to give the additional time needed for the state post, and he was eventually elected through a coalition of environmental and swing votes.

As for the all-important state commission, the general pattern of appointments was virtually all that the drafters of the Coastal Initiative could have hoped for.[18] Of the six public members, all except Roger Osenbaugh and, to a lesser extent, Bernard Ridder met the expertise requirement and could be expected to be generally sympathetic to the Initiative. Moreover, they included two moderate conservationists, Mel Lane and Richard Wilson, who were widely respected throughout the state. As might be expected, each appointing official sought geographic balance between Northern and Southern California and generally avoided (with the exception of Ellen Stern Harris) people identified as outspoken advocates during the Proposition 20 campaign.

It was in the selection of regional commission representatives, however, that the appointment process handed Proposition 20 proponents their greatest victories. The North Coast region, for example, sent a representative of its opposing majority in only 1 of the 4 years (and even then he was not a particularly articulate or outspoken opponent). The North Central and Central regions, as expected, sent strong advocates of coastal protection. But another surprise came from the South Central, which sent Ira Laufer, a local businessman who turned out to be one of the state commission's most articulate conservationists.[19] And, as already noted, the South Coast—the most evenly balanced of all the commissions—somehow managed to have an ardent environmentalist as its representative for the 1975–1976 period. Finally, the San Diego Commission—a majority of whose members apparently opposed Proposition 20—sent Jeffrey Frautschy, a marine biologist and a very moderate proponent.[20]

In considering the regional representatives, two points are worth making. First, contrary to our expectations and those of some of the framers, local elected officials constituted almost half of the original set of regional representatives, although by the crucial period of the adoption of the coastal plan in the summer of 1975 their number had been reduced to two.

Thus, the anticipated time constraints on local officials—most of whom already held a private job, local office, and regional commission membership—were apparently not a debilitating factor. Second, what seems to have happened among the set of potential regional representatives was that active opponents of the coastal commissions—such as John Mayfield in the North Coast and some of the San Diego commissioners—were simply unwilling to spend the 5 days a month involved in state commission meetings and travel, given the high probability that they would be a minority on that commission. In short, only people who either were strongly committed to successful implementation of the Initiative or were at least swing votes seemed willing to undergo the substantial time commitment that membership on both the regional and state commissions involved.

In sum, the process of appointing coastal commissioners seems to have been marked by (a) the selection of "public" members who generally met the expertise requirement, although often through previous experience on resource management agencies rather than by education; (b) distributing "public" members among coastal counties (to the exclusion of inland counties); (c) appointing local officials from coastal districts/towns where possible; (d) minimal concern with paying off political debts or appointing party stalwarts (at least among "public" members); (e) avoidance of people who had been outspoken advocates during the Proposition 20 campaign[21]; and (f) trying to reach some balance between the policy preferences of the appointing official(s) and the degree of support for Proposition 20 among the local electorate. Finally, appointments to the commissions were completed by the January 1973 start-up date, thereby avoiding the delay that has sometimes seriously impeded the genesis of new programs.[22]

It should also be mentioned that about three-quarters of the original appointees to the state and regional commissions retained their posts until the coastal plan had been adopted in the summer of 1975. This would suggest that most of the members were seriously concerned with fulfilling the task to which they had been assigned.[23] Moreover, this continuity of membership was an important factor in enabling the commissions to achieve their permit review and planning responsibilities within the very tight time limits imposed by the statute.

II. Personal Characteristics, Appointing Authority, and Commissioner Behavior—A Preliminary Look

Given this overview of the appointment process, it is now time to examine the outcome of that process in terms of the background characteristics, attitudes, and voting behavior of the commissioners. In particular, did the systematic differences between local elected officials and "public" members appointed by state officials that were anticipated by the proponents of the Coastal Initiative, and which had been observed on the Tahoe Regional Planning Agency (TRPA), actually emerge? And within these two general categories, were there differences by appointing official? For example, one might expect the appointees of Assembly Speaker Moretti—a strong supporter of the Coastal Initiative—to be more sympathetic towards the act's objectives than those of Governor Reagan (a consistent opponent of coastal legislation and Proposition 20), with the Senate appointees falling somewhere in between. With respect to local elected officials, there are two competing hypothesis: On the one hand, the slightly larger jurisdiction of county supervisors and their experience with the spillover effects of municipal development might make them less "parochial" than city councilmen; on the other hand, one might argue that city officials are generally more liberal and sympathetic to land use planning than their county counterparts.[24]

These questions are explored in Table 4-1, which compares the commissioners appointed by various authorities on a number of grounds. These include, first of all, a number of standard background characteristics that have been associated, both within the general public and among most political elites, with support for environmental protection and restrictive resource management: age (young), educational level (high), partisan affiliation (Democrat), and self-rated ideology (liberalism).[25] Second, the table provides three indicators of general attitudes toward coastal resource management: (1) vote on Proposition 20, (2) score on a scale measuring support for environmental protection, and (3) score on a similarly constructed scale representing support for local control and market allocation of resources.[26] Finally, the table compares the behavior of commissioners as seen in the percentage of "No" votes (i.e., votes to deny) for each category of respondents on all permit applications subjected to a public hearing in each of the three regions between April 1973 and June 1975.[27] It should be noted that the background and attitudinal data presented in Table 4-1 are drawn from a survey questionnaire distributed to members of the four commissions in the winter of 1974–1975, with an overall response rate of almost 80% and no

TABLE 4-1

COMPARISON OF COASTAL COMMISSIONERS BY APPOINTING AUTHORITY FOR STATE COMMISSION AND THREE REGIONAL COMMISSIONS[a]
(means, unless otherwise indicated)

Characteristic	"Public" members appointed by				Local officials appointed by				Total (35 of 44)
	Governor (7 of 9)	Senate Rules Committee (7 of 8)	Assembly Speaker (6 of 8)	Subtotal (20 of 25)	County Board of Supervisors (7 of 8)	Assoc. of municipalities (7 of 9)	Regional planning agency (1 of 2)	Subtotal (15 of 19)	
Background									
Age[b]	3.7	3.4	3.7	3.6	3.3	3.6	4.0	3.5	3.5
Education (% college graduate, BA/BS)	86%	71%	83%	80%	86%	43%	100%	67%	74%
Political party (% Democrat)	29%	50%	67%	47%	43%	17%	100%	36%	42%
Ideology[c] (liberalism)	3.4	2.7	3.8	3.3	2.3	2.7	4.0	2.6	3.0
Attitudes re: coast									
Vote on Prop 20 (% "yes")	71%	86%	100%	85%	43%	43%	100%	47%	69%
Environmental Protection Scale[d]	−.20	.63	.12	.16	−.63	−1.10	−.44	−.83	−.28
Localism/Market Solutions Scale[e]	.27	−.16	.16	.09	.68	.84	−.25	.69	.32
Voting behavior									
% "no" on permits[f]	15%	21%	16%	17%	8%	9%	28%	10%	14%

[a] Excludes regional representatives to state commission on the grounds that they may well not be representative of other regional commission members appointed by the same appointing authorities.
[b] Six-point scale: 1 = 18–29, 2 = 30–39, 3 = 40–49, 4 = 50–59, 5 = 60–69, 6 = 70 and over.
[c] Five-point scale from 1 = strong conservative to 5 = strong liberal.
[d] Factor score ranging from +3 (high support for environmental protection) to −3 (low support).
[e] Factor score ranging from +3 (high support for local control and market solutions) to −3 (low support).
[f] Only those permits on public hearing calendar of regional commissions, April 1973–June 1975: excludes (1) regional items on administrative and consent calendars and (2) all state commission voting.

significant variation in response among appointment categories (shown in paren-thesis under each column heading).

Looking first at the overall differences between "public" members and local officials on the four commissions, one notices that—with the exception of age, where the two groups were virtually identical—the "public" members ranked slightly higher than their local colleagues on the background correlates of en-vironmental concern. They were slightly better educated, had a higher percent-age of Democrats, and were somewhat more liberal. Not surprisingly, their attitudes toward coastal protection also differed. In particular, whereas 85% of the public members surveyed had voted for Proposition 20, only 47% of the local officials had done so. These differences were also reflected in the higher scores of public members on the Environmental Protection Scale (.16 vs. −.83) and lower scores on the Localism/Market Solutions Scale (.09 vs. .69). Finally, the public members voted to deny permit applications an average of 17% of the time, compared to a denial rate of 10% among local officials. In short, the general expectation of those involved in the 1970–1972 struggle over coastal legisla-tion—namely, that state-appointed officials would be more sympathetic toward coastal protection and therefore take a more restrictive view toward coastal devel-opment than local elected officials—was essentially confirmed, although the differences were perhaps not as great as some people might have expected.

Also, there was less difference than expected among the "public" commis-sioners appointed by the governor, the Assembly speaker, and the Senate Rules Committee. For example, while Governor Reagan not surprisingly appointed a higher percentage of Republicans than the Democratically controlled Senate Rules Committee and Assembly speaker, a substantial majority (71%) of his appointees nevertheless voted for Proposition 20 (compared to 86% of Senate appointees and 100% of those appointed by Speaker Moretti). Similarly, while Reagan's appointees scored somewhat lower on the Environmental Protection Scale and somewhat higher on the Localism/Market Solutions Scale than the Senate and Assembly appointees, their denial rates on development permits were quite close (15% vs. 21% and 16%). In short, there is considerable indication that Governor Reagan put aside his long-standing opposition to an active state role in land development review along the coast once Proposition 20 had passed, and generally appointed moderate Republicans like Mel Lane who were sym-pathetic to the Initiative's objectives. [28]

Among local government appointees there was very little internal variation (essentially ignoring the figures for the regional planning agency representatives because of the very small, and perhaps unrepresentative, number surveyed). Although county supervisors were considerably better educated and somewhat more Democratic than their municipal colleagues on the commissions, their degree of support for Proposition 20 was identical (43%) and their scores on the two attitudinal scales and their denial rates on permit applications were very similar. In short, while city councilmen may generally be more liberal than county supervisors on social issues, their views on state review of local governments' land use decisions were similarly antagonistic.

To what extent were there important differences between commissions along these lines? Table 4-2 compares "public" and local members in each of the three regional commissions examined, as well as state appointees and regional representatives on the state commission. As before, the response rate for each category is provided; a general knowledge of respondents and nonrespondents suggests no significant bias in any of the categories.[29]

Looking first at the three regional commissions, the expected differences between state appointees and local elected officials generally held in the North Coast, were most glaring in the South Coast, and essentially disappeared in the North Central. In the North Coast, for example, the "public" members were generally younger and more liberal—although less well educated—than their local colleagues. They had a much higher support rate for Proposition 20 (80% vs. 17%) and ranked considerably higher on the Environmental Protection Scale and comparably lower on the Localism/Market Solutions Scale. Nevertheless, they had only a moderately greater denial rate on permit applications (10% vs. 4%). This apparent anomaly will be addressed later in this chapter in the course of our discussion of the "group dynamics" of each commission.

In the South Coast the differences between "public" and local members were generally consistent and often striking. Thus, state appointees were considerably more Democratic and liberal than their local colleagues, had a higher support rate for Proposition 20, scored much higher on the Environmental Protection Scale and considerably lower on the Localism/Market Solutions Scale, and voted much more often to deny development permits (27% vs. 10%).

In contrast, there were essentially no significant differences (with the exception of age and education) between "public" and local members on the North Central Commission. They had virtually identical scores on partisanship, liberalism, support for Proposition 20, the two attitudinal scales, and percent "no" vote.

The source of the differences between the South Coast and North Central commissions clearly lay in their local members, as the public members of each had very similar scores on all the attitudinal measures and roughly comparable rates of denial.[30] In contrast, the local members in the North Central were much more liberal than their counterparts in the South Coast, had a higher support for Proposition 20, and showed considerably greater scores on the Environmental Protection Scale and comparably lower scores on the Localism/Market Solutions Scale; moreover, their denial rate was considerably greater (17% vs. 10%). This suggests that the differences between the North Central and South Coast commissions can be traced to the method by which local officials were selected: In the North Central Region most local governing bodies seriously took into consid-

TABLE 4-2
COMPARISON OF APPOINTMENT TYPES ON EACH COMMISSION

Characteristic	State commission			North Coast region			North Central region			South Coast region		
	Public members (5 of 6)	Regional reps. (6 of 6)	Total (11 of 12)	Public members (5 of 6)	Local members (6 of 6)	Total (11 of 12)	Public members (5 of 7)	Local members (5 of 7)	Total (10 of 14)	Public members (5 of 6)	Local members (4 of 6)	Total (9 of 12)
Background												
Age[a]	3.4	3.0	3.2	3.6	4.2	3.9	3.8	3.2	3.5	3.6	2.8	3.2
Education (% college graduate, BA/BS)	100%	83%	91%	40%	67%	54%	100%	60%	80%	80%	75%	78%
Political party (% Democrat)	20%	17%	18%	50%	20%	33%	60%	60%	60%	60%	25%	44%
Ideology[b] (liberalism)	2.8	3.5	3.2	3.0	2.2	2.6	3.8	3.4	3.6	3.6	2.0	2.9
Attitudes re: coast Vote on Prop 20 (% "yes")	80%	83%	82%	80%	17%	46%	100%	80%	90%	80%	50%	67%
Environmental Protections Scale[c]	.18	.42	.32	.02	-1.47	-.80	.18	.19	.19	.28	-1.16	-.36
Localism/Market Solutions Scale[d]	.29	-.51	-.15	.52	1.64	1.02	-.29	-.27	-.28	-.17	.92	.32
Voting Behavior % "no" on permits[e]	39%	41%	40%	10%	4%	7%	15%	17%	16%	27%	10%	19%

[a]Six-point scale: 1 = 18–29, 2 = 30–39, 3 = 40–49, 4 = 50–59, 5 = 60–69, 6 = 70 and over.
[b]Five-point scale from 1 = strong conservative to 5 = strong liberal.
[c]Factor score ranging from +3 (high support for environmental protection) to -3 (low support for environmental protection).
[d]Factor score ranging from +3 (high support for local control and market solutions) to -3 (low support).
[e]For regional commissioners, this involves all the hearing calendar items from April 1973 to June 1975; for state commissioners it involves a random sample of permits deemed to have raised a substantial issue in the period February 1973–June 1975.

eration the 60% support of their electorate for Proposition 20, while in the South Coast the overall electorate support of 53% (and only 51% in Orange County) was apparently much less persuasive.

Turning now to the state commission, there were few noteworthy differences between state appointees and regional representatives.[31] They were comparable in age, education, partisan affiliation, and vote on Proposition 20. In fact, contrary to expectations, the public members were slightly *less* liberal, scored slightly lower on the Environmental Protection Scale, and ranked slightly higher on the Localism/Market Solutions Scale than the regional representatives. Moreover, their voting behavior on permit appeals was almost identical: a 39% denial rate for state appointees, and a 41% rate for regional representatives.[32]

In short, the basic assumption of the proponents of Proposition 20—that local elected officials were less likely than state appointees to restrict development in the light of Proposition 20's objectives—was generally supported by the evidence. However, there is also need for some significant amendments to the basic proposition. First, it applied much less in regions, such as the North Central, in which local constituents had strongly supported Proposition 20. Second, it apparently did not apply at all when local officials were appointed to the state commission, except, perhaps, on those infrequent occasions when officials were voting on projects involving their own electoral constituency.[33]

On a more general level, our basic conclusion is that the policy predispositions of coastal commissioners were a function of both statutory efforts to structure the implementation process—i.e., by balancing state appointees and local officials on the regional commissions—and the vagaries of the appointment process. Of the latter, the most important were (a) the generally supportive nature of Governor Reagan's appointments, (b) the effect that strong popular support for Proposition 20 had on local appointments in the North Central Region, and (c) the tendency for regional representatives to the state commission to be either swing votes or strong supporters of the Coastal Initiative.[34]

It is now time to turn to the selection and policy preferences of the other group of officials directly involved in the implementation of the Coastal Initiative, namely, the professional staff of the commissions.

III. The Background, Selection, and Policy Preferences of Commission Staff

The 1972 California Coastal Zone Conservation Act said very little about the staffing of the commissions except to provide that the executive director of each commission be selected by its membership and be exempt from civil service. In addition, the rather limited funding provided the commissions precluded the creation of a large bureaucracy.

Despite this silence on the part of the enabling statute, the political situation surrounding the establishment of the commissions created a strong likelihood that the majority of its staff would be supportive of the Initiative's objectives. In particular, a number of studies of regulatory agencies have suggested that new agencies established after an intense political struggle are likely to be staffed primarily by representatives of the political forces that sponsored the agency's creation—particularly if the agency's enabling legislation clearly embodies the movement's objectives.[35] This was, of course, precisely the situation surrounding the coastal commissions.

There were three other factors that probably reinforced this tendency toward staffing by statutory supporters. First, the criteria developed for the civil service positions within the agency were strongly oriented toward people with a background in urban planning—a profession with a strong commitment to land use controls rather than deference to market allocation of resources.[36] Second, the commissions' status as one of the country's most innovative land use agencies and the young age distribution within the planning profession probably created selection pressures in favor of relatively young planners oriented toward public service.[37] Finally, the temporary nature of the commission's mandate and the likelihood that it would be embroiled in controversy probably discouraged applications from the stereotyped cautious bureaucrat.[38]

The background characteristics and policy orientation of the staff of the three regional commissions are provided in Table 4-3. Unfortunately, the state commission staff chose not to respond to the questionnaire from which these data were taken.[39] Nevertheless, comments from state commissioners and our own interaction with the state staff strongly suggest that they were very similar to the regional staffs, particularly those in the North Central and South Coast regions.

The data presented in Table 4-3 generally support our expectations. With the exception of the executive directors, the professional staff were young (generally in their 20s and 30s). With the exception of those in the North Coast, most had advanced degrees in planning and/or law. And, in what is surely one of the most striking figures to come out of this entire study, *all* of the professional staff in the three regions had voted for Proposition 20 (although one member of the North Coast staff admitted in retrospect that this had probably been a mistake).[40] On the other hand, the staff were, by and large, not environmental zealots. Their scores on the Environmental Protection and Localism/Market Solutions Scales in all three regions were rather close to the overall mean for the commissioners and staff surveyed.[41] They were almost evenly split between Democrats and Republicans and, on the whole, were moderately liberal in their general orientation.

Although the staff were surprisingly uniform in their general policy orientation, there was some interesting variation in their backgrounds and in the selection of the executive directors.

TABLE 4-3

CHARACTERISTICS OF THE PROFESSIONAL STAFF OF THE THREE REGIONAL
COMMISSIONS

Characteristic	Regional commission			Total (24 of 28)
	North Coast (11 of 11)	North Central (6 of 6)	South Coast (7 of 11)	
Background				
Age[a]	1.6	1.8	2.6	2.0
Education				
% college graduate, i.e., BA/BS	82%	100%	100%	92%
% with advanced degree	18%	83%	86%	54%
Political party (% Democrat)	46%	83%	33%	52%
Self-rated ideology (liberalism)[b]	3.3	4.0	2.5	3.3
Attitudes re: coast				
Vote on prop 20 (% yes)	100%[e]	100%	100%	100%[e]
Environmental Protection Scale[c]	.26	.10	.49	.29
Localism/Market Solutions Scale[d]	−.58	−.31	−.42	−.45

[a]Six-point scale: 1 = 18–29, 2 = 30–39, 3 = 40–49, 5 = 50–59, 6 = 70 and over.
[b]Five-point scale from 1 = strong conservative to 5 = strong liberal.
[c]Factor score ranging from +3 (high support for environmental protection) to −3 (low support).
[d]Factor score ranging from +3 (high support for local control and market solutions) to −3 (low support).
[e]In retrospect, one member of the North Coast staff would have voted against Proposition 20.

Unlike the regional commissions—which generally went through a rather elaborate search for their executive officers, relying on staff borrowed from one of the counties in the interim—the state commission at its initial meeting in January 1973 unanimously selected Joseph Bodovitz as its executive director.[42] Bodovitz was a former journalist in the Bay Area who had been BCDC's chief executive officer until he left it to come to the commission. The state staff was the largest of any of the seven commissions—about 15–20 professionals—and, with the exception of Bodovitz, was composed almost entirely of people with a background in planning and/or law, although a few members also had degrees in engineering or the natural sciences. Moreover, several—including the chief planner, Jack Schoop, and the assistant planner, Bill Travis—had previously held similar posts in BCDC.

Bodovitz felt rather strongly that, since the commissions were going to be basically involved in reviewing development permits and in preparing a general plan for the coast, what was needed was staff with skills and experience in land use law, permit work, and/or local planning—rather than, for example, natural resource management. Moreover, in his view, the commissions did not have the time or the resources to engage in any original research, nor were they charged with managing natural resources. If they needed expertise in these areas, they could go to the relevant state and federal agencies. What is important is that, as the chief executive officer of the state commission, his view of the proper training of commission staff was incorporated into the position descriptions for the civil service positions for not only the state but also the regional commissions.[43] While the vast majority of state and regional commissioners (and executive directors) interviewed agreed with him, this was not the case in the North Coast region.

The North Coast Commission, after a rather long and contentious search, selected Jack Lahr as its executive director in late March 1973. Lahr had grown up in the region, graduated from Humboldt State with a degree in wildlife biology, and then left the state to work for the Bureau of Land Management and the National Park Service, chiefly in land acquisitions. It appears that he was selected because of his knowledge of the region and timber management issues, his sensitivity to the problems of inverse condemnation, and a very disarming "country boy" style, coupled with a superb ability to mask his personal views.[44] In short, he could be expected to work well in what was essentially a rural region that had strongly opposed the Coastal Initiative and with a commission majority who reflected the views of their constituents. In selecting his staff, Lahr generally circumvented the civil service system. He felt strongly that the civil service position description was strongly oriented toward urban planners and therefore inappropriate in a rural area where natural resource issues, particularly timber and fisheries management, were crucial. Moreover, the monies provided for staffing the regional office were inadequate to hire, at civil service salaries, the number of people necessary to cover what was by far the largest of the six regions. As a result, he staffed the regional office almost entirely with "temporary" employees who were recent graduates (or, in some cases, graduate students) of Humboldt State University in some aspect of natural resources management.[45] This staffing pattern met with the approval of the regional commission, although, as we shall see, some problems were created by the staff's youth and general lack of experience. Lahr remained as executive director until April 1976, when he resigned to take a high executive position with the U.S. Bureau of Land Management.

The North Central region followed a more traditional path than the North Coast. After a rather extensive search, the regional commission hired Michael Fischer as its executive director in late March 1973. Although not a unanimous choice, he had the support of a substantial majority of commissioners.[46] A planner and political scientist by training, he had most recently been the deputy director of SPUR, a private community planning organization in San Francisco with a strong environmental orientation. As such, he obviously fit in well with the strong majority on the regional commission who supported Proposition 20. This staff, which was the smallest of any of those examined, had three other full-time professionals—all with degrees in planning and some experience in local government, although one person also had a degree in geology—and two part-time law students who worked on permits.[47] Fischer resigned his position in the spring of 1976 to become associate director of the state Office of Planning and Research and 2 years later replaced Bodovitz as executive director of the state coastal commission (under the 1976 Coastal Act).

Because the South Coast Commission was the most evenly balanced in its policy orientation and was subjected to the most intense pressures for development, the search for an executive director was, not surprisingly, rather contentious. After rejecting the choices of both the development-oriented and conservationist blocks, the commission selected Mel Carpenter, a relatively unknown retired naval officer who turned out to be an excellent administrator.[48] Because of its huge permit load, the South Coast had the largest of the regional staffs— between 14 and 17 professionals during the 1973–1976 period. Most had degrees in architecture or urban planning and 3–5 years experience with local governments. This was also the most bureaucratized of the regional staffs, with clear divisions between planning and permit personnel and with the latter clearly responsible for specific geographical areas.[49]

Given these overviews of the selection and policy preferences of the commissioners and staff in each of the four commissions examined, it is now time to put the pieces together in terms of brief portraits of the internal relations of each.

IV. INTERNAL COMMISSION DYNAMICS

The Coastal Initiative made very little direct effort to structure the internal dynamics of the state and regional commissions. The provision in Section 272743 that each commissioner elect its own chairman and appoint its own executive director did, however, serve to make each commission relatively autonomous. Moreover, the tight deadlines imposed by the statute—90 days for each permit application and essentially 2½ years to develop a plan for the entire 1100-mile coast—may have created some incentive to mute internal disagreements in the interests of meeting those deadlines, at least through the final decisions on the coastal plan in the summer of 1975. The relative insulation of the commissions from the governor, from the state legislature, and from partisan politics in general may also have reduced the incentives for political grandstanding and for ensnarling the commissions in the political ambitions of state officials.[50] On the whole, however, statutory drafters could only hope that the mixture of "public" members and local elected officials, the pressure of tight deadlines, and the general insulation from partisan politics would produce the same "chemistry" as had occurred on BCDC, in which a general sense of mission and the combination of diverse perspectives had resulted in a remarkably harmonious and hardworking agency that was able to meet its deadlines and generally to act in a manner consistent with statutory directives.[51]

This expectation was most clearly achieved at the state commission. All of the commissioners and staff interviewed agreed that this was a remarkably harmonious and dedicated group of people who respected each other's judgments, with no dominant individuals or cliques. As one commissioner remarked,[52] "It's a very unusual commission. This group works very well together. There is a high feeling of respect among commissioners for each other. When any member of the commission speaks strongly on an issue and really has something to say, the other eleven listen very carefully. No one tries to run with the ball that much—depends on the issue, the person who seems to have the most expertise."

Part of this effective interaction could be attributed to the general homogeneity of basic viewpoints, as the vast majority of commissioners and (in all likelihood) staff had voted for the Coastal Initiative and were deeply committed to see it work. Part could also be attributed to the remarkable leadership skills of Mel Lane, who was unanimously elected chairman at the initial meeting and was universally respected for his judgment, his integrity, and his ability to run meetings fairly and expeditiously. Similarly, Joe Bodovitz, the executive director, was highly respected by the commissioners, continued his excellent working relationship with Lane (as both had held similar positions at BCDC), and was able to get an enormous amount of work out of his staff, in part because of a general policy of involving all staff in policy decisions concerning staff recommendations on permits and on planning issues. Finally, part of the remarkable cohesiveness of this group of people could be attributed to the nature of commission meetings, which were generally 2-day affairs in which commissioners and staff were thrown together in the same hotel, with the result that a good deal of informed interaction occurred outside of commission meetings.[53] There were, of course, some tensions, particularly between planning and permit staff. In addition, a few commissioners expressed some resentment that the "BCDC crowd"—Lane, Bodovitz, and Schoop (the chief planner)—were running the show, although most commissioners felt this was somewhat inevitable given the tight dealines and, in any event, agreed with the general policies being pursued. On the whole, however, most commissioners viewed the state commission as probably the most stimulating and enjoyable group of people with whom they had ever worked, and this was confirmed by the remarkably low turnover on this commission (and staff, for that matter), despite the substantial inconveniences posed by the frequent meetings, travel, and general work load.

This general pattern was repeated in the North Central region, although to a slightly lesser extent. Here again, the commissioners got along well together and generally respected each other, although influential individuals and voting blocs were more discernible than at the state level.[54] Part of this could again be attributed to the general ideological homogeneity of the commission members, the vast majority of whom strongly supported Proposition 20; even the three or four more "conservative" members with a strong concern for property rights were genuinely respected by the majority and did not resort to invective or obstructionist tactics. Part of this could also be attributed to the leadership skills of the second chairperson, Margaret Azevedo, who was widely respected by most of the members for her judgment and ability to keep meetings from running past midnight (as they had frequently done under the first chairperson, Michael Wornum). Finally, the vast majority of the commissioners had enormous respect for the judgment, fairness, and tremendous amount of work of the staff and particularly the executive director, Michael Fischer. As one local government commissioner put it, [55] "The strength of the commission lies in the staff. . . . It reflects the quality of the director. . . . If you could take that staff and make a planning department for the city of San Francisco, you could do no better. It is a neat group of guys—and I am very critical of public servants in the large measure."

Relations within the staff were similar to the state level, with one person primarily responsible for planning, several others (the law students) working entirely on permits, and the remainder working on both. Unlike those of the state staff, however, staff decisions in the North Central were much more the prerogative of the executive director and the junior staff person(s) who had done the work on a particular issue than a result of group consultation.

The South Coast region departed from this general pattern of harmony. For one thing, this commission had relatively clear voting blocs, with three or four environmentalists, three or four development oriented commissioners, and four or five swing votes. [56] On balance, however, all of the leadership posts, including the three chairmen (Rooney, Bright, and Rosener) and the two state representatives (Hayes and Fay), shared moderate to relatively restrictive views of development. Moreover, each bloc had at least one articulate spokesman, with the result that there was probably more controversy and debate on this commission than on any of the others. While these internal divisions were kept within manageable levels during the first 2½ years, they occasionally disintegrated into petty bickering and badgering of both staff and applicants with the arrival of David Commons in the fall of 1975. Another salient feature of this region was its enormous permit work load, which required the commission to meet four to six times a month rather than once or twice as with the other commissions examined. As mentioned previously, this led to a relatively large and somewhat bureaucratized staff, although internal decisions were apparently the result of consultation among the senior permit (or planning) staff rather than being made by a hierarchically superior executive director. [57]

In many respects, however, the most interesting region was the North Coast. First, many commissioners remarked in interviews that there was a rather clear 8–4 split on the commission, with all of the local officials plus two of the state appointees constituting the localism/property rights majority. [58] These perceived attitudinal differences between local and (most) public members were confirmed by the survey data presented previously in Table 4-2. On the other hand, Table 4-2 also revealed very little difference in *voting behavior* between the two groups (1% vs. 5% denials), and, in fact, no member of the commission voted to deny permits more than 8% of the time. There are at least three possible explanations for this apparent dissonance between attitudes and behavior, all with some supporting evidence: Environmentally oriented members may have (1) been constrained by the strong, and sometimes virulent, opposition of most of the local populace to Proposition 20, and/or (2) doubted that the generally small (with the exception of timber harvesting) developments proposed in the North Coast posed any substantial conflicts with the Initiative's objectives, and/or (3) saved their "no" votes for those infrequent occasions when they could persuade one or two members of the majority to join them.[59]

Another interesting feature in this region was the relationship between the conservative majority on the commission and the considerably younger and more environmentally oriented junior staff. While most commissioners were apparently relatively satisfied with the staff, there were certainly occasions when a few of them—particularly the chairman, John Mayfield—publicly criticized the junior staff. This placed the executive director in a rather precarious position, as he served entirely at the pleasure of the commission majority. His solution was apparently to insist that staff carefully document the justification for any constraints on development, although he apparently very seldom changed the recommendation suggested by the permit staff.[60] Moreover, it is highly probable that the staff were motivated by some of the same concerns as the environmentally oriented commissioners for, as we shall see in Table 4-4, they seldom recommended denials. At any rate, the amount of tension between commission and staff (and within the commission as well) declined noticeably when John Mayfield—a very articulate, but also often acerbic conservative—was replaced as chairman by the much more diplomatic Donald Hedrick.[61]

This discussion of staff–commissioner relations in the North Coast region brings us to a topic that worried some of the framers of the Coastal Initiative, namely, the possibility that the full-time professional staff would come to dominate the part-time commissioners.[62] Commissioners received only travel expenses and a very modest *per diem* for the 2–6 days each month spent at commission meetings. And, the vast majority of commissioners held another full-time job—in the case of local elected officials, other governmental positions as well—and therefore could devote only a limited amount of time to their commission responsibilities.[63] In such a situation, the commission staff necessarily became their principal source of information on both permit and planning matters. Moreover, the staff largely determined the major issues to be discussed on particular permits and on general planning issues, as well as presenting a reasoned recommendation on both types of decisions.[64]

On the other hand, the commissioners were certainly not without resources. First, and most important, was the actual authority to make legally binding decisions, irrespective of any biases in the information presented by staff. Second, they had the authority to dismiss the executive director if displeased with staff performance. Third, the 2–3-month delay in assembling a permanent staff meant that it was the commissioners, rather than the staff, who were primarily responsible for designing the permit application form that, in effect, represented a checklist of topics to be investigated in each case. Fourth, all commissioners had access to information independent of the staff, primarily in the public hearings on all important permit applications and planning issues. Interviews also indicated that, while practically no commissioners regularly examined the permit files, many made their own site visits on important permits. Moreover, all but one were longtime residents of coastal counties and had a good knowledge of the coastal zone in their region; this would, however, be much less true of state commissioners, who would necessarily have only a superficial knowledge of much of the state's 1100-mile coast.[65] Finally, most of the commissioners were, by education and/or experience on other governmental bodies, quite knowledgeable about general issues of land use planning and resource management, as well as the dynamics of staff–commission relations. In short, there were probably fairly solid grounds for the framers of the Coastal Initiative to assume that the legal authority, experience, and general knowledge available to commissioners would be sufficient to enable them to overcome fairly severe time constraints in evaluating the information presented by staff and in setting the general policy direction of their commission.

The respective influence of staff and commissioners is examined in Table 4-4. For each commission, part A of the table provides several indicators of the general policy orientation of staff and commissioners. It clearly indicates that commissioners in the North Coast and, to a lesser extent, the South Coast were considerably less sympathetic to the objectives of Proposition 20 than were their staff. On the other hand, commissioners and staff in the North Central region and probably the state commission were all quite supportive.[66] If commissioners were able to establish the general policy orientation, one would thus expect the North Coast and South Coast regions to be considerably more permissive in their review of development permits than their colleagues in the North Central and State Commissions. On the other hand, if staff were able somehow to dominate commissioners, one would expect fairly similar decisions in all three regions, as staff in each region were remarkably homogeneous in their general policy orientations.[67]

Part B of Table 4-4 clearly supports the hypothesis of commissioner, rather than staff, control. Whereas 75% of the hearing calendar permits in the North Coast and 56% of those in the South Coast during the 1973–1975 period were approved as submitted by the applicant, this was true of only 11% of the permits in the North Central and 4% of the appeals actually voted on by the state commission. In short, this evidence suggests that the decisions of each commission were generally consistent with the basic policy orientation of its member-

Table 4-4

Comparison of the Attitudes and Behavior of Part-Time Commissioners and Full-Time Staff on Each Commission

	North Coast		South Coast		North Central		State	
	Staff	Comms.	Staff	Comms.	Staff	Comms.	Staff	Comms.
A. Attitudes								
1. % support Prop 20 in retrospect[a]	91%	36%	100%	56%	100%	100%	n.a.	91%
2. Environmental Protection Scale[b]	.26	−.80	.49	−.36	.10	.19	n.a.	.32
3. Location/Market Solutions Scale[c]	−.58	1.01	−.42	.32	−.31	−.28	n.a.	−.15
	Staff rec.	Comm. dec.	Staff rec.	Comm. dec.	Staff rec.	Comm. dec.	Staff rec.[e]	Comm. dec.[e]
B. Substantive nature of staff recommendation/commission decision[d]								
Approve as submitted	72%	76%	41%	56%[f]	15%	15%	5%	4%
Approve with minor conditions	12%	11%	28%	17%[f]	37%	35%	3%	3%
Approve with major conditions	10%	8%	8%	3%	31%	34%	52%	50%
Deny	6%	5%	22%	24%	16%	16%	40%	43%
	100%	100%	99%	100%	99%	100%	100%	100%

[a]This differs slightly from the item reported in earlier tables. Rather than asking respondents how they actually voted on Proposition 20 it asks them how, in retrospect, they would have voted. It is thus a better indicator of their attitudes once they had gained some experience administering the act.
[b]Factor score ranging from +3 (high support for environmental protection) to −3 (low support).
[c]Factor score ranging from +3 (high support for local control and market solutions) to −3 (low support).
[d]Based on a random sample of hearing calendar permits in the three regions, February 1973–June 1975.
[e]Based on a random sample of 166 permit appeals to the state commission, February 1973–June 1975.
[f]Statistically significant difference between staff recommendation and commission decision, p ≤ .05.

ship. Conversely, the North Coast staff recommended that 72% of the permits be approved as submitted and that only 2% be denied. While part of this seeming inconsistency with their rather strong environmental views can be attributed to a general feeling among many of the staff that most developments proposed in the North Coast were consistent with the objectives of the Coastal Initiative, staff were also well aware of the *laissez faire* policy orientation of the commission majority and this played an important background role in their deliberations.[68] A similar muting of personal policy preferences also occurred among the South Coast staff, although to a lesser extent. In the case of each commission, then, it is quite clear that staff operated within "a zone of tolerance" established by the commission majority. Or, as one state commissioner responded when asked about charges of staff domination,[69] "The staff is doing the job that we want them to do. If the staff were making different kinds of recommendations, we would disagree with them more often. The state commission is predisposed towards strong control over development as is the staff. I can guarantee you that if the staff were suggesting more development, they would be turned down more often. . . . It is the commission's philosophical orientation and its similarity to the staff's philosophical orientation that makes for this seeming dictatorship by the staff."

V. RELATIONS BETWEEN THE STATE AND REGIONAL COMMISSIONS

Although the 1972 Coastal Initiative clearly gave the coastal commissions a great deal of review authority over the actions of state and local agencies within the coastal zone (and particularly the 1000-yard permit area), it was much less explicit about allocating authority among the coastal commissions. The statutory provision for each commission to select its officers and its executive director gave each considerable internal autonomy. While this general independence was tempered by a strong state commission role in planning and permit review, the net result was a somewhat ambiguous degree of hierarchical control within the commissions. For example, the statute made the state commission ultimately responsible for the content of the coastal plan to be presented to the legislature while also indicating that the regional commissions should make recommendations; it was silent, however, on the manner in which the state and regional commissions should interact in this process.[70] Similarly, the statute clearly gave the state commission authority to hear appeals of regional commission permit decisions brought by "any aggrieved party."[71] On the other hand, the Coastal Initiative nowhere required the regions to accept state commission decisions as binding precedent. In short, while the statute made the state commission the ultimate arbiter of the content of the coastal plan and of any permit applications important or controversial enough to be appealed; it also allowed the regional commissions considerable autonomy and essentially left open the possibility that they would ignore the decisions of their nominal superiors at the state level.

In practice, however, a high degree of cooperation eventually evolved—albeit with a great deal of effort and some notable exceptions.[72] For one thing, the state commission and at least three of the regions were strongly committed to making Proposition 20 work. This essentially meant that regional diversity had to be recognized and that support in each of the regions had to be marshalled if the coastal plan was to pass the legislature. Thus, the state commission and staff made it a policy to give the regions whatever funds they requested (within overall budgetary constraints), to involve the regions at an early date in the planning process, and to seriously consider their recommendations. This involved monthly meetings between the chairmen and executive directors of the seven commissions, as well as meetings between the staff involved with specific plan elements. In addition, the regional representatives generally tried to articulate their region's rationale for both permit decisions and planning recommendations, although most did not feel bound to vote in accord with the preferences of the regional majority.[73] Even regions, like the North Coast, that were dominated by commissioners unsupportive of Proposition 20 had to seriously consider the general guidelines on planning and permit review developed by the state commission if they were to have any influence on the decisions ultimately adopted. Moreover, while there was some tendency in this region to ignore state decisions on permit appeals, most commissioners ultimately realized that this worked a real hardship on applicants, and thus there was some effort to work within state guidelines and at least to inform applicants of the likely consequences of appeal.[74]

Relations between the state and regional commissions on both planning and permit review will be discussed in greater detail in the next two chapters. In order, however, to illustrate some of the essential dynamics, state and regional commission decisions on a random sample of 166 permit appeals in the period between February 1973 and June 1975 are presented in Table 4-5. Part A simply indicates the regional commission decisions. For example, of the permits appealed from the North Coast region, the regional commission approved the permit application as submitted in 40% of the cases, while approving an additional 43% with conditions and denying the remaining 17%. The rest of the table deals with the difference that state review made. In order to understand this process, however, it is important to realize that the state commission had a two-stage review process: In the first part, the commission conducted an abbreviated hearing to decide if the appeal raised "a substantial issue," i.e., if it was important enough to subject to a full hearing. If not, the regional commission decision was simply affirmed. The percentage of appeals from each region that were deemed *not* to have raised a substantial issue are presented in part B of the table; as can be seen, this constituted about 54–64% of the cases from each region,

TABLE 4-5

REGIONAL AND STATE COMMISSION DECISIONS ON PERMITS APPEALED TO THE STATE COMMISSION, 1973–1975 (N = 166)

	Region						
	North Coast	North Central	Central	South Central	South Coast	San Diego	Total
A. Regional commission decision							
Approve as submitted	40%	0%	7%	14%	33%	50%	31%
Approve with conditions	43%	73%	41%	57%	28%	27%	36%
Deny	17%	27%	52%	29%	39%	23%	33%
Subtotal	100%	100%	100%	100%	100%	100%	100%
B. State commission decision on Substantial Issue							
% affirmed regional commission decision after abbreviated hearing	28%	62%	59%	64%	54%	54%	53%
C. State commission decision on the merits when subjected to a full hearing							
Approve as submitted	0%	0%	0%	0%	0%	17%	4%
Approve with conditions	55%	70%	64%	40%	60%	33%	53%
Deny	45%	30%	36%	60%	40%	50%	43%
Subtotal	100%	100%	100%	100%	100%	100%	100%
D. Cumulative decision of state and regional commissions							
Approve as submitted	7%	0%	4%	7%	12%	27%	13%
Approve with conditions	47%	69%	48%	50%	42%	31%	43%
Deny	47%	31%	48%	43%	46%	42%	44%
Subtotal	101%	100%	100%	100%	100%	100%	100%

[a] Following are the 95% confidence intervals ($p = .5$) for each of the regions: North Coast and North Central (± .09), South Coast (± .13), Central (± .14), San Diego (± .17), and South Central (± .38).

with the notable exception of the North Coast, where only 28% of the regional commission decisions were affirmed. The second stage of state review involved a full public hearing and vote by the state commission. State commission decisions subjected to this full review are indicated in part C of the table. Finally, the cumulative result of state and regional commission review—i.e. the regional commission decisions on appeals involving "no substantial issue" and commission decisions on permits subjected to a full hearing—are presented in part D of the table. For example, the end result of the appeals process was that 7% of the permits appealed from the North Coast were approved as submitted, 47% were approved with conditions, and the remaining 47% were denied.

A number of important conclusions emerge from this brief analysis. First, as should clearly be anticipated from the attitudinal data presented in Table 4-2 and as was expected by the framers of the Coastal Initiative, the state commission was generally less permissive in its review of development applications than the regions. On this sample of permit appeals, for example, 31% involved regional decisions to approve as submitted. By the end of the appeals process (comparing part A and part D of th Table 4-5), however, this was reduced to 13%, while the rate of denials rose from 33% to 44%. Second, the importance of the appeals process varied substantially from region to region. In the North Central and Central regions, for example, the appeals process had very little overall effect; the state commission denied a few permits that the regions had approved with conditions and, conversely, approved a few that they had denied, but the net result was very little change in development patterns. In contrast, the appeals process had a significant impact in the North Coast region. Whereas 40% of the permits appealed had been approved by the regional commission as submitted by the applicant, the net result of state commission review was that only 7% were ultimately approved as submitted; conversely, the rate of denials rose from 17% to 47%. State commission review had similarly major impacts in the San Diego region and somewhat lesser effects in the South Coast and South Central regions. In short, state review served precisely the function envisaged by the framers of Proposition 20: The state commission was an appeals board dominated by statutory sympathizers that served as a check on regional commissions with relatively permissive views toward coastal development.

V. CONCLUSIONS

One of the recurring difficulties confronting the effective implementation of innovative statutes is their assignment to one or more large existing bureaucracies, each of which has a number of conflicting mandates, inadequate resources, and surfeit of capable, but cautious, civil servants. In such a situation, implementation is delayed and often undergoes subtle changes as statutory directives are gradually incorporated into the agency's operating procedures by personnel who are often unenthusiastic about its objectives.

The Coastal Initiative generally managed to avoid this syndrome. Implementation was assigned to a single agency (or group of agencies), thereby avoiding the quagmire of interagency coordination. It was also assigned to a new agency, thereby avoiding the pitfall of integrating the new program into existing, and often conflicting, operating procedures. Moreover, as has often been noted with respect to new agencies, the professional staff came disproportionately from strong supporters of the statutory mandate. This in turn meant not only that the details of implementation were administered with a sympathetic eye but also that the staff were willing to work enormously hard in pursuit of statutory objectives.

The linchpin of the implementation process was, however, the appointment of coastal commissioners. The framers of the Coastal Initiative would have preferred a clear majority of state appointees on each of the commissions, on the grounds that they would generally be more supportive than local elected officials of the statutory objectives of protecting public access, wetlands, view corridors, and agricultural land/open space. But political reality obliged them to accept the compromise of an even split between state appointees and local officials on the regional commissions, with no direct local government representation on the state commission. There was also the hope that local officials would be less parochial when sitting on a regional or state board and that, as had happened with BCDC, they would work well with the state-appointed "public" members.

These expectations were generally realized. On the whole, state appointees were somewhat more supportive of the act's objectives than were local elected officials. This was particularly true in regions, like the North Coast, in which local officials mirrored their constituents' strong antipathy toward Proposition 20. But in regions like the North Central and, to a lesser extent, the South Coast in which the electorate had strongly supported the Initiative, local governing bodies generally sought to balance their own reservations about the Initiative with their constituents' support in making appointments to the regional commissions. A similar concern to balance personal and voter preferences may well have been important to Governor Reagan, who, although personally opposed to a strong state role in coastal management, nevertheless appointed a majority of statutory supporters to the commissions. In short, the overall balance of support among commissioners may have owed at least as much to the clear indication of voter support as to the relative proportion of state versus local appointees.

The Coastal Initiative incorporated a somewhat ambiguous stance toward hierarchical control within the overall agency (i.e., among the coastal commissions). On the one hand, it gave each regional commission considerable internal autonomy, with the result that each reflected a compromise between the views of its members (and their constituents) and statutory directives. In the long run, this expression of regional opinion and autonomy was probably advantageous to the development of support for the coastal plan in the 1976 legislature, as few groups could legitimately claim that they had been systematically excluded from the decision-making process. At the same time, however, the statute clearly assigned ultimate responsibility in both plan development and permit review to the state commission, i.e., the body most likely to be supportive of statutory objectives.

In short, the general effective structuring of the implementation process incorporated into the Coastal Initiative and the highly visible degree of voter and constituency group support represented in the Initiative campaign and throughout the implementation process meant that our third condition of effective implementation—namely, that implementing officials be committed to the achievement of statutory objectives and skillful in utilizing available resources—was essentially met. Most commissioners and staff were committed and the internal review process provided a check on recalcitrant regions like the North Coast. The skill of implementing officials in developing outside support, managing internal dissent, and maintaining morale is difficult to measure with any precision. But there can be little doubt that the leadership of Lane and Bodovitz at the state level was superb. Conversely, there were no cases in the three regions examined of clearly deficient skill and probably some cases, particularly by the executive directors in the North Central and North Coast, of above-average ability to work within resource and attitudinal constraints.

Given this general review of internal attitudes and processes, it is now time to turn to a more detailed examination of policy making within the commissions—first in permit review, then on plan development—as well as to the crucial question of the extent to which the policy outputs of the commissions conformed to statutory objectives.

NOTES

1. See, for example, Pressman and Wildavsky, *Implementation*, chaps. 3–6; Stephen Bailey and Edith Mosher, *ESEA: The Office of Education Administers a Law* (Syracuse: Syracuse University Press, 1968); Jerome Murphy, *State Education Agencies and Discretionary Funds* (Lexington, Mass.: D. C. Heath, 1974); Paul Berman and Milbrey McLaughlin, "Implementation of Educational Innovation," *Educational Forum* 40 (March 1976), pp. 345–370; and Christa Altenstetter and James Bjorkman, "Implementation of a Federal-State Health Program" (Paper presented at the 1977 Annual Meeting of the American Political Science Association, Washington, D.C., September 1–4, 1977).

2. For a discussion of this problem, see William Boyer, *Bureaucracy on Trial* (Indianapolis: Bobbs-Merrill, 1964), chap. 4.

3. Among existing state agencies, the most likely candidate for implementation of coastal legislation would have been the Department of Navigation and Ocean Development, whose major responsibility was the construction of boat marinas (i.e., a function in direct conflict with the wetlands protection mandate of the Coastal Initiative). For discussions of the resistance of most bureaucracies to programmatic change, see Kaufman, *The Limits of Organizational Change;*

Elmore, "Organizational Models of Social Program Implementation," pp. 199–216; and Downs, *Inside Bureaucracy*, chaps. 13, 18, 19.

4. Bernstein, *Regulating Business by Independent Commission*, chap. 3; Downs, *Inside Bureaucracy*, chaps. 2, 8, 9.

5. The assumption of differences between state-appointed and locally elected commissioners was based heavily upon the experience of the California members of the Tahoe Regional Planning Agency (Costantini and Hanf, *The Environmental Impulse and Its Competitors*, pp. 55–58). According to Section 27220 of the Coastal Initiative, "Each public member of the commission or of a regional commission shall be a person who, as a result of his training, experience, and attainments, is exceptionally well qualified to analyze and interpret environmental trends and information, to appraise resource uses in light of the policies set forth in this division, to be responsive to the scientific, social esthetic, recreational, and cultural needs of the state. Expertise in conservation, recreation, ecological and physical sciences, planning, and education shall be represented on the commission and the regional commissions."

6. Interview with Janet Adams, executive director of the Coastal Alliance, 2 December 1975, pp. 15–16.

7. Douglas and Petrillo, "California's Coast," p. 192.

8. California Public Resources Code, Sections 27300, 27423 (1973).

9. Douglas and Petrillo, "California's Coast," pp. 194–196.

10. California Public Resources Code, Sections 27320, 27420 (1973); Douglas and Petrillo, "California's Coast," pp. 214–217. The statutory emphasis on conducting the commissions' business in public was reinforced by an attorney general's ruling, in response to a suit by the Natural Resources Defense Council, requiring that a written summary of all *ex parte* communications be transmitted to other commissioners and to the commission staff (Healy *et al.*, *Protecting the Golden Shore*, p. 18.) Interviews with commissioners indicated that the vast majority strongly discouraged such *ex parte* contacts, particularly after that ruling.

11. The reasons for relying on appointees' knowledge of the reasons for their selections—rather than also interviewing the appointing authorities—were essentially practical. Interviews for this study were conducted in the winter of 1974–1975, or 2 years after the appointments had been made. By that time, Governor Reagan's appointments secretary (Ned Hutchinson) had died, and conversations with (former) Assembly Speaker Moretti indicated that he had only the vaguest recollection of what had occurred. Moreover, interviewing the members of the boards of supervisors and municipal associations in each county within the three regions was simply infeasible. We thus decided to rely entirely upon the appointees' memory and candor, and in many cases received surprisingly detailed and frank accounts. Nevertheless, the reader should be aware of the potential for measurement error.

12. Interviews with 11 of the 12 members of the North Coast Commission were conducted in the winter of 1974–1975.

13. The initial meeting of the commission on January 15, 1973, produced a 5–5 deadlock between May and Vaughn. At the next meeting on January 20, Bill Grader from Mendocino County was selected as chairman, and May from Humboldt County was chosen as regional representative on a 7–3 vote (see commission minutes). Interviews with several commissioners indicated that a crucial factor in the selection of May was that Vaughn was also from Mendocino County and a few members felt that the leadership posts should be distributed among the counties.

14. Peterson's election was essentially engineered by Dwight May, who suggested that since the chairman (Mayfield) was from Mendocino County and the vice-chairman (McClendon) from Del Norte, the regional representative should be from Humboldt County. This argument, plus some rather deft parliamentary maneuvering, produced a 9–1 vote for Peterson, who did not actively solicit the position (based upon commission minutes and interviews with several commissioners).

15. This discussion is based largely upon interviews with 12 of the 14 members of the North Central Commission in the winter of 1974–1975.

16. According to the annual reports of the commissions, the South Coast Commission processed about 40% of the permit load of the six regions, about twice the level found in any other region. Information in this section is based largely upon interviews with 10 of the 12 members of the commission in early 1975.

17. Following is the percent "yes" vote on Proposition 20 in each of the local jurisdictions with representation on the regional commission: Orange County (51%), Los Angeles County (55%), City of Los Angeles (61%), and City of Long Beach (43%).

18. This discussion is based largely upon interviews with all 12 members of the state commission in the summer of 1975.

19. Apparently the orientation of many commissioners in the beginning was largely unknown. But the supporters of Proposition 20 (who turned out to be a minority) decided to place their primary emphasis on the regional representative rather than on the chairmanship of the regional commission. And Laufer, although a supporter of the Initiative, was, as a businessman and former president of the local Chamber of Commerce, perceived to be sufficiently moderate as not to arouse intense opposition. He remained the regional representative, despite the subsequent reservations of many members, because the regional commission decided that its officers would serve for the life of the commissions (unlike the North Coast and North Central commissions, which held annual elections).

20. According to Frautschy, he was the only candidate. Although a supporter of Proposition 20, he was viewed as a moderate and turned out to be one of the most conservative members of the state commission.

21. For example, of the 15 activists (primarily county coordinators) in the Pro-20 campaign that we interviewed in conjunction with the campaign (see chapter 2, note 8), only Ellen Jonck from San Francisco and Phyllis Faber from Marin County were named to the commissions. This should not imply that the Pro-20 proponents were inactive or ineffective in the appointment process. According to Janet Adams, they made a major effort to nominate people and were often successful (Adams interview with Sabatier, p. 16). But it does suggest that the successful nominees were people like Mel Lane who were prestigious and had worked behind the scenes rather than the most visible activists. Adams also noted, parenthetically, that many of the activists who did not get appointed to the commissions subsequently used the Proposition 20 campaign organization to get elected to city and county governing bodies; examples included Bert Muhly from Santa Cruz, Bill Kortum from Sonoma County, Sandy Motley from Davis, and a lady from Ventura County.

22. In fact, the act demonstrated considerable concern with avoiding start-up delays, as Section 27221 required that all appointments be made by December 31, 1972, and Section 27255 required that the commissions hold their first meeting no later than February 15, 1973. For a discussion of the importance of this rapid start-up, see Douglas, "New Approach in California," pp. 15–16.

23. Another indication of commitment—even on the part of avowed opponents of Proposition 20 such as Bernard McClendon, Richard Brown, and John Mayfield—was their excellent attendance records, despite often having to drive 4–6 hours to commission meetings in the huge North Coast region.

24. For the former argument, see Rosener et al., "Environmentation vs. local Control," pp. 23–27, while the latter hypothesis was suggested by Gregory Jones, Sonoma County municipal representative to the North Central Commission, in an interview, 3 February 1975, pp. 1–2.

25. Partisan affiliation was coded in terms of Democrat, Republican, Independent; only two chose the third category. For reviews of the literature on the correlates of environmental concern, see J. Fred Springer and Edmond Costantini, "Public Opinion and the Environment," *Environmental Politics*, ed. Stuart Nagel (New York: Prager, 1974), pp. 195–224; James Coke and

Steven Brown, "Public Attitudes About Land Use Policy and Their Impact on State Policy-Makers," *Publius* 6 (1976), pp. 97–134; Riley Dunlap and Michael Allen, "Partisan Differences on Environmental Issues: A Congressional Roll Call Analysis," *Western Political Quarterly* 29 (1976), pp. 384–397; Fredrich Buttel and William Flynn, "The Politics of Environmental Concern," *Environmental Behavior* 10 (March 1978), pp. 17–36; and Daniel Mazmanian and Paul Sabatier, "Liberalism, Environmentalism, and Partisanship in Public Policy-Making: The California Coastal Commissions," *Environment and Behavior* 13 (May 1981), pp. 361–384.

26. The two attitudinal scales were developed from a survey questionnaire distributed to commissioners and staff on the four commissions at the conclusion of personal interviews in the winter and spring of 1974–1975. Sixty-five responses were received, for an overall response rate of 79%. Approximately 45 items in the questionnaire were devoted to general attitudes toward environmental protection, the proper scope and locus of governmental authority, and a variety of other public issues—many of them taken from the Costantini-Hanf study of the TRPA and the Coke-Brown study of land use in Ohio. Responses on these items were then subjected to a principal-components factor analysis with quatrimax rotation. Among the factors that emerged were one dealing with attitudes toward environmental protection and economic development and another with views toward the proper scope and locus of governmental authority on land use (and other) policies.

The first was labeled an Environmental Protection Scale: Its component items and their respective factor loadings are as follows: (1) "No further development should be allowed in any wild or natural areas, whether now under protection or not" (.75). (2) "If the number of people visiting a primitive area threatens to alter it substantially, the number should be reduced rather than adding facilities to handle the load" (.69). (3) "More emphasis should be placed on society's environmental rights and less placed on the individual's economic rights" (.67). (4) "One person's rights to a clean environment are not as important as another's rights to gainful employment" (−.65). (5) "I believe that plants and animals exist primarily for man's use" (−.63). (6) "Regulated mining operations can be operated in primitive areas without affecting their value" (−.60). (7) "Our national parks would better serve the public if motels and other vacation facilities were made available along access roads" (−.53). (8) "Aesthetic rather than economic factors must guide our use of natural resources" (.46).

The second scale was labeled Localism/Market Solutions. Its component items and their respective factor loadings are as follows: (1) "Decisions about development along the coast are best left to the economic market" (.74). (2) "The more governmental responsibility is placed at the local level, the better" (.71). (3) "The coast is a resource of importance to the people of the entire state and should be subjected to state oversight" (−.71). (4) "When environmental controls along the coast place heavy burdens on individual property owners, special exceptions should be made" (.68). (5) "Government planning almost inevitably results in the loss of essential liberties and freedoms" (.62). (6) "Individuals with the ability and foresight to earn and accumulate wealth should have the right to enjoy that wealth without government interference and regulation" (.57). (7) "While in the past, local governments in California have been primarily concerned with increasing their tax base through economic development, today they are doing an adequate job of balancing economic development with considerations of environmental quality" (.46). (8) "Lobbyists and campaign contributions may be strictly regulated to prevent special interests from corrupting the political process" (−.45). As might be expected, the two scales are negatively correlated (−.79); moreover, as we shall see in the next chapter, they were both excellent predictors of commissioner voting behavior. For additional discussion of the derivation of the two scales and their relationship to the background items presented in Table 4-1, see Mazmanian and Sabatier, "Liberalism, Environmentalism, and Partisanship."

27. Because of the all-inclusive definition of "development" under the Coastal Initiative, only about 25% of permit applications before the regional commissions were deemed sufficiently important or controversial to be subject to a full public hearing; the remainder were voted on *en masse* by

the regional commissions, although they could be moved to the hearing calendar at the request of two commissioners. The data reported here involved all hearing calendar items voted on in the three regions between April 1973 and December 1975; they were gathered with the cooperation of Judy Rosener of the University of California, Irvine.

28. This was also the view of Mel Lane in his informal relations with Reagan (see interview with Lane, 2 December 1975); for additional evidence, see Healy *et al.*, *Protecting the Golden Shore*, p. 212, note 42.

29. In the North Coast, the only nonrespondent was McHugh, an Assembly appointee who voted consistently with the conservative majority; thus, the responses given in Table 4-4 are probably slightly more Pro-20 than they should have been, especially among the "public" members. In the North Central, there were four nonrespondents: two public members (Jonck and Zankich) and two local officials (Wornum/Giacomini and Theiller). In each category, however, they were split between strong environmentalists and members of the commissions' conservative minority; thus, the overall sampling error was probably negligible. In the South Coast, the situation was complicated by the turnover in the two governor's slots and in the local planning agency representative; suffice it to say that the only bias may have been to portray local members as slightly more in favor of coastal protection than they actually were. Finally, the only nonrespondent among state commissioners was Ellen Stern Harris, the commission's most outspoken environmentalist; thus, that commission was probably slightly even more environmental than it is portrayed. All of these *caveats* apply, however, only to the items taken from the attitudinal survey; there is no "sampling" bias in the permit data, as the entire population of votes during the 1973–1975 period were counted.

30. The reader may well question whether the denial rates of 15% and 27% for public members in the North Central and South Coast, respectively, are, in fact, "very similar." However, the figures for the North Central are somewhat understated because of the tendency in this region, as we shall see in the next chapter, to approve Projects with major conditions—in many cases amounting to *de facto* denial—rather than to deny them outright.

31. Selection of the regional representatives to be surveyed presented some problems because of the turnover in the North Coast, Central, and South Coast regions. We could have averaged the responses from occupants of the various positions (weighted by the time served as regional representative), but that seemed needlessly complicated. Instead, we chose to use the responses of the occupant in the winter of 1974–1975, when the remaining commissioners were surveyed, i.e., Peterson in the North Coast, Andresen in the Central, and Fay in the South Coast. In the first two cases, this probably produced no bias, as Peterson was probably somewhere between Vaughn and May, while Andresen was probably similar to Harry in her overall orientation. In the South Coast, on the other hand, Fay was more environmentally oriented than Hayes (although moreso in his attitudes than in his voting behavior on the state commission). In short, the responses of the regional representatives presented in these tables probably contain a slight "sampling" error.

32. This difference is not statistically significant, i.e., it may simply be an artifact of sampling error. Voting on permit appeals at the state level is not comparable to voting on hearing calandar items at the regional level because (a) the permits appealed were generally the most important and the most controversial of the regional decisions, and (b) the state commission used a two-stage voting procedure by which only those appeals judged to have raised a "substantial issue" (about 47% of the total) were actually subjected to a roll call vote by the entire commission. The data presented here are from a random sample of 166 permit appeals (77 of which involved a substantial issue) voted on by the state commission between February 1973 and June 1975. For a more complete discussion, see Paul Sabatier, "State Review of Local Land-Use Decisions: The California Coastal Commissions," *Coastal Zone Management Journal* 3 (1977), pp. 255–290.

33. The permit appeals data from the sample of 166 cases in 1973–1975 (*supra*, note 32) could not be used to investigate the voting behavior of local officials on projects from their own electoral constituencies because the number was so small as to be swamped by possible sampling error.

34. At the most aggregate level, whereas 69% of the respondents from our four commissions reported having voted for Proposition 20, 64% of the 56 commissioners (out of 84) from all seven commissions responding to a *Los Angeles Times* poll in early 1973 had done so (Healy, *Land Use and the States*, p. 74). Moreover, informal conversations with various coastal officials suggest that the Central Coast Commission was very similar to the North Central's, while the South Central and San Diego commissions were probably slightly more conservative than the South Coast (but much less so than the North Coast). But detailed evidence on the appointment process and the attitudes (and voting behavior) of "public" and local officials in these regions is simply not available.

35. See, for example, Bernstein, *Regulating Business by Independent Commission*, chap. 3; Downs, *Inside Bureaucracy*, chaps. 2, 9, 13; and Pendleton Herring, *Public Administration and the Public Interest* (New York: Russell and Russell, 1936), chap. 7 on the FTC and chap. 9 on the FPC. This was apparently not, however, the case with the California Energy Conservation and Development Commission. We would suggest that the reason lay in the genesis of the Energy Commission in legislative compromise rather than a vigorous political movement, as well as the general appointment process whereby each of the five full-time commissioners apparently staked out "turfs" with the line staff.

36. Essentially, the civil service position description required people with a background in land use law and/or planning and preferably some experience in local government (interview with Joseph Bodovitz, executive director of the state commission, 25 October 1975, pp. 2–5). For discussions of the ideology of planners, see Melvin Weber, "Comprehensive Planning and Social Responsibility," *Journal of the American Institute of Planners* 29 (November 1963), pp. 232–241; Martin Rein, "Social Planning: The Search for Legitimacy," *Journal of the American Institute of Planners* 35 (October 1969), pp. 233–244; and Kenneth Topping, "Ideology and Urban Planning" (Unpublished paper, California State University at Los Angeles, March 1972).

37. Letter from Bob Brown, executive director, North Central Regional Commission, 16 July 1979. There is also the possibility that differences in public/private salary scales may have led to the recruitment of younger staff [on the general phenomenon see Gary Wamsley and Mayer Zald, *The Political Economy of Public Organizations* (Bloomington: Indiana University Press, 1973), pp. 46–51], but this hypothesis was rejected by at least one knowledgeable observer.

38. In Downs's terms, the institutional insecurity of the commissions and a number of other factors tended to attract policy advocates rather than "conservors" (Downs, *Inside Bureaucracy*, chap. 5, pp. 8–9).

39. Concerted efforts were made to obtain the cooperation of the state staff in the late summer of 1975 and again late in 1976. These proved unsuccessful because of (1) the substantial time constraints under which the state staff operated; (2) the institutional insecurity that pervaded the state staff before, during, and after the 1976 legislative session and the consequent fear that any data reported would jeopardize the fate of the coastal legislation and, in turn, the jobs of the staff; and (3) the contention of the executive director that staff attitudes—as opposed to their behavior—were not a proper subject of investigation.

40. Technically, three of the staff did not actually vote for Proposition 20 because they were not in the state at the time. But they admitted that, in retrospect, they would have voted for the Initiative and thus are counted as supporters.

41. The scores on both of these scales are standardized (z) scores. Thus, the mean for the entire population of commissioners and staff surveyed is necessarily zero. Recall from Table 4-5 that the overall means for commissioners on the Environmental Protection Scale and the Localism/Market Solutions Scale were $-.28$ and $.32$, respectively. In short, commissioners as a whole were somewhat more "conservative" than staff as a whole, although overall staff scores were quite similar to those of commissioners on the North Central and state commissions.

42. Interviews with state commissioners indicate that Bodovitz was really the only person considered for the job. Because of his experience with BCDC, he had been asked to give an orientation session to the commissioners at their first meeting. He so impressed them that they offered him

the job on the spot. Instrumental in his selection were his outstanding record at BCDC, his articulateness, and his ability to work well with Mel Lane, who had been elected chairman. Apparently the high regard for Bodovitz's performance at BCDC extended even to the Reagan administration (interview with Roger Osenbaugh, 27 August 1975).

43. Interview with Bodovitz, 25 October 1975, pp. 2–3.

44. This is based on interviews with several commissioners, notably Grader, Mayfield, May, Hedrick, Benioff, Brown, Vaughan, and Rusher. The other major candidate was Jack Frazier, who subsequently became the chief staff person in the California Department of Fish and Game. Frazier was strongly opposed by Mayfield, however, and the commission eventually turned to Lahr, who several people mentioned was seen as a compromise candidate. Many of the people who were not originally impressed with him came to really appreciate the judgment and intelligence behind his self-effacing exterior. During the selection process, staff for the commission was provided by Sam Winston of the Humboldt County Planning Department.

45. "Temporary" refers to a non-civil service position. Only 3 out of the 10 professional staff were civil service, and one of those (Bob Lagle) left after about 1 year to take the planner's position in the Central Coast region (interview with Jack Lahr, 24 September 1974).

46. Interviews with several commissioners. Fischer was strongly supported by the chairman and several members of the personnel selection committee and apparently emerged as almost a consensus candidate fairly early, although a few members were worried about his environmentalist background.

47. Interview with Mike Fischer, 21 October 1974.

48. Interviews with several commissioners in late 1975.

49. Interviews with Carpenter, 10 June 1975, and with Gordon Craig, chief permit analyst, 15 December 1977.

50. In contrast, this was apparently a significant problem for the California Energy Commission, whose chairman (Richard Maulin) was Governor Brown's campaign manager, while another member (Bob Moretti) had just been defeated by Brown in the gubernatorial primary.

51. Swanson, "Coastal Zone Management from an Administrative Perspective," pp. 91–97.

52. Interview with Ira Laufer, 26 August 1975, pp. 7–8. As part of each interview, commissioners were asked to indicate the most influential members on their commission. One of the remarkable things about the state commissioners was their lack of consensus on who was influential, although Lane, Laufer, and Wilson probably had the most "votes."

53. The state commission generally held two 2-day meetings each month, alternating between northern and southern California (usually San Francisco and Los Angelos). Moreover, it was Bodovitz's policy that all staff whose work would be discussed attend each meeting. While this certainly reduced staff work time, it did promote interaction with commissioners. In this context, it was not unusual for staff who had been unable to get their way in staff meetings to present their argument directly to sympathetic commissioners. Bodovitz apparently felt that this untidy arrangement was a necessary price for commitment—that staff would work the incredible hours they did only if they felt their views were being fully considered.

54. Interviews with commissioners indicated fairly widespread agreement that the commission had three or four moderate conservatives (Theiller, Owen, Zankich, and Molinari/Tamaras), a like number of strong environmentalists (Jonck, Egger, Grote, Faber, and Lundborg), and a large number of swing votes who generally sided with the latter. There was also a pretty fair consensus that the most influential members were Azevedo (even before she became chairman), Mendelsohn, and, to a lesser extent, Lundborg, Faber, and sometimes Owen or Theiller. All of the members agreed that, with the exception of Molinari, there was little if any distinction made between "public" members and local elected officials.

55. Interviews with Gregory Jones, 3 February 1975, p. 6. Even the commissioner who was most critical of the staff because of its alleged domination of the commission admitted that Fischer was highly competent and was, in fact, carrying out the wishes of the commission majority (interview with Robert Theiller, 10 February 1975).

56. The presence of such voting blocks was mentioned by several commissioners interviewed. In fact, one commissioner and a lobbyist presented virtually identical "lineups": Nowell, Rubley, Phillips, and, to a lesser extent, Holmes and Warshaw as the development bloc; Rosener, Bright, Fay, and, to a lesser extent, Rooney as the environmentalists; and Wilson, Hayes, and Schmit as the swing votes.

57. Interviews with Carpenter and Craig.

58. Several also mentioned, however, that the coalitions became more flexible (1) on issues involving the preservation of agricultural land, in which Galletti (a farmer) would sometimes join the minority and (2) after Peterson replaced Rusher as the Humboldt County supervisor in January 1975.

59. Several of the commissioners mentioned the latter two points in interviews; there are also some data from the survey questionnaire that weakly support the second hypothesis, but the question is too complex to discuss here. Nobody explicitly mentioned the first point, but it is difficult to imagine how they could not have been affected, at least subconsciously. For an excellent discussion of the subtle pressures to limit overt disagreements in small groups, see Sidney Verba, *Small Groups and Political Behavior* (Princeton: Princeton University Press, 1961), pp. 22–29.

60. In particular, Lahr and several of the staff mentioned in interviews their strongly held feelings that the developments proposed in the North Coast were much less in conflict with Proposition 20's objectives than those proposed—and often approved—in the South Coast. But several junior staff also tacitly admitted that they were well aware of the views of the commission majority, particularly chairman Mayfield, and that they ultimately served at the pleasure of the commission. Further evidence for this latter point comes from the fact that the most outspoken environmentalist on the staff was the only one with civil service protection (Dick Laursen), although there is no evidence that Lahr ever overtly threatened junior staff if they did not put aside their personal views in making recommendations.

61. Mayfield was criticized by many environmentalists appearing before the commission for trying to intimidate them and limit discussion. Moreover, several commissioners were privately critical of the chairman for his lack of tact. In his defense, Mayfield argued—quite correctly, in some cases—that environmentalists were not always as pertinent and succinct in their comments as they should have been.

62. Douglas and Petrillo, "California's Coast," p. 192. For a much more extensive discussion of this topic, see Paul Sabatier and Daniel Mazmanian, "Relationships Between Part-Time Policy-Makers and Full-Time Staffs: The California Coastal Commissions," *Administration and Society* 13 (Fall 1981), pp. 207–248. Among the major studies of staff–board relations are Mayer Zald, "The Power and Functions of Boards of Directors: A Theoretical Synthesis," *American Journal of Sociology* 75 (July 1969), pp. 97–111; L. Harmon Zeigler and M. Kent Jennings, *Governing American Schools*, (North Scituate, Mass.: Duxbury, 1974), chaps. 8–13; William Boyd, "School Board–Administrative Staff Relationships," *Understanding School Boards*, ed. Peter Cistone (Lexington: Lexington Books, 1975), pp. 102–129; and Ronald Loveridge, *City Managers in Legislative Politics* (Indianapolis: Bobbs-Merrill, 1971). For a discussion of information processing in the South Coast Region, see Mark Rosentraub and Robert Warren, "Information Utilization and Self-Evaluating Capacities for Coastal Zone Management Agencies," *Coastal Zone Management Journal* 2 (1976), pp. 193–222.

63. Of the 43 commissioners interviewed, all but about 6 held at least part-time jobs. Moreover, only one of the local elected officials (Hayes) was paid well enough that he did not have to also hold down a job in the private sector. During this period, commissioners were reimbursed for expenses and were paid $50 for each full day of commission meetings, although not for the time spent preparing for meetings; under Section 30314 of the 1976 Coastal Act this was changed to provide up to $100 for preparation.

64. For an excellent discussion of the importance of the issue definition in determining the important resources and actors in any conflict, see E. E. Schattschneider, *The Semi-Sovereign People* (New York: Holt, Winston, 1960), p. 68. For example, to the extent that staff permit reports

focused on technical issues such as the impacts of a proposed development on water quality— rather than on broad policy questions of the proper distribution of costs and benefits—their own influence, and that of expert commissioners, would be enhanced. In addition, there is some evidence from the literature on small groups that the person who casts the initial "vote"—in this case, the staff recommendation—is likely to have additional influence, as subsequent voters are reluctant to dissent in a face-to-face situation where disagreement puts a strain on relationships that have to be maintained; see A. Paul Hare, *Handbook of Small Group Research*, 2nd ed. (New York: Free Press, 1976), p. 29.

65. This discussion of commissioner and staff resources is based upon personal interviews, where the capacity of commissioners to obtain information independent of staff was a major topic. Bob Mendelsohn, the North Central's representative to the state commission, did an excellent job of illuminating the greater difficulties confronting state commissioners: "The regional commissioners are much closer, physically, to the staff [than are state commissioners] and can interact more with them. They can also go out and make their own investigations and they do. They are also subject to much more interaction with applicants because everybody is from the same region. That is much less the case at the state level. Regional people also know their coast very well, so they don't need the staff to tell them what it is like. Whereas when the state staff is telling me something about San Diego, I have to believe them because I have no reference points."

66. See note 39 concerning the lack of survey data from the state staff. Nevertheless, frequent interaction with staff and conversations with state commissioners indicate that the staff was probably at least as environmentally oriented as the state commissioners.

67. This argument assumes a high correlation between attitudinal indicators and voting behavior. As we shall see in the next chapter, this was certainly the case for commissioners; for example, r = .71 between commissioners' scores on the Environmental Protection Scale and their % No Vote.

68. This is based upon interviews with staff as well as the more extended analysis reported in Sabatier and Mazmanian, "Relationships Between Part-Time Policy-Makers and Full-Time Staff."

69. Sabatier interview with Bob Mendelsohn, 14 March 1975, pp. 4–5. The term *zone of tolerance* is taken from Boyd, "School Board–Administrative Staff Relationships," p. 118.

70. California Public Resources Code, Sec. 27320 (1973). For a more general discussion of this and other ambiguities in state–regional relations, see Douglas and Petrillo, "California's Coast," pp. 194–196.

71. This authority was expanded when the state commission very early interpreted "aggrieved party" to mean anyone who had participated, either in person or in writing, at the regional hearing [California Administrative Code, Title 14, Division 5.5, Sec. 13903 (1974)]. The effect of this interpretation was to clearly give environmentalists and neighbors who had participated in the regional hearing standing to bring an appeal without having to prove that they would be directly and materially harmed by the proposed development. It thus assured that virtually all important regional commission decisions would be appealed by either applicants or opponents.

72. This cooperation required, however, a great deal of effort on the part of Bodovitz in particular. Moreover, there were still cases of (1) inconsistent planning and permit area designations and (2) disputes over the settlement of litigation, most notably involving the Bodega Harbor subdivision in the North Central Region. See Douglas and Petrillo, "California's Coast," p. 195.

73. Interviews with almost all of the regional representatives (May, Peterson, Mendelsohn, Vaughn, Andresen, Laufer, Fay, and Frautschy). The vast majority indicated that they were selected to exercise their own judgment (rather than simply to reflect regional opinion), although they also felt a responsibility to articulate the regional commission's viewpoint.

74. This is based upon interviews with staff and commissioners over several years and represents our summary judgment. Nevertheless, the situation was ambiguous enough that someone else might come to a different conclusion.

5

PERMIT REVIEW BY THE COASTAL COMMISSIONS
Decisions and the Factors Affecting Them

Previous chapters have examined the history and content of the 1972 Coastal Initiative, the external factors affecting the commissions' decisions, and the attitudes and interaction of commissioners and staff at the regional and state level. This chapter takes the analysis one important step further by focusing on two crucial questions: First, to what extent were the permit decisions of the coastal commissions in fact consistent with statutory objectives to, e.g., protect wetlands, scenic areas, and physical access to the coast? Second, of the several factors identified in our conceptual framework, which appear to be the most important in affecting the commissioners' permit decisions?

We shall begin by briefly reviewing the process of permit review, including a broad overview of the nature of permit decisions by the regional and state commissions and changes in the stringency of those decisions over time. The bulk of this chapter will deal with an analysis of the extent to which permit decisions conformed to statutory objectives, both in the aggregate (at least with respect to the state commission) and in three specific areas that were important sources of controversy during the 1972 Initiative campaign: The protection of wetlands and open space around Humboldt Bay (North Coast region), the provision of public access and the minimization of the cumulative effects of development at the Sea Ranch (North Central region), and the provision of public access in highly developed areas of Los Angeles (South Coast region). The final section presents a causal model of the factors affecting commissioner voting behavior during the 1973–1975 period.

In general, the analysis will suggest that the commissions' permit decisions were, by and large, consistent with statutory objectives and that the most important factors affecting those decisions were community needs and resources, commissioners' general policy orientation, and the participation of environmental groups and other opponents in the permit review process.

I. Permit Review: A Descriptive Overview

Within their jurisdiction, the commissions had final review over all developments proposed by private parties and state and local agencies. Their substantive jurisdiction included not only (1) the erection of structures but also (2) the laying of pipes and utility lines, (3) the disposal of any wastes, (4) any grading, dredging, or filling, (5) logging, and (6) changes in permissible intensity of use such as rezoning and lot splits.[1] In short, their authority was probably greater than that of any other state land use agency—most of which have small geographic jurisdictions and/or review over a much more limited range of proposed developments.[2] In fact, the commissions' legal authority was probably matched in scope only by federal land management agencies, particularly the U.S. Forest Service.[3]

Recall that the coastal commissions' permit review process began only *after* a project had obtained approval of the relevant local governments and other state agencies and that it was a *de novo* review. Thus, an analysis of their decisions represents a magnificent opportunity to examine the ability of a state government agency with certain characteristics to alter the allocation of critical resources from what probably would have taken place in the normal system of control by the economic market and local governments. At the very least, every condition imposed or application denied by the commissions represented a clear effect of their superimposition on the antecedent process of local review.

Moreover, there is some evidence that a simple comparison of developments as approved by local governments and as approved (or denied) by the commissions underestimates the commissions' impact on the overall magnitude and quality of coastal development. A longitudinal study of housing construction in Los Angeles and Ventura counties indicates that the number of building permits declined in the permit zone during the 1972–1976 period (compared with pre-1972 figures), while that in the area just beyond the permit zone increased, thereby suggesting a displacement of some development from the permit zone.[4] Given the low rate of outright denials by the coastal commissions (about 3%), this displacement can probably be attributed to the anticipatory response of developers: Rather than risk the more stringent review process within the half mile nearest the coast, they either moved inland or withheld their application within the permit zone in the hope that the commissions' mandate would be abolished or at least drastically altered by the 1976 legislature. In addition, virtually all observers agreed that the quality of proposed developments in the coastal area increased noticeably during the 1973–1976 period—again, probably an anticipatory response of applicants to the commissions' policies and to other recently enacted environmental protection laws.[5] Finally, as noted previously in Chapter 3, there were a few communities in Southern California that changed their development review criteria—e.g., off-street parking and bluff setback requirements—in response to coastal commission decisions.[6] In sum, the permit review figures presented throughout this chapter probably underestimate the commissions' overall impact on coastal development because they do not include the effects on changes in design and quality of development made in anticipation of commission decisions, or the shift of potential developments to areas outside the permit zone because of anticipated difficulties with the commissions.

Our primary focus throughout this chapter is on understanding the commissions' review of permits applied for (i.e., their direct affect on development) as a traditional regulatory function by an agency operating within a specific legal and political environment. But one must also understand the role of permit review within the overall implementation of the Coastal Initiative. Unlike most state land use statutes, the Initiative combined the dual functions of permit review and planning by utilizing the former not only to preserve planning options but also to stimulate the formulation of plan policies in the context of concrete controversies.[7] Moreover, the permit process provided a base of 3 years' experience by which the 1976 legislature could evaluate the commissions' performance in deciding whether to renew or alter the commissions' statutory mandate.

A. The Permit Review Process

Once an applicant had obtained all other local and state permits, he submitted a form to the relevant regional commission describing the nature and location of the proposed project (and, in the South Coast, its cost) and its probable effects on habitat, water quality, public access, and a number of other topics; in large projects an environmental impact report was also required by California law. After a preliminary analysis, the regional staff assigned the application to one of three calendars:

1. The administrative calendar involved developments costing less than $10,000 or repairs and maintenance worth less than $25,000 to existing facilities.
2. The consent calendar involved those items that, in the view of the staff, involved no significant issues in terms of the Initiative's objectives.
3. The public hearing calendar was reserved for potentially significant or controversial projects.

The application was then assigned to the staff person familiar with that area or type of project for more detailed analysis. This normally involved discussions with the applicant and personnel from the relevant state and local agencies, as well as a site visit. The staff then prepared a report and recommendation for the regional commission. At the commission meeting, the administrative and consent items were voted on *en masse*, although any two commissioners could review an item from the consent calendar and schedule it for a public hearing. Discussion of the hearing calendar items normally involved presentations by the staff and the applicant, as well as testimony from neighbors, environmentalists, or government officials who wished to speak. In most cases, the permit was voted on at the same meeting. Approval required a majority of the commission membership (not simply those members present) and a two-thirds majority in projects involving dredging/filling, restrictions on physical or scenic access to the coast, or effects on water quality, fisheries, or agricultural land.[8]

This was normally the end of the process. About 4% of all regional permit decisions were, however, appealed to the state commission by the applicant and/or "any aggrieved party" (meaning anyone who had testified in person or in writing at the regional hearing). This very liberal interpretation of legal standing provided ready access to opponents of development, and in fact, about two-thirds of appeals were brought by environmental groups or neighborhood associations.[9] The decision process at the state level involved two distinct stages. First, the state staff contacted regional staff, the applicant and appellants, personnel in the relevant state and local agencies, and occasionally other actors in preparing its report and recommendation concerning whether or not the appeal posed "a substantial issue."[10] The public hearing involved brief presentations by state staff, the applicant and appellants, and other interested parties. As at the regional level, the hearing was rather informal. If the commission decided that the appeal raised no substantial issue, the regional commission decision was final. In the approximately 50% of all appeals that the commission decided raised a substantial issue—thereby triggering the second stage of the process—additional testimony was taken, and a final vote was deferred until the subsequent meeting to enable the staff to conduct additional analyses and to make a substantive recommendation. At the second meeting, discussion was normally dominated by staff and commissioners. The commission then voted, using the same rules as at the regional level.

About 25% of state commission decisions (or about 1% of all applications) were subsequently appealed to the courts by the applicant or other "aggrieved party."[11] As previously indicated, however, the courts (at least at the appellate level) seldom overturned the commissions' decisions, although appellants could sometimes hope for a negotiated settlement.

B. *The Permit Decisions*

A general—though, as will be noted, somewhat misleading—overview of the commissions' permit review activity from February 1973 to December 1975 is presented in Table 5-1. The first two columns indicate the distribution of permits among the regional commissions. Of the 15,025 permits acted upon

TABLE 5-1
PERMIT REVIEW ACTIVITY OF COASTAL COMMISSIONS
FEBRUARY 1973–DECEMBER 1975[a]

Commission	Permit applications		Permit decisions	
	Number	% of total	% approved	% denied
Regional commissions				
North Coast (Del Norte Humboldt, Mendocino Counties)	1,140	7.6%	98.8%	1.2%
North Central (Sonoma, Marin, San Francisco)	599	4.0%	94.5%	5.5%
Central (San Mateo, Santa Cruz, Monterey)	2,089	13.9%	97.2%	2.8%
South Central (San Luis Obispo, Santa Barbara, Ventura)	2,231	14.8%	99.1%	0.9%
South Coast (Los Angeles, Orange)	5,961	39.7%	95.4%	4.6%
San Diego (San Diego City)	3,000	20.0%	97.0%	3.0%
Subtotal all regions	15,025	100%	96.8%	3.2%
Appeals to state commission				
Raising *no* substantial issue	293	49.7%	51.8%[a]	48.2%[b]
Raising a substantial issue	296	50.3%	59.1%	40.9%
Subtotal	589	100% (3.9% of all reg. decisions)	55.5%	44.5%

[a]Sources: California Research, *State Coastal Report*, April 1976, p. 17, and unpublished data compiled by the state commission. In fact, this was virtually the only type of aggregate permit data published by the commissions in their annual reports.
[b]These were the regional commission decisions simply affirmed by the State Commission.

during these 3 years, about 40% were in the South Coast region and almost 75% were in the three southern regions from San Luis Obispo to the Mexican border. The second two columns indicate the regional and state commission decisions on a straight approve–deny basis. As can be seen, only 3.2% of permit applications were denied by the regional commissions, ranging from a high of 5.5% in the North Central to a low of 0.9% in the South Central. In contrast to this miniscule (and relatively invariant) percentage of denials at the regional level, the state commission denied about 41% of the cases deemed to have raised a substantial issue during the 1973–1975 period. On the basis of these figures— which were essentially the only type of aggregate permit data published by the commissions—one would probably conclude (a) that the regional commissions had very little effect on the pattern of coastal development and (b) that there was an enormous gulf in policy orientation between the state commission and even the most environmentalist regional commissions (e.g., the North Central).

These figures are, however, quite misleading for at least three reasons. First, as indicated previously, they do not include the effect of the commissions on the number and quality of developments *proposed* within the permit zone. While very difficult to quantify, these essentially anticipatory effects were certainly not insignificant. Second, the regional data do not distinguish among the three permit calendars. They thus lump projects having an insignificant effect on coastal resources—e.g., the addition of a few rooms to a single-family residence, the repair of a highway culvert, or the construction of a small apartment house a half mile from the coast in highly urbanized Los Angeles—with those projects on the regional hearing calendars that had potentially significant effects on statutory objectives. Given the very broad substantive and geographic scope of the commissions' authority, a large percentage of the proposed developments did not affect coastal resources and thus did not in any way merit denial. Third, the figures on both regional and state commission decisions fail to indicate that "permit approvals" covered a wide spectrum, from permits approved as submitted by the applicant to those approved with minor conditions to those approved with major conditions (normally over the strong objections of the applicant). In fact, as we shall see later when discussing the Sea Ranch, permits approved with major conditions were sometimes perceived by the applicant as *de facto* denials.

A more accurate view is obtained if attention is restricted to the approximately 25% of all permit applications that were deemed potentially controversial or serious enough to be placed on the hearing calendars of the regional commissions. This is done in Table 5-2, which indicates the decisions of the three regional commissions on a 4-point scale, from "approve as submitted by the

TABLE 5-2
PERMIT DECISIONS OF REGIONAL COMMISSIONS ON HEARING CALENDAR ITEMS,
1973–1975

	Regional commission		
Commission decision	North Coast (N = 273)	North Central (N = 177)	South Coast[a] (N = 1348)	
Approve as submitted by applicant	76%	15%		(56%)
Approve with minor conditions	11%	35%	79%	(17%)
Approve with major conditions	8%	34%		(3%)
Deny	5%	16%	21%	(24%)
	100%	100%	100%	

[a]Unfortunately, the coding of the complete *census* of South Coast permits (N = 1348) did not distinguish among types of approval. Thus, the figures from a separate analysis of random *sample* of 100 hearing calendar permits during roughly the same period are provided in parenthesis.

applicant" through "approve with minor/major conditions" to "deny." From this analysis it can be seen that the North Coast Commission was the most lenient, approving 76% as submitted by the applicant while denying only 5%. The North Central was the most stringent, approving only 15% as submitted while denying 16%, with the vast majority of decisions involving conditional approvals. And the South Coast occupied a middle ground, approving 56% as submitted while also denying the highest percentage, 21–24%.

Thus far we have examined the overall record of permit review by the commissions throughout the 1973–1975 period. But any analysis of regulatory policy making must address the supposed "cycle of decay," whereby agencies gradually become "captured" by regulated groups over time as public support for stringent regulation declines while the regulated continue to actively present their case before the agency and its sovereigns.[12] It has been argued, however, that such a change in policy orientation can be forestalled if (1) the statute provides relatively clear objectives and coherently structures the implementation process so as to enable supportive agency officials to "ride out" temporary crises and if (2) the groups that originally supported establishment of the agency continue to play an active role in its implementation.[13] While the 3-year time span of our data does not permit a thorough test of the "cycle of decay" hypothesis, the commissions certainly suffered a decline in diffuse public support and an increase in opposition from the construction industry and other regulated groups because of the energy crisis of the winter of 1973–1974, a deepening recession (as statewide unemployment rose from 7.0% in 1973 to 7.9% in 1975), and a decline in employment in the construction industry from 345,000 in 1973 to 303,000 in 1975. Thus, one of the major pressures for a relaxation of stringent regulatory controls was certainly present. Moreover, the Coastal Initiative did provide relatively clear objectives and did coherently structure the implementation process, while data to be presented shortly indicate that environmental groups and other opponents of development continued to play an active role in permit review. In short, within the limited time frame of our data base, we have a relatively clear test of the ability of a well-structured statute and a supportive constituency to overcome substantial declines in diffuse public support and in the economic well-being of the regulated industry.

Table 5-3 presents time series data for the three regional commissions and the state commission on, first, the nature of hearing calendar decisions and, second, some measures of the presence of a constituency supportive of stringent regulation. Looking first at the commissions' decisions in part A of the table, it can be seen that there was a general tendency for the three regional commissions to become somewhat more stringent in 1974 than in the first year of their operation and then to relax slightly in 1975. In each case, however, the figures suggest a more stringent review process in 1975 than in 1973.

The situation was somewhat different on appeals to the state commission.

TABLE 5-3

CHANGES IN REGIONAL AND STATE COMMISSION DECISIONS OVER TIME ON HEARING CALENDAR PERMITS

	Commission and year											
	North Coast			North Central			South Coast			State commission[a]		
	1973 (N = 92)	1974 (N = 71)	1975 (N = 109)	1973 (N = 74)	1974 (N = 74)	1975 (N = 29)	1973 (N = 488)	1974 (N = 521)	1975 (N = 338)	1973 (N = 43)	1974 (N = 75)	1975 (N = 46)
A. Commission decision												
Approve as submitted	82%	73%	73%	30%	4%	3%				10%	15%	12%
Approve w/minor cond's.	13%	7%	11%	39%	27%	45%	84%	74%	79%	3%	7%	7%
Approve w/major cond's.	3%	11%	9%	16%	53%	35%				19%	41%	48%
Deny	2%	9%	6%	15%	16%	17%	16%	26%	21%	68%	37%	34%
	100%	100%	99%	100%	100%	100%	100%	100%	100%	100%	100%	101%

B. Participation of supportive constituency
  1. Percentage of permits w/opposition

By anyone	18%	55%	35%	34%	49%	62%	28%	32%	34%	—	—	—
By statewide env. group	7%	49%	21%	20%	30%	24%	6%	2%	2%	—	—	—
By local env. or homeowners' group	4%	13%	4%	11%	22%	28%	9%	12%	7%	—	—	—

  2. Permit appellant

Environmental group	—	—	—	—	—	—	—	—	—	29%	28%	42%
Neighborhood assoc.	—	—	—	—	—	—	—	—	—	8%	3%	6%
Applicant	—	—	—	—	—	—	—	—	—	44%	40%	26%
Other (e.g., individual neighbors)	—	—	—	—	—	—	—	—	—	19%	28%	26%
										100%	99%	100%

[a] This involves the direct and indirect decisions of the state commission, i.e., its own decisions on appeals involving a substantial issue and its reaffirmation of the regional commission decision on appeals deemed not to have raised a substantial issue.

Note that the percentage of cases that were either directly or indirectly approved as submitted by the applicant remained very stable (at about 12%) over time. But the percentage of denials dropped dramatically from 68% in 1973 to about 35% in the succeeding 2 years. In short, the pattern of state commission decisions changed from heavy reliance on denials in 1973 to a much greater emphasis on conditional approvals in 1974–1975. Thus, there is a possibility that the decline in public support may have caused the commission to rethink its exceedingly stringent initial orientation. Moreover, the timing suggests that the controversy over the San Onofre nuclear power plant—initially denied in December 1973, and subsequently approved with conditions in February 1974—may have played a major role in this reorientation. On the other hand, the quality of development proposals may have increased substantially once it became obvious that clearly deficient projects would be denied. In all probability, both of these factors—i.e., a more moderate policy by the commission and improved project proposals— were involved in the change in state commission decisions over time.

Part B of Table 5-3 indicates one of the (potentially) major reasons the regional commissions and, to a leasser extent, the state commission did not succumb to the "cycle of decay" during this period—namely, the continued presence of supportive regulatory constituency. As can be seen, the percentage of permits with some opposition to granting the permit (as submitted or with only minor conditions) *increased* in all three regions between 1973 and 1975; in fact, in both the North Coast and North Central regions it essentially doubled. The data further suggest that organized statewide environmental groups (principally the Sierra Club) and local environmental or homeowners' groups—rather than unaffiliated individuals—were the principal sources of opposition in the North Coast and North Central regions and that they increased or at least maintained their level of activity over time. In contrast, organized groups were a less viable force in the South Coast region, with most of the opposition apparently coming from unaffiliated individuals. Finally, if one looks at another measure of the presence of a supportive constituency—namely, the ability to appeal regional decisions to the state commission—one finds that the percentage of appeals initiated by environmenal groups increased from 29% in 1973 to 42% in 1975, with homeowners' groups playing a much lesser role and 20–25% of appeals coming from unaffiliated individuals.

In sum, then, these data certainly indicate that the recession of 1974–1975 and the decline in diffuse public support did *not* result in more permissive permit regulation by the regional commissions, although the case is more ambiguous at the state level. The data also tend to support our general argument that an agency with a stringent mandate can weather temporary crises if it has a statute with a clear policy orientation and if it continues to benefit from an active supportive constituency. Finally, it suggests that the coastal commissions were applying more stringent review criteria in the protection of coastal resources than had the local agencies that had previously approved these applications. But this still leaves open the extent to which these more stringent decisions actually conformed to the objectives of the Coastal Initiative.

II. Consistency of Permit Decisions with Statutory Objectives

The fundamental objective of the 1972 California Coastal Zone Conservation Act was "to preserve, protect, and, where possible, to restore the resources of the coastal zone for the enjoyment of the current and succeeding generations" (Section 27001). While this gradiloquent language was sometimes interpreted as calling for a moratorium on coastal development during the 1973–1976 interim period, this extreme view was denied by the Initiative's principal sponsors, was clearly inconsistent with the extensive permit review provisions in the act, and was explicitly rejected by commission officials as legally dubious and politically suicidal.[14] Instead, the Coastal Conservation Act—particularly the section specifying the conditions under which a two-thirds vote would be required and/or permits should be subject to "reasonable terms and conditions"—sought to ensure that (1) public access to publicly owned or used beaches, or other recreation areas, not be reduced and, in fact, be increased to the maximum extent possible; (2) the scenic resources of the coast—and particularly the line of sight from the coastal highway toward the sea—be protected; (3) wildlife habitat areas—particularly wetlands and commercial fisheries—be protected; (4) adverse effects upon water quality, including the disposition of human and other wastes, be minimized. In addition to these major objectives, there was also language indicating that[16] (5) dangers of flooding, erosion, and geologic hazards (e.g., earthquakes) should be minimized; and (6) the conversion of agricultural land should be given considerable scrutiny. Finally, it must be recalled that one of the major objectives of the interim permit review process was to (7) preserve planning options pending decisions on the content of the coastal plan to be presented to the 1976 legislature.[17] It should also be remembered that the act nowhere indicated that economic hardship or technological feasibility were to be considered in permit review.

Any effort to compare the extent to which the permit decisions of the coastal commissions actually complied with these objectives encounters, however, some serious difficulties. First, while the general thrust of the above objectives is clear, they are not entirely unambiguous; precisely what, for example, constitutes "minimum adverse effects to scenic resources?" Second, it is in principle exceedingly difficult—and in practice essentially impossible—to ascertain the precise impacts on these statutorily protected values of each of the 15,000 developments for which coastal permits were sought during the 1973–1976 interim period.

These difficulties notwithstanding, we shall first attempt to get an overall estimate of the extent to which the commissions either denied or imposed relevant conditions in the pursuit of statutory objectives. We shall then examine in greater detail the specific cases of (1) the protection of wetlands and agricultural land around Humboldt Bay in the North Coast Region, (2) the frustrating pursuit of public access at Sea Ranch in the North Central region, and (3) the attempt to obtain some access in urbanized areas within the South Coast region. While certainly not exhaustive, these three areas were certainly among the most important in their respective regions and involve most of the principal objectives of the Coastal Conservation Act.

A. An Overview

Any aggregate analysis of the conformity of permit decisions with statutory objectives necessarily involves some simplifying assumptions. Thus, instead of trying to estimate the actual impacts of proposed developments, we shall assume that certain impacts would probably have occurred in instances where the issue was raised by two or more individuals (whether staff, commissioners, or outsiders) in the staff reports or the public hearings on specific permits. And instead of examining all permits or even the random samples of hearing calendar items in the three regions examined in this study, we shall restrict our attention to the random sample of permits appealed to the state commission from all six regions during the 1973–1975 period. The assumption here is that the appellate process was sufficiently open, and the costs of appeal sufficiently low, that almost all important or controversial regional commission decisions were appealed by the applicant and/or opponents to the state commission. [18]

Thus, Table 5-4 provides the final decision of the coastal commissions— i.e., the state commission decision on appeals raising a "substantial issue" and the regional commission decision in cases of "no substantial issue"—when various issues pertaining to statutorily protected values were raised on a random sample of 166 permit appeals during the 1973–1975 period. More specifically, for each issue it indicates the extent to which the final commission decision addressed that issue by either denying the permit or imposing relevant conditions. For example, when public access to the beach was raised as an issue by two or more participants, on what percentage of the cases did the commissions either deny the permit or require some form of access dedication or other condition relevant to providing access? In addition, the first column of the table indicates the percentage of appeals in which various issues were raised.

The data suggest that—with the exception of habitat protection and preserving planning options—the commissions either denied the permit or imposed relevant conditions about 75% of the time that any of the seven statutorily related values was a major issue in the permit hearings. Note, however, that the

TABLE 5-4

PERCENTAGE OF PERMIT APPEALS IN WHICH FINAL DECISION OF COASTAL COMMISSIONS "ADDRESSED" ISSUES RELATING TO VARIOUS STATUTORY OBJECTIVES (N = 166)[a]

Issue	% of appeals on which issue raised %	N	A. Relevant condition(s) imposed	+	B. Permit denied	=	Percent of cases in which issue "addressed" (Total of A + B)
A. Issues directly related to statutory objectives							
1. Public access to beach	21%	35	8%	+	58%	=	66% of 35
2. Effects on scenic resources and/or view corridors	24%	40	43%	+	48%	=	91% of 40
3. Wildlife habitat	9%	14	n.c.[b]	+	37%	=	37% of 14
4. Effects on water quality	11%	18	51%	+	26%	=	77% of 18
5. Geologic hazards/erosion	11%	18	36%	+	44%	=	80% of 18
6. Conversion of agricultural land	11%	19	11%	+	59%	=	70% of 19
7. Preserve planning options	19%	31	n.c.	+	54%	=	54% of 31
B. Other issues							
1. Transportation, e.g., parking	20%	34	16%	+	61%	=	77% of 24
2. Cumulative impacts on land use	23%	38	12%	+	60%	=	72% of 38
3. Consistency with existing development	20%	34	12%	+	72%	=	84% of 34
	Totals more than 100% because of multiple issues on many permits.						

[a] This is based upon a content analysis of the state commission files on a random sample of 166 permit appeal decisions made from February 1973 through June 1975.
[b] n.c. = not coded because could not identify conditions relevant primarily and exclusively to this issue.

manner of addressing the issue varied, with public access and the conversion of agricultural land being associated with outright denials while the protection of scenic resources and water quality impacts were often dealt with by approving the project with relevant conditions, e.g., landscaping and the undergrounding of utility lines for the former, better septic systems for the latter. Even the issues where this analysis suggests that the commission had a comparatively poor record—habitat protection (37%), the preservation of planning options (54%), and, to a lesser extent, public access (66%)—can at least partially be attributed to methodological difficulties in developing categories of "relevant" conditions. For example, we did not code transportation conditions as relevant to public access because we could not ascertain whether this was generally the case; nevertheless, there were certainly instances in the South Coast and other regions where parking restrictions were definitely imposed to improve public access to the beach. Moreover, as we shall see shortly, the commissions did an excellent job of protecting wildlife habitat in and around Humboldt Bay. In short, on these three issues we have almost certainly underestimated the extent to which relevant conditions were imposed.[19]

Table 5-4 also presents a similar analysis for three other issues—transportation, cumulative impacts on land use, and consistency with existing development—that were probably indirectly related to statutory objectives. This was clearly the case, for example, in the parking restrictions just discussed. Similarly, cumulative impacts on land use and consistency with existing development were related to the commissions' efforts to restrict development as much as possible to existing urban areas, thereby protecting the scenic resources of the rural coast (among other things). But these issues—all of which were similarly "addressed" by the commissions about 75% of the time they were raised—are also traditional preoccupations of planners. This suggests that the professional norms of commission staff may well have supplemented statutory directives in serving as decision guides to the commissions.

In sum, this analysis should be treated with some caution. There is no way of knowing whether the raising of an issue by two or more participants always signaled a potentially significant impact on coastal resources. Similarly, we have almost certainly underestimated the percentage of cases in which relevant conditions were imposed for some issues, primarily habitat protection and public access. Conversely, just because a relevant condition was imposed did not mean that the adverse impact had been entirely mitigated. These *caveats* notwithstanding, this analysis indicates that—at least in the case of those permit applications that were important or controversial enough to be appealed to the state commission—the commissions either imposed relevant conditions or simply denied the permit application in about 75% of the cases in which issues relating to statutory objectives were raised in the staff reports or at the public hearings.

While this analysis of aggregate data provides an indication of the general conformity of permit decisions with statutory objectives, it does not give a sense of how the commissions handled specific cases and their effects on specific areas of the coast. For this we turn to three interesting, though not necessarily representative, cases: Humboldt Bay, the Sea Ranch, and public access on the South Coast.

B. The Protection of Wetlands and Agricultural Land around Humboldt Bay (North Coast Region)

Humboldt Bay was the largest and most productive wetland under the jurisdiction of the coastal commissions, second in the state only to the enormous expanse of San Francisco Bay/Suisun Marsh. Like most wetlands in the state, it had shrunk in size—from perhaps 27,000 acres a century ago to 16,000 acres in 1973—because of siltation and the reclamation of salt marshes and intertidal flats. The bay and its surrounding agricultural land provided an important link for migratory waterfowl (particularly the Pacific black brant), as well as major nesting areas for egrets, herons, cormorants, and numerous other species. In addition, the bay constituted a major fisheries resource, providing a nursery area for English sole and several crab species, rearing areas for herring and anchovy, lifelong habitat for large numbers of oysters, mussels, and clams, and transportation for salmon and steelhead trout on their way to spawning areas on the Eel River and other tributaries. In fact, the bay and its tributaries annually provided waterfowl hunting for about 25,000 hunter-days, sports fishing for about 3,500 angler-days (exclusive of marine fishing), by far the largest oyster fishery in the entire state, and the largest commercial fishery north of Los Angeles.[20]

But the area around Humboldt Bay was also the largest urban and industrial center on the California coast north of San Francisco. Eureka, with a 1970 population of 24,000, was the administrative and commercial hub of Humboldt County, as well as the major port and mill center for the North Coast lumber industry. And, on the northern shore of the bay, Arcata (1970 population of 9000) was the home of Humboldt State University. Although the area had historically grown at a very slow rate, a modest (5%) population increase in the 1970–1974 period subjected the 8 miles of marshes and marginal agricultural lands between Eureka and Arcata on the east side of the bay to considerable pressures for development between 1972 and 1976. For a map of the area and the location of the major projects during this period, see Figure 5-1.

The principal permit applications affecting Humboldt Bay during the 1973–1976 period are briefly reviewed in Tables 5-5A and 5-5B.[21] Indicated are the nature of the proposed development, the principal issues raised at the regional commission hearing, the regional staff recommendation, and the regional commission decision. In those cases that were subsequently appealed, the comparable information at the state commission level is provided. Table 5-5A deals with those permits involving primarily issues of wildlife habitat, while Table 5-5B deals with permits along the east side of Humboldt (Arcata) Bay between

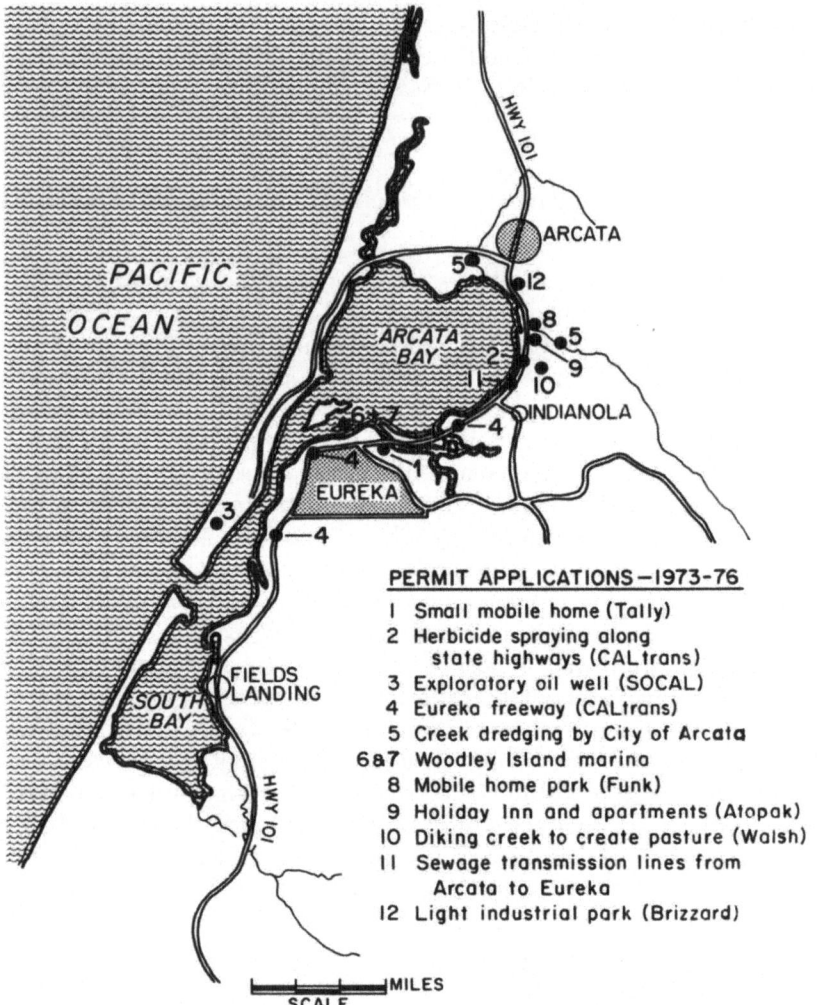

FIGURE 5-1. Permit applications in and around Humboldt Bay, 1973–1976. 1 = small mobile home (Tally); 2 = herbicidal spraying along state highways (Caltrans); 3 = exploratory oil wel (SOCAL); 4 = Eureka freeway (Caltrans); 5 = creek dredging by City of Arcata; 6 and 7 = Woodley Island marina; 8 = mobile home park (Funk); 9 = Holiday Inn and apartments (Atopak); 10 = diking creek to create pasture (Walsh); 11 = sewage transmission lines from Arcata to Eureka; 12 = light industrial park (Brizzard).

Eureka and Arcata involving issues of not only wildlife habitat but also the protection of agricultural land from urban sprawl.

Although some of the most valuable habitat area in the area was part of a National Wildlife Refuge in 1973—particularly the heron egret rookery on Indian Island—much of the rest was either in private ownership or under the jurisdiction of a variety of public agencies. Virtually all of the underdeveloped portion of the bottomlands around the bay, particularly those portions subjected to seasonal flooding, provided important foraging and protective habitat during at least part of the year.[22] Thus, one of the responsibilities of the coastal commissions was to somehow protect as much of this habitat as possible while avoiding the inverse condemnation of private property and also permitting needed development in this economically depressed region.

Table 5-5A indicates seven permit applications involving potential effects on wildlife habitat, other than those in the bottomlands east of Arcata Bay to be discussed in the next subtable. Three of them—the mobile home at Eureka Slough, the exploratory oil well, and the dredging by the City of Arcata—were resolved entirely at the regional level in discussions involving the regional staff, the applicant, and the California Department of Fish and Game.

Two others involved the California Department of Transportation (Caltrans), both of which were eventually appealed to the state commission. In the case of herbicide spraying along the roads to serve as a fire retardant, Fish and Game did not get involved because they could not demonstrate adverse impacts on wildlife, and the regional staff recommended a straight approval. The permit was nevertheless appealed by the North Coast Environmental Center because of alleged effects on habitat and on human health. When the state commission and staff indicated considerable reservations about the spraying program, Caltrans simply withdrew its permit rather than risk a denial. The other Caltrans project involved the Eureka Freeway, a proposal strongly supported by local government and business leaders to reroute and upgrade Highway 101 in order to reduce serious noise and traffic congestion (chiefly from the constant stream of logging trucks) in and around Eureka. The project was approved by the regional commission with conditions designed to mitigate adverse impacts on wildlife, although the proceedings were clouded by equivocation on the part of Fish and Game. The regional decision was appealed by both the Sierra Club (concerned about habitat loss) and local residents concerned about the loss of housing. The state staff then negotiated additional conditions in both areas with Caltrans. For example, the agency was prohibited from using dredge spoils in its embankments and, most importantly, was required to purchase 61 acres of wildlife habitat (to be deeded to Fish and Game) to replace the 18 acres that would be destroyed by the project.[23]

Similar in many respects to the Eureka Freeway was the Woodley Island Marina, as both illustrate the efforts of the regional and state commissions to balance the economic needs of the community against the strong statutory mandate to protect wetlands. As originally proposed, the marina would have included 400 slips and numerous support facilities on the southern half of Woodley Island, just across from downtown Eureka. Not even the Sierra Club questioned the need for additional commercial fishing berths, although there was considerable discussion over the size of the project and the inclusion of numerous berths

TABLE 5-5A

PRINCIPAL PERMIT DECISIONS AFFECTING HUMBOLDT BAY, 1973–1976: CASES FOCUSING ON THE PROTECTION OF WILDLIFE HABITAT[a]

Applicant (Permit No.)	Type and location	Regional commission		State commission on appeal	
		Permit hearing: Staff rec., outsiders, issues	Commission decision	Hearing: Appellant issues, staff rec.	Commission decision
1. Don Tally (73-066)	Mobile home park near Eureka Slough on northern outskirts of Eureka	Staff recommended approval with minor condition of installing fence along slough; Fish and Game persuaded applicant to abandon proposed filling of marsh and stick to high ground	Approved 11–0 as recommended by staff (6/14/73)	No appeal	
2. Calif. Dept. of Transport. (73-069)	Renewal of permit to use herbicides along highways in 3 counties	Staff rec. approval, opposed by Sierra Club and local envir. group; Fish and Game did not participate because could not demonstrate adverse impacts on wildlife	Approve 8–1 with one abstention (5/8/75)	Appealed by local envir. group b/c of alleged danger to habitat and human health	Application withdrawn in anticipation of adverse comm. dec.
3. Std. Oil of Calif. (75-192)	Build an exploratory oil well on Samoa Peninsula	Staff recommended approval with conditions; no opposition because concerns of Fish and Game voluntarily met by applicant	Approved staff recommendation 10–0 (5/8/75)	No appeal	
4. Calif. Dept. of Transport. (75-186)	Construct 6.4 mi. of freeway from Elk River through Eureka to 1	Staff rec. approval w/conditions, project strongly supported by local gov-	Approved 9–1 with conditions rec. by staff (4/16/75)	Appealed by SC and homeowners; staff rec. approval w/cond's. that	Approved staff rec. 10–2 (6/17/75)

	mi. no. of Eureka Slough	ernment and Chamber of Commerce; Opposed by envir. groups and some displaced home-owners; Fish and Game sought some habitat mitigation measures—which Caltrans and comm. staff agreed to—but then sought additional ones that were not incorporated into (revised) staff recommendation		Caltrans acquire additional off-site habitat for F&G and additional replacement housing
5. City of Arcata (75-219)	Dredging of 6 creeks, with utilization of silt at sewage treatment plant	Fish and Game (and Sierra Club) objected to dredging on 3 creeks, which applicant then deleted, staff recommended approval of revised application	Approved revised application 11–0 (3/18/76)	No appeal
6. Humboldt Harbor Dist. (75-231)	400-slip marina, support facilities (restaurant, coffee shop, parking, offices) and road on So. half of Woodley Island across from downtown Eureka	Staff rec. approval of 214-slip marina, and some support facilities, but deletion of restaurant and offices; also required that northern half of island be kept in wildlife habitat, separated from marina by chain link fence, Fish and Game and Sierra Club essentially concurred, project strongly	Commission approved development, deleting staff restrictions on number of slips, restaurant, wire fence, and off-site habitat, but retaining general restriction that half of island remain wildlife habitat (11–1, 6/10/76)	Appealed by Sierra Club, staff rec. denial because of excessive damage to wetlands, tho acknowledged need for harbor for *commercial* fishing boats	Denied 4–6 (8/31/76)

aSource: Minutes of the North Coast Regional Commission and the state commission.

(continued)

TABLE 5-5A (Continued)

Applicant (Permit No.)	Type and location	Regional commission		State commission on appeal	
		Permit hearing: Staff rec., outsiders, issues	Commission decision	Hearing: Appellant issues, staff rec.	Commission decision
		supported by commercial fishermen, chamber of commerce, and local governments, opposed by Humboldt State economist on economic grounds; staff subsequently deleted some conditions, whereupon Sierra Club recommended denial, Fish and Game equivocal			
7. Humboldt Harbor Dist. (76-369)	Revised application for Woodley Is. marina after state denial, now includes 228 slips, restaurant and offices; product of advisory committee with broad-based membership, which found this to be the best of several alternatives	Staff rec. approval with cond's. for (1) permanent protection of remainder of island as wildlife refuge, (2) fence separating marina from wildlife refuge, and (3) purchase of off-site wildlife habitat, strong support from fishermen and most envir. groups, tho Sierra Club concerned about need for restaurant	Approved staff rec. 10–0 (12/8/76)	No appeal	

for pleasure boats. The regional staff (supported by Fish and Game) originally recommended approval of a scaled-down project—deleting the noncommercial berths, the restaurant, and offices—with the added conditions that the northern half of the island remain in its natural state, separated from the marina by a chain link fence, and that the Harbor District purchase off-site wildlife habitat to compensate for that destroyed by the project. Many of these conditions were resisted by the District for legal and/or economic reasons, and Fish and Game rescinded its recommendation for the chain link fence. After considerable discussion, the North Coast Commission in June 1976 deleted most of the conditions, although it did retain the proviso that half the island be retained in its natural state.[24] Not surprisingly, the permit was appealed by the Sierra Club. When the Harbor District and the state commission staff were unable to agree on a scaled-down project—perhaps because of the uncertainty of the coastal bill then pending in the legislature—the staff recommended denial because of the excessive damage to wetlands. This was accepted by the state commission on August 31, 1976.

After passage of the 1976 Coastal Act, but before it went into effect, the Harbor District convened a broadly based advisory committee (including environmental groups) to examine a number of alternatives. Then in December 1976 it submitted a new application for a 228-slip marina, restaurant, parking, and a few offices, with the northern half of the island remaining in its natural state. The regional staff recommended approval with the conditions that (1) the U.S. Fish and Wildlife Service become the effective manager of the northern half of the island (with a veto over any additional development), (2) the chain link fence be constructed, and (3) 15–20 acres of off-site wildlife habitat be purchased by the District and returned to tidal action to offset habitat loss from construction of the marina. These conditions were reluctantly accepted by the applicant and unanimously approved by the regional commission. Although the Sierra Club had some reservations about the need for a restaurant and the lack of restriction to commercial vessels, no appeal was filed and thus the regional decision was final.[25]

The second group of permits dealt not only with the preservation of wildlife habitat but also with the protection of marginal agricultural land on the east side of Humboldt Bay between Eureka and Arcata (see map in Figure 5-1). Because of their low productivity (due to poor soils and seasonal flooding) and accessibility to Highway 101, these lands were subjected to considerable development pressures, at least by the standards of the North Coast.

Of the five projects listed in Table 5-5B, the first two occurred very early in the life of the commissions and were symptomatic of the changes that the Coastal Initiative was to make in even the North Coast Region. Located across the road from each other adjacent to Jacoby Creek about 2 miles south of the built-up portion of Arcata, they involved the construction of (1) a 200-unit mobile home park, (2) a 100-unit Holiday Inn and 150-unit apartment complex, and, inciden-

Table 5-5B
Principal Permit Decisions Affecting Humboldt Bay, 1973–1976:
Cases Involving Habitat, Ag Land, and Urban Sprawl Between Arcata and Eureka[a]

		Regional commission		State commission on appeal	
Applicant	Type and location of development	Permit hearing: Staff rec., outsiders, issues	Commission decision	Hearing: Appellant, issues, staff rec.	Commission decision
8. Funk and Winston (73-028)	200-unit mobile home park just off Hwy 101 between Eureka and Arcata near Jacoby Creek	Staff rec. denial because of effects on wildlife habitat; opposed by neighbors, local envir. group, and Fish and Game Depts. because of impacts on wildlife, leapfrog development, loss of open space and agricul. land, conformity with Arcata general plan and/or traffic congestion; on other hand, would represent major addition to tax base	First voted that, despite grazing 35–40 animals per year, did not constitute agricul. land; then voted 7–5 to approve as submitted (5/24/73)	Appealed by Sierra Club and local envir. groups; staff rec. denial because of habitat loss, loss of agricul. land, and growth-inducing impacts	Denied 1–9 (9/5/73)
9. Atopak (73-008)	100-unit motel (Holiday Inn) and 150 apt. units between Eureka and	Essentially same as above	Same as above	Same as above	Denied 2–9 (9/5/73)

	Description	Staff recommendation	Commission action	Appeal	Final action
10. Dan Walsh (75-432)	Arcata near Jacoby Creek. Construct a dike to reclaim 20 acres of salt marsh as pasture (between Eureka and Arcata)	Staff rec. denial because of effects on wildlife habitat. Sierra Club and Fish and Game urged denial	Approved permit 9–2 largely on grounds that it was agricul. land (7/10/75)	Appealed by Sierra Club and Fish and Game because of loss of wildlife habitat, staff rec. denial	Denied 2–9 (9/3/75)
11. Humboldt Wastewater Authority (76-271)	Construct sewage transmission lines on Samoa Penin. and from Arcata to Eureka	Staff recommended approval with conditions to protect agricul. and wildlife habitat between Eureka and Arcata; Fish and Game and Sierra Club generally concurred; some local residents objected to growth-inducing impacts	Commission approved development, deleting some staff conditions (8–4, 5/13/76)	Appealed by SC; staff rec. approval with conditions to limit hookups between Arcata and Eureka; Sierra Club urged deletion of East Bay Interceptor	After defeating motion to delete Arcata Eureka line, approved staff rec. with conds. 8–3 (7/6/76)
12. Brizzard Co. (75-230)	Light industrial park on 100-acre site on southern outskirts of Arcata subject to flooding	Staff rec. approval with mitigation measures for flood control and deeding of 17 acres of salt marsh habitat to Fish and Game (which concurred); Sierra Club still opposed	Approved staff recommendation 10–0 (4/8/76)	Appealed by Sierra Club because of loss of wildlife habitat, moreover, not included in sewer hookup in East Bay Interceptor decision (#76-271)	Application withdrawn (10/4/76) after staff indicated would probably recommend denial

aSource: Minutes of the North Coast Regional Commission and the state commission.

tally, (3) a sewer interceptor from Arcata. These projects had previously been turned down by the Humboldt County Board for Supervisors, whereupon the City of Arcata annexed and then rezoned the land.[26] In short, these projects represented precisely the sort of aggressive pursuit of an expanded tax base by some local governments that had been partially responsible for passage of the Coastal Initiative. At any rate, the regional staff—supported by Fish and Game and local environmental groups—recommended denial because of damage to the anadromous fishery in Jacoby Creek, the loss of agricultural land (used to graze about 37 head of cattle and sheep per year), and the growth-inducing impacts of the proposed developments on wildlife habitat in the entire area (much of which was subject to seasonal flooding). After four hearings, the North Coast Regional Commission by identical 7–5 votes first ruled that the area constituted neither agricultural land nor salt marsh (thereby precluding the need for 8 votes, or ⅔ of the commission's membership) and then approved the two developments as submitted on May 24, 1973. These decisions were quickly appealed by the Sierra Club and denied by the state commission for essentially the same reasons noted by regional staff.[27]

This decision may have had some value as precedent, for no projects were proposed in the East Bay area for the next 2 years. Then, in the summer of 1975 Daniel Walsh submitted an application to construct a small dike to reclaim as pasture 20 acres that had become inundated by the construction of a similar dike on his neighbor's land some years previous. The regional staff—supported by Fish and Game and the Sierra Club—recommended denial on the grounds that this was essentially a salt marsh. The regional commission disagreed, approving the permit on the grounds that it was viable agricultural land (although only ½ mile and almost certainly less viable than the Atopak and Funk properties, which it had previously found did *not* constitute "agricultural" land). The Sierra Club and Fish and Game then appealed, whereupon the state commission denied the permit in September 1975.[28]

The Walsh case illustrates a number of interesting points. In what was a clear conflict between two statutory objectives, the state commission ruled that wildlife was more important than marginal agricultural land. This was certainly a reasonable reading of the priorities in the Coastal Initiative. Moreover, it was sensitive to the problem of the gradual loss of wildlife habitat through such seemingly insignificant actions as reclaiming 20 acres of seasonally flooded land. Nevertheless, the denial by the state commission raised the question of inverse condemnation of private property: If Walsh couldn't use his land as pasture, what could he use it for? Simple fairness would suggest that an owner ought to obtain sufficient return at least to pay his taxes.[29] In fact, it was precisely this "dilemma of the small project"—the cumulative impacts of individually insignificant developments versus inverse condemnation and sympathy for the individual citizen harassed by arrogant bureaucracy—that continually plagued the coastal commissions, as we shall see again at Sea Ranch.

The final two cases illustrate the potential importance of public works projects in stimulating cumulative changes in land use. The first involved the construction of sewage transmission lines along the Samoa Peninsula and from Arcata to Eureka to remedy local problems with overloaded septic tanks. The regional staff recommended approval with conditions to minimize the growth-inducing impacts of the Arcata-Eureka line. This was supported by Fish and Game and the Sierra Club, though opposed by some local residents who felt the staff recommendation was not sufficiently restrictive. Several of the conditions were, however, opposed by the sewage treatment district (in part because of difficulties entailed in obtaining federal funds for the project) and then deleted by the regional commission on an 8–4 vote. In bringing an appeal, the Sierra Club and some local residents urged the deletion of the Arcata-Eureka line. The state staff, however, recommended approval with conditions designed to eliminate hookups except from existing single-family residences or new developments approved by the coastal commissions. After rejecting a motion to delete the Arcata-Eureka line, the state commission accepted the staff recommendation.[30] The second permit involved a 100-acre industrial park on the southern outskirts of Arcata that was subject to seasonal flooding. Following the recommendation of its staff, the regional commission approved the permit with conditions pertaining to flood control and requiring the deeding of 17 acres of salt marsh habitat to Fish and Game. Upon appeal, the Sierra Club raised the issue not only of habitat loss but also of the apparent exclusion of this parcel from hookup to the Arcata-Eureka sewer line. Upon learning that the state staff was going to recommend denial, the development company withdrew its application in October 1976.[31]

These permit cases suggest, first and foremost, that the coastal commissions did a very good job of protecting wetlands and agricultural land in and around Humboldt Bay during the 1973–1976 period. Of the 12 permit applications, 6 were either denied or withdrawn. Even those that were approved involved very little *net* loss in wildlife habitat, in part because of restrictions on the development process (i.e., the prohibition against using dredging soil on the Arcata Freeway), in part because the commissions generally required applicants to purchase off-site habitat (and return it to tidal action) comparable in size to what had been destroyed. In addition, the commissions brought a considerable amount of habitat owned by private owners or development-oriented public agencies under the legal protection of fish and game agencies. This was most notable, of course, in the Woodley Island case, where the entire northern half of the island was transferred from the Harbor Department to the U.S. Fish and Wildlife Service. As a result, both local fish and game officials and environmental group leaders felt there was no appreciable loss in wildlife habitat around Humboldt Bay during the 1973–1976 period.[32]

This success would not, however, have been possible without the strong appellate process built into the 1972 Coastal Act. In fact, in 8 of the 12 cases the state commission imposed more stringent review criteria than the North Coast Regional Commission, denying 6 that had been approved by the region and imposing additional conditions on 2 others. Moreover, it is quite likely that the threat of appeal obliged the Harbor District to finally accept conditions on the Woodley Island Marina that it (and probably the regional commission) deemed unnecessary and/or undesirable. The appellate process would not have worked, however, without the active participation of the Sierra Club, local environmental groups, and, to a lesser extent, the California Department of Fish and Game, for it was they who formally brought the appeals and who presented a counterweight to applicants before the state commission. It is also possible that the appeals process would not have worked as well without the willingness of the regional staff to make the case for wetlands protection and to recommend specific measures to achieve that objective. In fact, there were 4 cases (Funk, Atopak, Woodley Island, and the Arcata-Eureka interceptor) in which the recommendation of the regional staff was overturned by the regional commission but then largely upheld at the state level. In these cases, the regional staff essentially provided much of the detailed knowledge of the local situation, which was then utilized by their state colleagues.

In short, then, the permit review process incorporated into the Coastal Initiative seems to have worked quite well in protecting wetlands and agricultural land in and around Humboldt Bay. We now turn to the ability of the coastal commissions to achieve their most important objective(s)—physical and scenic access—at what was certainly one of the most controversial areas on the entire coast.

C. The Frustrating Struggle to Obtain Additional Access and Deal with Cumulative Impacts at the Sea Ranch

A planned residential community, the Sea Ranch stretches for 10 miles along both sides of the coastal highway in northern Sonoma County about 110 miles north of San Francisco (see map in Figure 5-2). In 1968 the county approved a subdivision map and rezoning for 5200 units, which, at full buildout, would make this the largest coastal community between San Francisco and Eureka. By February 1973, 1770 lots with the necessary infrastructure (e.g., streets and utility hookups) had been sold and 300 houses constructed. About 1375 residences (involving about 1750 people at the Sea Ranch at any time) were expected by 1980, with 4000 residences (and 10,000 people) shortly after the turn of the century. [33]

The Sea Ranch was distinctive among communities along the northern coast not only for its size but also for a number of features incorporated into the overall development plan designed to minimize the obtrusive features of the development and to meld it as much as possible into natural landforms. Thus, houses and condominiums were clustered, with the extensive "commons" areas planted in natural vegetation; individual houses were normally finished in un-painted woods and had to be approved by a design review committee composed of architects; and a general effort was made to avoid putting houses on exposed promontories, with many of them actually built into the rolling hills leading from the highway down to the ocean (see photographs in Figure 5-2). Not only was this probably the best designed rural development along the entire coast, but the restrictions placed on homeowners and the developer's advertising tended to attract residents who took pride in their efforts to preserve natural habitat in the meadows and the rocky tide pools along much of the oceanfront.[34]

The Sea Ranch was thus a large, well-designed planned community. It was also a *private* development occupying 10 miles of the most potentially accessible shoreline in northern Sonoma County. As part of the 1968 county permit, the developer had deeded two rather poorly marked trails near the northern end of the property, as well as a 150-acre public park at the northern extremity overlook-ing the Gualala River and approximately a mile of ocean shoreline.[35] Although it was clearly understood at the time that these dedications were in lieu of periodic access points through the ranch, the settlement was vehemently op-posed as insufficient by some county and state organizations.[36] In addition to physical access, by the early 1970s scenic views of the ocean from the coastal road were gradually being reduced by houses (particularly in a few units where the original subdivision layout did not meet Sea Ranch's normal quality) and, even more, by a row of trees planted along the western edge of the highway that would eventually form a relatively solid wall for much of the 10 miles.

There is little question that, had the Sea Ranch come before the North Central Regional Commission in 1973 as an entirely new development, substan-tial changes would have been required to bring it into conformance with the policies of the 1972 Coastal Act. Additional access dedications would have been required. The "green wall effect" would have been reduced, and the cumulative impacts of the entire development on coastal resources would have been more adequately addressed: How many septic tanks could be constructed before affect-ing water quality and bluff stability? How much water could be drawn from the Gualala River before harming fish habitat? And how much of the limited capaci-ty of narrow, winding Highway 1—the only access route for 30 miles in either direction—should be allocated to Sea Ranch residents and visitors (assuming that substantial congestion would discourage the general public from visiting this part of the coast or at least dramatically impair the pleasure of those who did so)?[37] Given such considerations, nothing like the 5200 units originally proposed (or even the 4000 subsequently envisaged by the developer) would likely have been approved.

But the coastal commissions were not faced with a new project in 1973. Instead, they were confronted with a steady stream of individual applicants (26 in

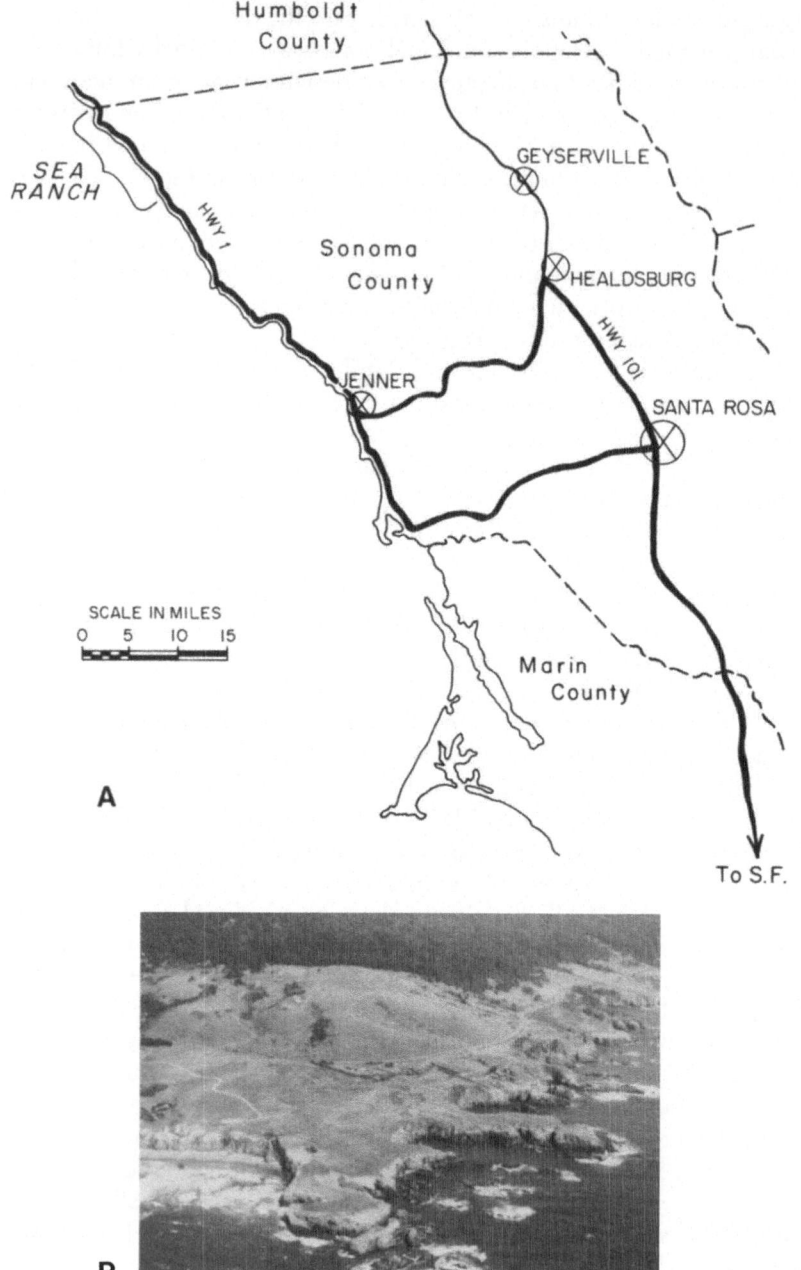

FIGURE 5-2.   Sea Ranch. A. Map of the Sonoma coast, showing principal access routes to Sea
Ranch. B. Aerial view of a portion of the coast at Sea Ranch. C. View of the ocean across the coastal
terrace from the highway. Note row of young trees. D. Restrictions on public access.

C

D

FIGURE 5-2. (*Continued*)

the first 8 months), each seeking to build a single-family residence on a lot purchased in a subdivision approved before the passage of Proposition 20. These individuals had no control over the future size of the development. That was in the hands of the developer, Oceanic California. Individual applicants could do virtually nothing to assure physical access to the ocean or to remove the trees. For they represented only a miniscule percentage of the 1770 voters (1 for each lot) in the Sea Ranch Association, the lot owner's group that controlled the "commons" areas constituting about 70% of the land seaward of the highway.[38] The problem was further compounded by the symbolic significance of the Sea Ranch: Probably more than any other single development (as least in Northern California), it had crystallized voters' perception that the coast was being "lost" to private developers by the actions of local governments and thus that a statewide coastal agency was needed.[39]

Given this situation, the North Central Regional Commission had a number of options in the spring and summer of 1973[40]:

1. It could have denied all permits during the 1973–1976 period, or at least until the coastal plan had been approved by the commission in late 1975. While such a temporary moratorium was advocated by many environmental groups and was probably permissible under California law, it would have aroused a firestorm of protests from potential applicants and considerable opposition in the legislature—as almost everyone acknowedged that, with a few exceptions, the 100–150 houses expected to be constructed during the interim period would themselves have very little effect on coastal resources.

2. The commission could have reviewed each permit application on its individual merits, deferring consideration of the overall issues of access and the impacts of eventual buildout until completion of the coastal plan. This was the solution advocated by Sea Ranch residents and the approach initially followed in a somewhat similar situation by the North Coast Regional Commission.[41] But it would have been widely perceived as a capitulation by environmental groups and would undoubtedly have resulted in litigation. Moreover, this strategy would have deferred the really crucial question of how to convince Oceanic and the Sea Ranch Association to implement overall controls until the very uncertain period after 1976, while every additional step in development reduced the commission's leverage with Sea Ranch authorities.

3. Finally, the commission could have used the threat of denying individual permit applications as a means of extracting concessions on overall problems from the developer and the homeowners' association. This essentially involved telescoping the planning process by having the commission decide what it wanted in terms of access, overall buildout, and mitigation measures of such buildout (e.g., on water quality and quantity). It would then make the approval of individual permits contingent upon agreement of the developer and the lot owners' association with these "overall conditions." Given that people who had

purchased lots but not yet constructed their houses made up the overwhelming majority of votes in the Sea Ranch Association, it was probably assumed that they could persuade the association to resolve overall issues so they could build.[42] In effect, individual applicants would be held "hostage" to extract concessions from the developer and the association.

After routinely processing about 30 Sea Ranch applications (including three denials because of view corridor obstructions), in the summer of 1973 the North Central Commission adopted the third strategy for dealing with the issues raised by the Sea Ranch and five other preexisting but incompletely built-out subdivisions within its jurisdiction.[43]

The formal process of adopting overall conditions began with a series of staff reports and public hearing, culminating in the formal staff recommendation of October 31, 1973. It advocated the following:

1. The dedication of two additional public access easements from the high-way to the beaches at the southern end and midpoint of the 10-mile stretch. This would also include parking areas for cars along the highway at these and previous access points.
2. A program of selective tree-removal to recapture viewsheds from the coastal highway to the ocean. This would be paid for by Sea Ranch authorities.
3. A program of periodic monitoring of septic tanks.
4. Height and bulk restrictions on houses west of the highway to protect viewsheds from the highway to the ocean.
5. Deferral of a decision concerning permissible buildout pending further study.

Moreover, the staff recommended that final action on four Sea Ranch applications before the commission be deferred until an agreement was reached on the overall conditions.

While the Sierra Club, Coastal Alliance, COAAST, and other environmental groups did not disagree with many of the specific recommendations made by the regional staff, they expressed considerable concern over the effects of ultimate buildout and questioned whether permits should not be denied until a decision on the permissible ultimate size of the development had been made.[44] The reaction by Sea Ranchers was to express (a) serious reservations about the effect of public access on fragile tide pools and bluff erosion, (b) rejection of any effort to restrict development rights without compensation, and (c) outrage at the entire "hostage" strategy behind the overall conditions; for example, Oceanic's attorney observed that the staff recommendation "constituted the longest and most verbose extortion note he had ever had the pleasure of reading."[45] Nevertheless, Sea Ranch authorities accepted the commission's offer to meet with an *ad hoc* committee of commissioners in an effort to find an equitable solution.

Over the next several months these deliberations provded general agreement on a number of issues, including septic tank monitoring, tree trimming, restrictions on house size to protect viewsheds, and, most surprisingly, on public access—with a system of 100 visitor-passes per day to selected areas replacing the two designated access points originally proposed by staff.

Nevertheless, the negotiations eventually floundered on two issues[46]: The first was ultimate buildout. Oceanic requested binding approval of 3900 units (with some hint of a willingness to come down as low as the 2500 units that it contended had already been subdivided), while the coastal commission was reluctant to accept the probable impacts of that large a development. For example, a detailed staff report indicated that, were Highway 1 to remain unchanged, the traffic volume generated by 1600 additional homes (or 1900 overall) would reduce travel on the scenic coastal highway to bumper-to-bumper, stop-and-go conditions even with the absurd assumption that no visitors were permitted on the public road; if Sea Ranchers were restricted to one-third of existing highway capacity, only 500–600 additional homes could be permitted.

The second stumbling block involved the form of the agreement between the coastal commissions and Sea Ranch authorities. The latter wanted a binding contract in the form of a blanket permit that would be the sole criterion for evaluating permit applications, that would be incorporated into the coastal plan to be presented to the 1976 legislature, and that could be rescinded by the Sea Ranch parties (but not the commission) were the agreement to be revised by the legislature. In a very strongly worded opinion, the attorney general's office informed the commission that such an agreement would be a violation of the Coastal Initiative and would abrogate the commission's duty to review future development. Faced with this advice, the commission on February 21, 1974, requested the staff to develop a revised set of overall conditions incorporating the areas of substantive agreement that had been reached during the negotiations.

Thus, on April 4, 1974, the North Central Regional Commission (by a 9–3 vote) adopted its overall conditions for the Sea Ranch.[47] Much to the disappointment of both Oceanic and the Sierra Club, the commission did not establish a level of permissible buildout. While implying that 4000 units was unacceptable, it nevertheless argued that the expected 100–150 additional units during the 1973–1976 period would not have significant negative cumulative effects and thus that decisions concerning ultimate buildout could be deferred, pending further study, until the adoption of the coastal plan. The regional commission's decision also sought to allay the fears of Sea Ranchers by acknowledging that all owners of single-family lots had some development rights, and by clearly stating its intent that the overall conditions would not be changed during the planning period and that the access, view protection, and septic tank conditions were intended to be permanent. Within these general parameters, the overall conditions imposed several requirements.

1. The Sea Ranch Association and the county parks department were to institute a pass system for public access, with the actual number of daily passes to be determined by the coastal commission after study of the effects of public use on beach and tide pool areas. Moreover, permittees would be subject to internal Sea Ranch regulations concerning protection of wildlife habitat. Recognizing the untested nature of such a system, the association was also required to dedicate the two access easements mentioned in the original staff recommendation.

2. Oceanic and the Sea Ranch Association were required to institute a detailed tree-trimming program (involving removal of approximately 2000 trees) at their expense in order to protect views of the ocean from the highway.

3. In place of the existing system of individual septic tank permits from the county health department, the Sea Ranch Association would not be required to obtain a more comprehensive waste discharge permit from the regional water quality control board (a state agency), subject to coastal commission approval.

4. Individual permit applicants were required to conform to a detailed set of site, height, and bulk requirements very similar to those proposed in the October staff recommendation (although waivers could be applied for in individual cases). Essentially limiting houses to 1000–1500 square feet of floor area, these were designed to minimize obstruction of view corridors from the highway. In additions, construction would be prohibited on a few very sensitive lots.

That same evening, the regional commission put the Sea Ranch overall conditions into operation by approving seven applications that had accumulated during the month of negotiations. But in addition to individual size and height limitations, each permit was subjected to the condition that "construction not begin until the commission finds that satisfactory progress has been made toward the initiation or accomplishment of overall conditions for the Sea Ranch."

As a finding of "satisfactory progress" was in turn clearly contingent upon actions of the Sea Ranch Association concerning establishment of the trail system, dedication of the access easements, signing of an agreement concerning tree removal, and application for a waste discharge permit—all of which were beyond the authority of individual lot owners—applicants justifiably argued that these supposed approvals amounted to *de facto* denials. Not surprisingly, they appealed the regional commission's decision to the state commission (as did the Natural Resources Defense Council).

The state staff and commission, confronted with essentially the same arguments as their regional counterparts, arrived at a somewhat different resolution. On the one hand, the Natural Resources Defense Council (representing the Coastal Alliance) and the Sierra Club urged an outright denial in order to preserve planning options with respect to e.g., the desirability of a blufftop trail, the feasibility of septic tanks, the adequacy of the water supply from the Gualala River, and the cumulative effects of development on Highway 1 and other resources. On the other hand, the Sea Ranch Association complained bitterly about holding individual applicants as hostages, about the lack of guarantees that the 1700 lot owners would be allowed to build, and—clearly foreshadowing a court challenge—about the constitutionality of placing the burden of proof on applicants. In response to these diametrically opposed views, the state staff recommended—and the commission subsequently approved—a policy that accepted a slightly modified version of the overall conditions recommended by the regional commission but then rejected the "hostage" aspect by allowing individual applicants to proceed—in the absence of agreement between the commission and the Sea Ranch Association on overall conditions—by placing a $1500 "environmental deposit" with the commissions, which would be utilized by the regional commission to help defray the costs of cutting trees, providing access, and monitoring septic tanks and water diversion from the Gualala River.[48]

Like many compromises, however, this one satisfied virtually no one. While it did allow a total of 86 lot owners to construct residences between June 1974 and December 1976, other lot owners rejected what they perceived to be "stinking, illegal graft."[49] The Sea Ranch Association was similarly unmollified because of the cloud of uncertainty placed on the property of its members by the commission's refusal to guarantee buildout of at least the 1968 subdivided lots and by the commissions' abrogation of the 1968 access agreement between Sea Ranchers and the county.[50] Environmental groups saw the $1500 deposit as a thinly veiled sellout because it allowed development to continue with little assurance that the overall conditions would be effectively addressed. And even the regional staff foresaw difficulties because the Sea Ranch Association would probably never give the coastal commissions access to the Sea Ranch to perform the tree trimming and the commission did not have eminent domain authority to condemn and purchase accessways; thus, effective use of the deposit would require new legislation and extensive cooperation from other agencies.[51]

Not surprisingly, a raft of litigation resulted from this and other decisions pertaining to the Sea Ranch.[52] A suit brought by the Natural Resources Defense Council directly challenging the state commission's decision as a violation of the 1972 Coastal Initiative was rejected by the state court of appeals in February 1976 on the grounds that the 15 permits granted at that time would not have a substantial adverse environmental effecct and that the commissions did not have to consider the cumulative effects of full buildout during the interim permit process. A few months later a federal appeals court rejected the claim of the Sea Ranch Association that the entire development should be exempt from the permit provisions of the 1972 Coastal Act, thereby removing another challenge to the coastal commissions' authority and, indirectly, to their substantive decision. And, to indicate the bitterness that this entire controversy posed to Sea Ranch residents, a number of them unsuccessfully sued the Sea Ranch Association to prohibit it from granting any public access through the commons areas. But the major suit filed by Oceanic and the Sea Ranch Association concerning the constitutionality of the 1972 and 1976 Coastal Acts was not decided until 1981. (This, together with other events after 1976, will be discussed in Chapter 9.)

As of 1976, then, the ultimate fate of Sea Ranch was still uncertain. It is thus very difficult to come to any precise conclusions about the extent to which Sea Ranch permit decisions during the 1973–1976 period conformed with statutory objectives. The general cloud of uncertainty that the commissions posed on the development—together with higher interest rates, construction costs, and fuel costs—effectively reduced the annual buildout rate from 60 per year during 1970–1972 to about 30 per year during 1973–1976. Moreover, the commissions soon put a moratorium on the creation of new lots. Thus, the total increase in houses was probably sufficiently small to retain planning options. In addition, the restrictions placed on septic tanks and on the size and location of building minimized negative effects on water quality and on view corridors. But, as of December 1976, the commissions had not yet obtained what they deemed to be adequate public access, nor had they succeeded in removing the trees impairing views of the ocean from the coastal highway.

This is not to say that the coastal commissions did not conscientiously attempt to promote these statutorily protected values. The regional commission's basic strategy of developing overall conditions for preexisting subdivisions and then imposing a *de facto* moratorium on new construction until the homeowners' association agreed to abide by the conditions was an innovative and extremely aggressive approach. In fact, the state commission (and subsequently at least one previously supportive regional commissioner) decided that the hostage strategy posed intolerable hardships on individual applicants, given prior county approval of their lots and the minimal effect on coastal resources of constructing a relatively small number of widely dispersed houses. [53]

The fundamental point, however, is that the coastal commissions were confronted with an extremely difficult situation. Even had they been able to obtain the Sea Ranch Association's agreement to a satisfactory set of overall conditions relating to physical and scenic access and septic tanks (as the North Central Commission very nearly did in 1974), the *quid pro quo* would have been permission to allow houses on the approximately 1200 subdivided lots owned by individuals on which houses had not been constructed prior to February 1973. If the regional staff's analysis of the impacts of buildout on the coastal highway was at all correct, however, such a decision would have impaired the value of any access easements obtained at Sea Ranch by making it difficult for visitors to reach the area because of the crowded conditions on Highway 1 for the 30 miles between Jenner and Sea Ranch. The same analysis suggested that highway capacity would cease to be a limiting factor only if development could somehow be forestalled on 400–600 of the 1200 unbuilt lots. Any attempt, however, to permanently deny development on that many lots would have raised a storm of protest in the legislature and might have been overturned by the courts. But public purchase of development rights was also not a viable solution because of the enormous cost (perhaps $24 million) involved.[54] In short, the commissions inherited a situation from Sonoma County that made it extremely difficult—if not impossible—to provide the public access mandated by the 1972 Coastal Initiative.

This emphasis on the intractability of the problem at Sea Ranch—rather than lack of effort on the part of the commissions—is confirmed by the success of the general strategy of the North Central Commission in obtaining access at all over preexisting subdivisions within its jurisdiction.[55] But none of these approximated the Sea Ranch situation of a huge development with at least 1200 sold-but-unbuilt lots and with such limited highway capacity to the area.

D. Providing Beach Access in Los Angeles County: Malibu and Redondo Beach

Probably no single issue was as important in the passage of the 1972 Coastal Initiative as the need to provide greater access to the beaches for the 7 million residents of Los Angeles and the 1.5 million residents of Orange County. While almost half of the 116 miles of coast in this region was under public ownership, recreational demand was enormous—with estimates of as many as 2 million people at the beaches on peak summer weekends.[56]

There were at least four components to "the access problem" in the South Coast region:

Limited Physical Access to the Beach/Tidelands. This was particularly severe in the 26-mile stretch of Malibu coast in northern Los Angeles County from the Ventura County line to Santa Monica. While there were a few public beaches in this enormous expanse, most of the area was occupied by a wall of expensive houses between the Pacific Coast Highway and the ocean, which essentially precluded the public any vertical access from the highway to the beach; in addition, many of the owners had placed fences down to the water's edge, thereby precluding even lateral access along the wet sand. There were similar problems, although on a smaller scale, in some of the walled communities in and around Newport Beach in Orange County.[57] In effect, then, the public was largely precluded access to the publicly owned tidelands for approximately a fifth of the coast in this region. It was these restrictions that made "Don't Lock Up the Beach" such an effective slogan during the Coastal Initiative campaign.

Parking. Because most public beaches had only limited parking facilities, many users were forced to park in adjacent neighborhoods. In many places, this was becoming increasingly difficult as single-family residences were being replaced by higher density uses such as apartments, office buildings, and commercial establishments. As a result, "permanent" residents were filling the on-street parking capacity, leaving very little for beach users. While this problem was perhaps most acute in Marina del Rey—a recreational harbor area that was becoming dominated by very high-density apartment and office complexes—it was also a serious problem throughout much of the South Coast region from Santa Monica to Newport Beach. For example, Redondo Beach was a rapidly growing municipality of 60,000 people along the central Los Angeles coast whose beach-front area was rapidly being transformed from low-density residences (with many senior citizens) to high-rise apartments, much of it as part of a redevelopment project.[58]

Visitor Facilities. Several areas, most notably the Malibu coast, had a shortage of restaurants, campgrounds, and other services for beach users.

View Corridors. This too was particularly a problem in Malibu and some of the wealthy Orange County communities where a row of houses between the coastal highway and the beach obscured public views of the water (see Figure 5-3).

We cannot hope to address in a comprehensive fashion these multiple facets of access throughout the South Coast region. Instead, we shall focus on access issues in Malibu and Redondo Beach as they were raised on appeals to the state commission.

Although beaches in the southern third of Malibu were too narrow to provide recreation for large numbers of people, Malibu nevertheless represented one of the most important disputes over access in the entire South Coast region, in part because it occupied so much of the coast, in part because it raised the issue so clearly of whether wealthy residents should be able to block public access to the state-owned wet sand beaches. In the period between February 1973 and June 1975, 14 appeals involving proposed developments between the coastal highway and the ocean (all but 1 involving single-family residences) were acted upon by the state commission (see Table 5-6). Of these 14, the commission either denied or imposed vertical access conditions on 10, while imposing lateral access conditions on an additional 2 projects. Since most of the denials explicitly involved attempts to preserve options for public acquisition, the commissions would seem to have been doing a good job of using their permit review authority to provide public access to the coast. In addition, 2 other Malibu projects involved visitor facilities: In the case of a proposed 200-unit private recreational vehicle park, the state commission reversed the regional commission denial on the grounds that the land would best be used for such a facility rather than for housing, although it also imposed conditions to improve the design of the project

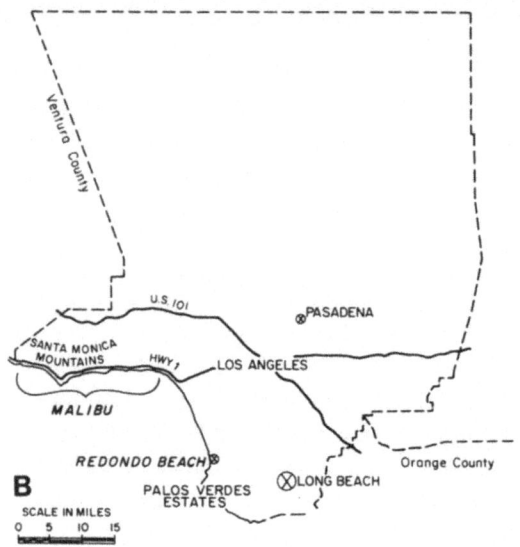

and to replace 75 recreational vehicle sites with tent campsites.[59] In another case, the state commission approved the conversion of apartments to retail shops on the condition that the owner provide additional parking, dedicate a portion of his land for a park, and require the shops be oriented toward beach users.[60]

This brief review of appeals would suggest that the coastal commissions did a good job of enhancing public access in the Malibu area. But the record also raises a number of *caveats* to such a conclusion. First, the state commission minutes and conversations with regional commission officials reveal that the regional commission did not aggressively pursue access until early 1974, when it began requiring lateral access conditions on all permits and vertical access easements approximately every half mile.[61] Second, the appeals process did not appear to work very well in this area. Note that all but 2 of the 16 appeals listed in Table 5-6 involved denials by the regional commission, with the remaining 2 involving conditional approvals. In fact, all of the appeals were brought by permit applicants. If the South Coast Commission had not been aggressively promoting public access in Malibu during 1973, then why weren't the Sierra Club and other environmental groups appealing its decisions? Part of the explanation was the huge permit load in this region, as well as legitimate concerns about the desirability of promoting access in this area, part of which was characterized by narrow beaches and narrow lots (thus making access easements difficult to obtain). Nevertheless, the failure to initiate any appeals in this area certainly fueled the suspicions that environmental groups were willing to tolerate wealthy enclaves that impeded access—a suspicion that was to come back to haunt environmentalists during the 1976 legislative session. Third, it is worth remembering that through the permit process the commissions could only deny permits in order to preserve acquisition options or require access easements as a condition of development. The effective provision of public access was in turn contingent upon future actions of other agencies and/or the legislature concerning the maintenance of accessways and the provision of funds for public purchase. As we shall see in Chapter 7, while some lots were subsequently designated for purchase as part of the 1976 Coastal Parks Bond Act, as of June 1979, few, if any, of the accessways were actually open to the public because of the unwillingness and/or inability of local parks departments to assume maintenance and liability responsibilities.[62] In short, the record of the coastal commissions during the 1973–1975 period in providing access at Malibu may not have been as good as the data in Table 5-6 suggest, and many of their efforts may have been subsequently undermined by the inability or unwillingness of other actors (veto points) to perform needed functions.

←————————————————————————————————

FIGURE 5-3.   Beach access in the South Coast region. A. The demand for beach access on a hot summer day (*Los Angeles Times* photo). B. Map of Los Angeles County. C. Restrictions on the supply of beach access: a wall of houses between the coastal highway and the beach (Newport Beach).

TABLE 5-6
PERMIT APPEALS INVOLVING PUBLIC ACCESS AT MALIBU
FEBRUARY 1973–JUNE 1975[a]

Applicant (Appeal no.)	Nature of proposed development	Regional commission decision	Appellant	State commission: Issues at hearing	State commission: Decision	State commission: Date
Cases involving houses between highway and ocean						
1. Corrodi (18-74)	1 SFR[b]	Deny	Applicant		Affirm regional denial	3/20/74
2. Higgins Trust (94 & 113-74)	2 SFR	Deny	Applicant	1. Geol. stability 2. Parking 3. Septic tanks	Affirm regional denial	11/20/74
3. Gordon (140–74)	1 SFR	Deny	Applicant	1. Geol. stability	Affirm regional denial	9/18/74
4. Stephens (244-74)	1 SFR	Approve w/conds. 1. Vertical access 2. Lateral access	Applicant	1. Access to beach	Affirm regional decision	11/20/74
5. Ventress (299-74)	1 SFR	Deny	Applicant	1. Consistency of reg. comm. decisions. 2. Intrusion on beach	Affirm regional denial	2/4/75
6. Frumkes (11-75)	1 SFR	Deny	Applicant	1. Possible acquisition	Deny	3/25/75
7. Feldman (11-75)	1 SFR	Deny	Applicant	1. Consistency of reg. 2. Public access 3. View corridors	Approve w/cond's. 1. Vertical access 2. Lateral access	3/25/75
8. Nicholas (33-75)	1 SFR	Deny	Applicant	1. Change in reg. com. policy 2. Geol. stability 3. Intrusion on beach 4. Septic tank	Approve w/cond's. 1. Lateral access 2. Bulkhead stability 3. Septic tank	3/25/75

Case	Project	Regional decision	Appealed by	Objections	State commission decision	Date
9. Oliver (37-75)	1 SFR	Deny	Applicant	1. Geol. stability	Approve w/cond's. 1. Reduce size of house 2. Bluff set back	4/22/75
10. Apollo (66-65)	1 SFR	Deny	Applicant	1. Possible acquisition	Affirm reg. com. denial	4/22/75
11. Bucklin (75-75)	Addition to existing SFR	Approve w/cond's. 1. Lateral access	Local envir.	1. Sufficient access	Affirm regional decision	5/13/75
12. Beck	1 SFR	Deny	Applicant	1. Possible acquisition 2. Septic tanks 3. Consistency of reg. comm. dec.	Deny	6/3/75
13. Farrida	1 SFR	Deny when wouldn't agree to cond's.	Applicant	1. Bluff stability 2. Septic tank	Approve w/cond's. 1. Lateral access 2. Septic tank	6/13/75
14. Bernstein (120-75)	6-unit apt.	Deny	Applicant	1. Intrusion on beach 2. Septic tanks	Affirm regional denial	6/17/75
Cases involving possible visitor facilities						
1. Adamson Co. (163-73)	200-space RV park	Deny	Applicant	1. Houses vs. visitor facilities 2. Poor design 3. Traffic	Approve w/cond's. 1. Reduce to 125 RV spots and 75 tents 2. Design features	11/28/73
2. Segal (273-74)	Conversion of apartments to retail shops	Deny	Applicant		Approve w/cond's. 1. Parking 2. Park 3. Shops to serve beach areas users	12/18/74

[a]Source: Minutes of state commission.
[b]SFR = single-family residence.

Unlike that in Malibu, the major problem in Redondo Beach was not private developments restricting physical access from the coastal highway to the beach. For with the exception of the King Harbor area, most streets opened directly onto the beach. Instead, the problem was finding a parking place—particularly given the trend in converting the near-beach area to high-density development. Much of the conversion was being done by the Redondo Beach Redevelopment Agency. In early 1973 it applied for an exemption from coastal commission authority on vested rights grounds; this was denied by the commissions and then appealed to the courts. After a *district* court decision exempting over half of the 1153 units originally requested, the Redevelopment Agency and the state commission entered into direct negotiations pending an appeal. The commission's record in the appellate courts was such, however, that the agency accepted an out-of-court settlement permitting it to construct 542 dwelling units (and 140 senior citizen units) on the condition that it also deed a 3-acre park along the waterfront to the city.[63] In short, the state commission eventually obtained a 50% reduction in density and a park.

The other appeals from Redondo Beach during the 1973–1975 period involved regional denials of multiunit residences, which were all reaffirmed by the state commission. As a result, after several years of conflict with the regional commission, the city modified its zoning along the beachfront to restrict the conversion of single-family residences and small apartments to much higher density residential developments, as well as revising its building code to require two off-street parking spaces per residential unit.[64] In sum, while the coastal commissions' permit review authority did not enable them to construct additional parking facilities for beach users, their decisions prevented much of the on-street parking traditionally employed by beach users from being swallowed up by higher density residential and commercial developments.

E. Conclusion

Let us return briefly to our overall argument. In the first two chapters we hypothesized that the decisions of a regulatory agency would be consistent with statutory objectives if (1) the statute provided clear and consistent objectives; (2) it incorporated a sound causal theory identifying factors impeding the achievement of those objectives and gave implementing agencies jurisdiction over those factors; (3) the statute coherently structured the implementation process by assigning implementation to supportive agencies within a hierarchically integrated process; (4) the appointment process resulted in leaders of implementing agencies who were committed to statutory objectives and were reasonably skillful; (5) the agency was supported by constituency groups and a legislative/executive fixer, with the courts neutral or supportive; and (6) the priority of statutory objectives was not significantly undermined over time by changing socioeconomic conditions.

Chapters 2–4 indicated that these conditions were generally met in the case of the coastal commission during the 1972–1976 period. The 1972 California Coastal Zone Conservation Act mandated a consistent "tilt" in favor of environmental protection and public access over economic development; it assigned implementation to new agencies with final jurisdiction over all development within the 1000-yard permit zone; and it established decision rules, such as the two-thirds voting requirement and liberal rules of standing, which certainly favored statutory supporters. The commission had staff who strongly supported statutory objectives and commissioners who, with the exception of the North Coast region, generally did so. Environmental groups actively participated in the commissions' deliberations (although somewhat less so in the South Coast region), and the appellate courts consistently upheld the commissions' decisions, while the Coastal Initiative largely insulated the commissions from the need for a "fixer" during the 1973–1975 period. The only pessimistic note for effective implementation was the decline in public support for environmental protection and the slump in the construction industry during 1974–1975 as a result of the nationwide recession.

Given that most of the conditions were met, one would expect permit decisions of the commissions during the 1973–1976 period to have been consistent with statutory objectives. The data presented here have generally confirmed this expectation, although a definitive judgment is difficult because of the lack of precision in statutory objectives and the essentially impossible task of estimating the effects of 15,000 permit decisions on statutorily protected values.

Turning first to the overall stringency of the commissions' decisions, we saw that both the North Central and state commissions imposed conditions on or denied over 80% (23% of all) of the permit applications important enough to be granted a public hearing. The percentage of conditional approvals/denials on hearing calendar items dropped to about 45% in the South Coast and only 25% in the North Coast. (If attention is restricted to simply *major* conditions or denials, the combined figures are 82% for the state commission, 51% for the North Central, 27% for the South Coast, and 11% for the North Coast.) There is considerable evidence, however, that the appellate process served to correct most of the cases in which the North Coast Commission probably violated statutory objectives. This was certainly the conclusion of our detailed review of permits involving Humboldt Bay. Moreover, the aggregate data on permit appeals presented in Chapter 4 (Table 4-9) indicated that over 90% of the permits appealed from this region ultimately resulted in conditional approvals or denials. On the other hand, the appellate process may not have been as successful in correcting the errors of the South Coast Commission. While the aggregate data in Table 4-9 indicate that only about 10% of permits appealed from this region were eventually approved as submitted, our analysis of public access at Malibu suggests that no regional approvals were appealed by environmental groups or other opponents during 1973, when the regional commission was applying criteria that it subsequently felt to be inadequate to provide lateral and physical access at Malibu.

With regard to the commissions' performance on specific statutory objectives, the data presented in Table 5-4 indicate that—at least on those permits appealed to the state commission—the commissions either denied or imposed relevant conditions on at least 70% of the permits involving possible geologic hazards or erosion, potential effects on scenic resources or view corridors, possible impacts on water quality, or the conversion of agricultural land. As for the remaining statutory objectives to provide public access to the beach, protect wildlife habitat (especially wetlands), and preserve planning options, the case studies of Humboldt Bay, Sea Ranch, and Malibu/Redondo Beach suggest that the commissions probably did a much better job than indicated by the (admittedly problematic) figures in Table 5-4. They certainly did an excellent job of protecting wetlands in Humboldt Bay; they were clearly concerned with preserving planning options (including possible acquisitions) at Sea Ranch and Malibu; and they certainly attempted to provide public access at Sea Ranch, Malibu, and Redondo Beach.

With respect to the crucial access question, however, their leverage was limited by prior local government approvals—which created enormous difficulties for the commissions at Sea Ranch—and by the absence of statutory authority for the commissions to acquire and/or manage access ways. This, in turn, meant that the actual *provision* of access required the cooperation of several autonomous entities, i.e., constituted a very loosely integrated implementation system. In sum, the commissions' permit decisions during the 1973–1975 period appear generally to have been consistent with statutory objectives.

III. A MODEL OF COMMISSIONER VOTING

Turning from decisions rendered to the driving forces behind them, Figure 5-4 conceptualizes the principal relationships among the most probable factors affecting permit review by the coastal commissions. Consistent with the work of Hofferbert, Katz and Kahn, and other social scientists, the diagram distinguishes (1) basic background factors (community needs and resources, and public opinion), (2) fairly stable characteristics of the coastal commissions (statutory attributes and the basic policy orientation of agency officials), and (3) those factors specific to individual permit cases (the probable impacts of the proposed develop-

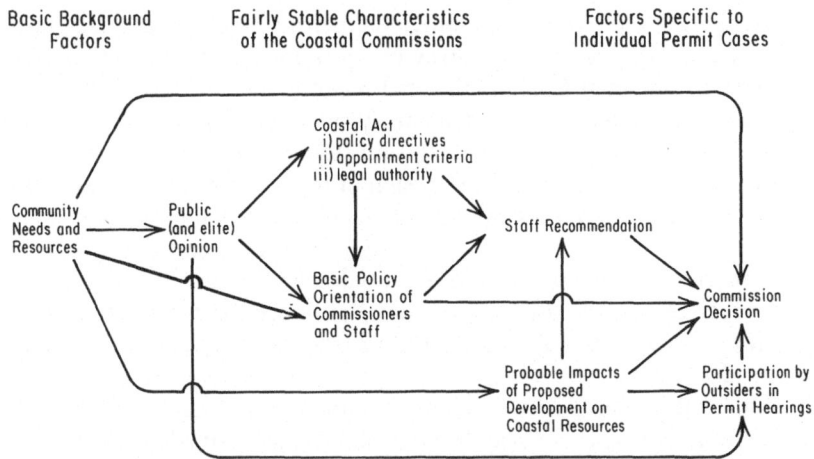

Basic Background Factors   Fairly Stable Characteristics of the Coastal Commissions   Factors Specific to Individual Permit Cases

FIGURE 5-4.   Causal model of factors affecting permit decisions.

ment, the staff recommendation, outsider participation, and, finally, the commission decision).[65]

The background factors presumably operate primarily through the fairly stable characteristics of the commissions. For example, the higher educational and income levels and greater population density and recreational needs of residents in the North Central region presumably led to greater public support for coastal protection than in the North Coast region. This, in turn, resulted in the selection of commissioners (particularly among local elected officials) whose basic policy orientation was more sympathetic to the objectives of the Coastal Initiative. But community needs/resources and public opinion can also affect specific permit decisions through affecting commissioner's perceptions of community needs and attitudes on individual permit cases, as well as by affecting the reservoir of possible participants with various points of view at the public hearings.

Turning to the relatively stable characteristics of the commissions, some of the statutory variables—most notably, the appointment criteria and general policy directives—presumably affected the basic policy orientation of commissioners and staff. For example, state appointees generally ranked higher on our environmental protection scale than local elected officials; moreover, there is some reason to think that the statute's environmentalist policy directives may have led to the selection of staff and commissioners who were more environmentally oriented than may have been the case with an entirely different statutory mandate. The statutory characteristics and general attitudes of agency officials then influenced commission decisions, both directly by commissioner votes and indirectly via the staff recommendations.

Among the factors specific to individual permit cases, the probable impacts of the proposed development obviously interacted with the value orientation of commissioners and staff to affect the staff recommendation and commissioner voting. They also affected the extent of outsider participation at the hearing, which in turn may have influenced commissioner voting.

Ideally, quantitative indicators of each of these factors would be developed and then integrated into a testable quantitative model. Unfortunately, some of the variables—such as the "force" of statutory directives and the probable impacts of proposed developments—are exceedingly difficult to quantify over a large number of permit cases. In addition, of course, statutory characteristics were constant throughout all cases and thus unsuitable for explaining variation in permit decisions.

On the other hand, it is possible to examine many aspects of the general model depicted in Figure 5-4 and thus to obtain an approximate indication of the relative importance of most of the factors involved. Specifically, we propose to examine the effects of several subsets of variables in explaining the voting behavior of the 30 regional commissioners in our study.

This is done in Table 5-7. The table presents the results of a correlation and regression analysis of the effects of a variety of predictor variables on the denial rate of individual commissioners in the three regions, i.e., the percentage of (all) hearing calendar items on which the commissioners voted to deny the permit application over the 1973–75 period.[66] The following are the variables (and indicators) incorporated into the analysis:

Needs and Resources of Commissioner's Community/Constituency.[67] Preliminary analysis of a number of such indicators revealed that the best predictors of commissioners' voting behavior were (a) community population (1970), (b) the percent of the community's population within the planning jurisdiction of the coastal commissions, and (c) per capita acres of park and forest land suitable for recreation within the community.[68] The first is presumably related to support for open space, scenic vistas, and additional coastal recreational opportunities— in short, an alternative to the congestion of city life. The second is an indication of the percent of a community's activities subject to the jurisdiction of the coastal commissions; presumably those commissioners whose constituents are *least* subject to regulatory control by the commissions will be most likely to deny development applications. And the third measure is a straightforward indicator of the supply of existing outdoor recreational opportunities and thus inversely related to the need for coastal recreation sites.

Public Opinion Within the Commissioner's Community/Constituency. Here we employ the percentage of voters who supported Proposition 20 in

TABLE 5-7

EFFECT OF A VARIETY OF PREDICTOR VARIABLES ON THE PERCENT VOTE TO DENY
OF REGIONAL COMMISSIONERS ($N = 30$)

| | Association with commissioners' % no vote | | | |
| | Single correlation (r) | Regression analysis | | |
Independent (predictor) variables		Standardized regression coefficient	R^2 added	Total R^2
Needs and resources of commissioner's community				
1970 population	.65[b]	.41[b]	.42	.42
Percent of population within commission's planning jurisdiction	−.61[b]	.01	.15	.57
Acres of park and forest per capita	−.39[a]	.07	.05	.62
Public opinion within comissioner's community				
Percent vote for Prop. 20	.53[b]	.16	.01	.63
Appointment criteria and process				
Local (rather than "public") official	−.40[b]	−.03	.01	.64
Commissioner's basic policy orientation				
Concern with coastal degradation	.69[b]	.32[b]	.15	.79
Environmental Protection Scale	.71[b]	.01	.04	.84
Commissioner's responsiveness to staff				
% no when staff rec. denial−% no when staff rec. approval	.39[a]	−.09	.01	.85
Commissioner's responsiveness to public opposition in permit hearings				
% no with opp.−% no without opposition	.76[b]	.50[b]	.05	.90

[a,b]Significance levels of .05 and .01, respectively. Recall, however, that we are not dealing with a random sample of commissioners but rather with an incomplete census of 30 out of 36 commissioners in the three regions.

November 1972. The greater the public support for coastal "protection," the higher the commissioner's denial rate.

Appointment Criterion and Process. This is simply a two-cell variable indicating whether or not the commissioner is a local elected official (rather than a "public" member appointed by state officials). It is thus a straightforward measure of one of the means by which Proposition 20 sought to structure the implementation process. But, as we saw in Chapter 4, it is also a measure of the entire appointment process which was a result of the interaction of statutory criteria, the policy preferences of appointing officials, and the degree of local support for Proposition 20. At any rate, the assumption is that local elected officials will have lower denial rates than "public" members.

Commissioner's Basic Policy Orientation. Here we use two indicators: (a) the commissioner's score on the Environmental Protection Scale and (b) his/her score on a Coastal Degradation Scale, designed to measure the commissioner's perception of the seriousness of a variety of potential coastal problems (e.g., the destruction of scenic resources, the shortage of public access to the beach).[69] Both are presumably positively correlated with denial rates.

Commissioner's Responsiveness to Staff. This is simply the difference in a commissioner's denial rate when the staff recommended denial and when it recommended approval. It is thus a straightforward measure of the overall influence of staff recommendations on his/her voting behavior.[70]

Commissioner's Responsiveness to Public Opposition in Permit Hearings. Similarly, this is the difference in a commissioner's denial rate when there was some public opposition and when there was none.[71]

In short, the statistical analysis incorporates all of the variables in our model except the probable impacts of proposed developments and those aspects of the statute concerning the influence of policy directives (e.g., to protect public access) on commissioners' voting behavior.

The first column in the table reports the zero-order correlation (Pearson r) between each predictor variable and the denial rates of the 30 regional commissioners. The relationships are all in the expected direction: For example, commissioners from urban constituencies had higher denial rates than their rural colleagues; commissioners with a relatively large percentage of their constituents under the jurisdiction of the commissions were less willing to deny permits than their colleagues whose constituents were less subject to coastal regulatory controls; and local elected officials generally had lower denial rates than "public" members, although the relationship was only moderately strong ($r = -.40$). In fact, viewed in isolation, the best predictors of commissioners' voting behavior are their responsiveness to public opposition ($r = .76$) and the two measures of their basic policy orientation (Concern with Coastal Degradation, $r = .69$; Environmental Protection, $.71$), while two of the community needs indicators (population and percent of population within the commissions jurisdiction) and community public opinion also had correlations of at least $.50$ in magnitude.

While such bivariate relationships are useful in identifying good predictors of commissioners' voting behavior, they cannot address more complicated causal relationships. For example, what effect do commissioners' basic policy orientations have on their voting *once the needs and the resources of their communities have been taken into account?* In order to explore such questions—and to provide an additional exploration of our more complex general model presented in Figure 5-4—the latter three columns in the table present the results of a hierarchical regression analysis in which variables were introduced in the following order: (1) community needs and resources, (2) public opinion, (3) the appointment criterion distinguishing local from public members, (4) the commissioner's basic policy orientation, (5) his/her responsiveness to staff, and, finally, (6) his/her responsiveness to public opposition. The sequence is pretty straightforward. For example, responsiveness to staff comes before responsiveness to public opposition because the staff recommendation preceeded the public hearing in the permit process. The only real ambiguity concerns the placement of the appointment criterion. It clearly had to precede commissioners' basic policy orientation, as commissioners had to be appointed before their views became relevant. It was placed after community needs/resources and public opinion on the grounds that it is basically a statutory variable and thus antedated by the basic political character of the community (particularly given our measure of public opinion, i.e., the November 1972 vote on the Coastal Initiative).

Such a causal model reveals that the three measures of community needs and resources employed were in fact the driving force behind the entire system— accounting for .62 of the .90 variance explained. In fact, once community needs and resources are taken into account, public opinion adds very little to the model. Conversely, commissioners' policy orientation still proves to be quite important, adding .20 to the variance explained. After that, staff recommendations adds very little, while public opposition still adds .05—not at all trivial, given its place in the temporal sequence. The model as a whole proved to be quite powerful, predicting 90% of the variance in denial rates among regional commissioners. In fact, a greatly simplified version involving only 1970 community population, the perceived coastal degradation aspect of a commissioners' basic policy orientation, and the responsiveness to public opposition was able to predict 88% of the variance.[72]

One final note: In addition to the straight hierarchical regression, we did a limited path analysis in order to separate out the direct versus the indirect effects of the community need/resource variables. Much to our surprise, it revealed that approximately 58% of the effect of 1970 community population on commissioners' vote was *direct*, i.e., unmediated by the other variables such as public opinion and commissioners' basic policy orientation.[73] In other words, in many instances commissioners were apparently responding to their perceptions of community needs unfiltered by their own policy predilections or public opinion.

What can we conclude from this analysis? First, it seems quite clear that community needs and resources are certainly one of the fundamental driving forces in agency decision making and thus in policy implementation (see this chapter's Appendix for additional analysis of this point). This suggests that any statute will have only a limited ability to affect local decision making in hostile environments. The basic strategy, then, should be to incorporate an appeals process that transfers final authority to agency subunits responsive to more supportive communities. As we saw earlier in the chapter, the permit appeals process incorporated into the Coastal Initiative apparently did this quite well (at least with respect to the North Coast region). The importance of community needs and resources also confirms our belief that one of the conditions of successful implementation is that the relative priority of statutory objectives not be undermined over time by changing socioeconomic conditions.

Second, the analysis confirmed the importance we have placed on the role of supportive interest groups in policy implementation. This is particularly crucial in any permit or licensing process in which regulatory officials are constantly and directly confronted with the pleas of the people being regulated; without a counterweight of articulate spokesmen for regulation, the normal human tendency is to succumb to protestations of hardship.

Third, our focus on a single statute has not permitted a very precise assessment of the importance of statutory variables in permit review. We have, however, been able to examine the importance of one structural variable, the appointment of local government versus public members to the commission. The assumption of the framers of the Coastal Initiative that "public" members would be more supportive of statutory objectives than local elected officials was confirmed in practice, although this was certainly not as important a factor in explaining variation in commissioner voting behavior as one might have thought.

In the final analysis, however, a full testing of our model—particularly given the importance attached to statutory variables—would require a prodigious effort to obtain estimates of all the variables for a substantial number of agencies with different statutes, probably in the same policy area. To the best of our knowledge, such an analysis has been attempted only once and, even then, dealt with only about half of the variables.[74] The reasons are straightforward: To obtain accurate estimates of the independent and particularly the dependent variable (agency policy decisions), one would essentially have to replicate this type of study—including surveys of agency officials on random samples of agency decisions, and developing *relevant* indicators of public opinion and community needs and resources—for 50 or more agencies!

Pending the results of such an analysis, we shall have to be content with our tentative conclusions (1) that community needs and resources appear to be the principal driving force guiding agency decision making, but that the policy orientation of agency officials and the participation of outsiders exercise important independent effects; and (2) that the ability of a statute to structure the implementation process through the selection of supportive officials, the consistency of its policy directives, the "bias" of its decision rules, and the establishment of a hierarchically integrated process almost certainly has a strong—but very difficult to measure—effect on the stringency of agency decisions and their consistency with statutory directives.

APPENDIX: EXPLAINING INTERREGIONAL VARIATION
IN COMMISSION DECISIONS

Table 5-8 presents data on most of the variables in our overall model for the North Coast, North Central, and South Coast regions:

Community Needs and Resources. The first three indicators (population, per capita income, and median education) have all been found to be modestly related to public support for environmental protection and land use controls. Conversely, unemployment should be inversely related to such support. The last two indicators (county population per mile of coast and per capita park/forest land) probe different aspects of the need for coastal recreation opportunities and thus public support for controls of nonrecreational coastal development: The greater the population per mile of coast, the greater the demand for the coast as a scenic and recreational resource; conversely, the greater the existing per capita land available within the region for recreational purposes, the less the demand for coastal recreational opportunities.

Public Opinion. Regional vote on Proposition 20 provides a straightforward measure of public support for coastal protection. While it would have been desirable to include subsequent indicators of public opinion during the 1973–1975 period, none were available on a regional basis.

Basic Policy Orientation of Commissioners and Staff. Reported here are (a) the percent of each group who supported Proposition 20 in retrospect and (b) their scores on the Environmental Protection Scale (with a maximum of $+3$ and a minimum of -3).

Outsider Participation at Public Hearings. Here we use the percent of hearing calendar permits on which there was opposition expressed, as what we are really interested in is the continued participation of the constituency that had originally supported Proposition 20.

Staff Recommendation. Reported here are the staff recommendations on all hearing calendar items in the 1973–1975 period on a 4-point scale from 1 = approve as submitted by applicant to 4 = deny.

Commission Decision. This too involves all hearing calendar permits during the 1973–1975 period, reported on a 4-point scale.

In fact, the only variables in our model for which indicators are not provided in Table 5-8 are, first, the policy directives and the legal authority provided by the

TABLE 5-8

INTERREGIONAL COMPARISON OF COMMUNITY NEEDS AND RESOURCES, PUBLIC
OPINION, POLICY ORIENTATION OF COMMISSIONERS AND STAFF, OUTSIDER
PARTICIPATION AT PUBLIC HEARINGS, STAFF RECOMMENDATIONS, AND, FINALLY,
COMMISSION DECISIONS ON HEARING CALENDAR PERMITS

	Region		
Characteristics	North Coast	North Central	South Coast
A. Community needs and resources			
1. Population in coastal counties (1974)[a]	176,000	1,132,000	8,617,000
2. Per capita income (1973)[b]	$4,423	$7,157	$5,782
3. Median years education (1970)[b]	12.2	12.5	12.4
4. % unemployed[c] a. 1973	9.0%	7.5%	6.2%
b. 1975	15.0%	11.5%	9.3%
5. City population per mile of coastline ('74)[a]	613	8,028	74,284
6. Public park and forest land per capita (1975)[d]	7.48 ac.	0.11 ac.	0.09 ac.
B. Public opinion			
Vote on Prop. 20 (Nov. 1972)[a]	34%	60%	53%
C. Policy orientation of commissioners and staff			
1. % vote for Prop. 20 in retrospect[e]			
a. Commissioners	36%	100%	56%
b. Staff	91%	100%	100%
2. Environmental Protection Scale[e]			
a. Commissioners	−.80	.19	−.36
b. Staff	.26	.10	.46
D. Outsider participation at public hearings			
% permits with opposition[f]	34.6%	42.0%	30.7%
E. Staff recommendation			
Rec. on hrg. calendar items, 1973–75[e]			
Approve as submitted	72%	15%	41%
Approve with minor conditions	12%	37%	28%
Approve with major conditions	10%	31%	8%
Deny	6%	16%	22%
F. Commission decision			
Decision on hrg. calendar items, 1973–75[e]			
Approve as submitted	76%	15%	56%
Approve with minor conditions	11%	35%	17%
Approve with major conditions	8%	34%	3%
Deny	5%	16%	24%

[a]Figure 1-1.
[b]Table 3-1.
[c]County Supervisors Association of California, *County Fact Book*, 1975 and 1977.
[d]State Lands Commission, *Public Land Ownership*, 1975.
[e]Table 4-8. Recall that the entire census of hearing calendar items were used in the two northern regions ($N = 277$ and 173, respectively), while information on permit conditions was available only for a sample of 99 items in the South Coast.
[f]Table 3-3.

1978 Coastal Initiative and, second, the probable impacts of development. The former is omitted because the statute was constant across all three regions and thus incapable of explaining interregional variation in permit decisions, while the latter is not included because of the lack of feasible empirical indicators.

The model assumes that community needs and resources are the driving force behind the entire permit process through their effects on public opinion, the basic policy orientation of commissioners (and perhaps staff), and the extent of outsider opposition at the permit hearings. In addition, community needs and resources have some direct effect on permit decisions as commissioners respond to perceived needs, e.g., for recreational opportunities, unmediated by any of the intervening variables. On the basis of this reasoning, one would clearly expect the *least* stringent permit review in the North Coast region. It had the lowest population, per capita income, and educational levels of any of the three regions; the highest unemployment levels; and the least need for coastal recreational opportunities. This prediction is certainly confirmed by the data on permit decisions, as the North Coast had by far the highest percentage of outright approvals and the lowest percentage of denials.

Moreover, the data confirm the model's predictions concerning several of the intervening variables. For example, the North Coast had by far the least support among the general public and regional commissioners for coastal protection of any of the three regions. On the other hand, agency staff in this region were as supportive of the Coastal Act as their colleagues in the other two regions. Nevertheless, the data presented here and in the previous chapter suggest that staff recommendations on permit items were less a function of staff members' beliefs concerning the desirability of environmental protection than of their perceptions concerning community needs and resources and the prodevelopment policy orientation of their bosses on the commission. Finally, the extent of outsider opposition to permit approval at the public hearings was quite comparable to that of the other two regions—despite the low level of support for coastal protection among the public as a whole. This is, in fact, a tribute to the role of environmental groups (particularly the Sierra Club) in maintaining an effective presence in a region where the general public and several commissioners were quite hostile to their point of view.

Turning now to the other two regions in our study, an examination of community needs and resources would lead one to expect slightly more stringent commission decisions in the South Coast than in the North Central, mainly on the grounds of a lower unemployment rate and a much higher ratio of population per mile of coast. This was not, however, the case, as the North Central had a much lower percentage of permits approved as submitted (although also a slightly lower percentage of denials). The link between community needs and commission decisions was similarly reversed throughout the intervening variables: It was the North Central rather than the South Coast that had the greater degree of public and especially commissioner support for coastal protection, as well as the higher precentage of permits with opposition. The explanation for this mild reversal in rank between the two regions probably lay in the fairly long history of support for land use restrictions in the Bay Area as well as the significantly greater economic consequences of restricting coastal development in the South Coast: Although the *percentage* of county population in the permit area was very similar in the two regions (4–5%), the *number* of people—and almost certainly the amount of construction activity—within the coastal permit area was much greater in the South Coast than in the North Central. As a result, the construction industry—to say nothing of the concentrations of economic power associated with the ports of Long Beach and Los Angeles and the commercial complex of Marina del Rey—had a much greater stake in the South Coast than did their counterparts in the North Central, and therefore much greater incentive to influence the vote on Proposition 20, the appointments to the regional commission, and the voting behavior of regional commissioners.

NOTES

1. *California Public Resources Code,* Section 27103 (1973). Excluded from the commissions' jurisdiction, however, were (1) development in and around San Francisco Bay under the jurisdiction of BCDC, (2) repairs to single-family residences of under $7500, (3) maintenance dredging operations of the U.S. Army Corps of Engineers, and (4) activities on federal military installations.
2. Of the other state land use programs with wide geographical authority, Hawaii's dealt only with the conversion of agricultural land and Florida's only with major projects with regional impacts. The only institutions with authority comparable to that of the California coastal commissions were the state and regional boards in Vermont under Act 250 (1970) and the Adirondack Park Commission in New York [see Robert Healy, *Land Use and the States,* 2nd ed. (John Hopkins University Press, 1981) chap. 3–5; and Nelson Rosenbaum, *Land Use and the Legislatures,* (Washington, D.C.: Urban Institute, 1976)]. Among coastal zone management agencies, the California commissions had the greatest geographical scope and the most autonomy from local governments [Federal Office of Coastal Zone Management, "State Coastal Zone Management Activities," (December 1976)].
3. Paul Culhane and Paul Friesema, "Land Use Planning for the Public Lands," *Natural Resources Journal* 19 (January 1979), pp. 43–74.
4. Robert Kneisel, "The Impact of the California Coastal Commissions on the Local Housing Market" (Unpublished Ph.D. dissertation, University of California, Riverside, 1979).
5. See Peter Joseph Douglas and Petrillo, "California's Coast: The Struggle Today, A Plan for Tomorrow," *Florida State Law Review* 4 (1976), p. 333. Other laws that probably contributed to the overall improvement in the quality of proposed developments were the California Environmental Quality Act and the 1973 California Forest Practices Act.
6. As noted in chapter 3, Newport Beach incorporated the commissions' requirements on off-street parking, Laguna Beach revised its density requirements, and Redondo Beach revised its entire building code to bring about greater conformity with coastal commission decisions.
7. For an excellent discussion of this topic, see Robert Healy *et al., Protecting the Golden Shore: Lessons from the California Coastal Commissions* (Washington, D.C.: Conservation Foundation, 1978).

8. *California Public Resources Code*, Section 27400 (1974).
9. Paul Sabatier, "State Review of Local-Land Decisions: The California Coastal Commissions," *Coastal Zone Management Journal* 3 (1977), pp. 258–259, 290.
10. This discussion is based on interviews with Bodovitz and with the senior permit staff at the state level, particularly Frank Broadhead. The formal regulations of the state commission stated that a finding of "substantial issue" was contingent upon procedural irregularities or an unsubstantiated decision at the regional level and/or an application that involved issues of statewide concern or effects on the coastal plan being prepared. An analysis of a random sample of 166 appeals during the 1973–1975 period revealed that most cases of "substantial issue" dealt with regional decisions to approve a project, often with no or minor conditions. Ibid, pp. 263–269.
11. This approximation is based on periodic reports from the attorney General's office concerning litigation involving the commissions.
12. This argument is most clearly presented in Marver Bernstein, *Regulating Business by Independent Commission* (Princeton: Princeton University Press, 1955).
13. See chapter 1.
14. The argument that the act imposed a moratorium was made by opponents of Proposition 20 during the 1972 campaign and by some supporters after its passage. It was, however, explicitly rejected by proponents during the campaign and subsequently by commission officials. For the ballot arguments for and against Proposition 20 mailed to all registered voters, see California Secretary of State, *Propositions and Proposed Laws, Together with Arguments, General Election,* November 7, 1972, pp. 53–55. For an interesting discussion of this point, see Judy Rosener, "Hightide: A Coastal Commissioner's Perspective," *Coastal Zone Management Journal* 4 (1978), pp. 1–6.
15. All four of these objectives were mentioned both in Section 27401 requiring a two-thirds vote of the commission and in Section 27403 requiring that permits be subject to "reasonable terms and conditions to ensure. . . ."
16. Geologic hazards were mentioned only in the section dealing with reasonable conditions, while agricultural land was mentioned only in the section dealing with a two-thirds vote. The latter is particularly difficult to interpret as a substantive standard because it simply requires that "no permit shall be issued for any of the following without a two-thirds affirmative vote: . . . (e) Any development which would adversely affect . . . agricultural uses of land which are existing on the effective date of this division" (California Public Resources Code, Section 27401, 1974).
17. Healy *et al.*, *Protecting the Golden Shore*, pp. 71–75.
18. The act and commission regulations gave authority to bring an appeal to anyone who had participated in person or in writing at the regional meeting. In part because of this extremely liberal provision for legal standing, over 60% of appeals were brought by *non*applicants. Moreover, the costs of appeal were generally minimal: the time required to complete the forms, plus time and travel expenses to one (or more) commission meetings in San Francisco or Los Angeles. (The latter could sometimes be onerous, particularly from the North Coast where plane connections were poor.) Lawyers were certainly not required and were often not employed by appellants. Commission staff we interviewed in the North Coast and North Central regions could not think of any important permits that had not been appealed, while state commission permit staff estimated that at least 80% of all important regional decisions had be appealed.
19. The basic problems were, first, that specific conditions often related to several issues and, second, that the categories of conditions used in the content analysis of commission files were based upon the suggestions of commission staff rather than upon an explicit linkage with statutory directives. For example, conditions relating to erosion or water quality controls were often designed in part to protect aquatic habitats but were not coded here as "relevant" conditons because they often related to quite different issues. Similarly, some of the categories of conditions—for example, "make permit subject to actions of other agencies" or "require studies prior

to construction"—were essentially procedural and thus could not be related to substantive issues. Nevertheless, they were sometimes used for habitat protection and/or public access. For a somewhat similar analysis of the conformity of local zoning decisions with plan policies, see Daniel Mandelker, *The Zoning Dilemma* (Indianapolis: Bobbs-Merrill, 1971), chap. 4.

20. California Department of Fish and Game, *The Natural Resources of Humboldt Bay*, Coastal Wetlands Series #16 (Sacramento: State Printing Office, December 1973).
21. The cases were selected by starting from a list provided us by Fish and Game and then by adding a few cases from our own review of the regional commissions' minutes during the entire period.
22. Fish and Game, *Natural Resources of Humboldt Bay*.
23. *North Coast Regional Commission, Minutes*, March 13 and April 16, 1975; State Commission, *Minutes*, May 13 and June 17, 1975.
24. The Harbor District argued (1) that the restaurant was needed to provide revenues to help finance the project and (2) that any restriction to commercial fishing boats would violate the terms of a grant from the California Department of Navigation and Ocean Development (North Coast Regional Commission, *Minutes*, April 12 and June 10, 1976).
25. North Coast Regional Commission, *Minutes*, December 8, 1976.
26. In addition, the Arcata city attorney represented one of the applicants (Atopak). This discussion is based upon the North Coast's *Minutes* of April 11 and 27, and May 10 and 24, 1973, as well as interviews with several commissioners.
27. The Sierra Club probably did not appeal because Chapter 8 of the 1976 Coastal Act (which would go into effect the next month) permitted visitor facilities and recreational boating berths in harbors. Both applicants subsequently submitted claims of exemption from the permit provisions of the Coastal Initiative on the grounds that substantial monies had been expended prior to the establishment of the commissions. In a repeat of the actual permit case, the regional commission voted 7–5 to grant the exemptions but was then reversed on appeal to the state commission. The latter's decision was subsequently upheld by the courts in the *Avco* case. See State Commission, *Minutes*, October 3, 1973, January 23, 1974, and August 31, 1976, pp. 6–7.
28. North Coast Regional Commission, *Minutes*, May 8 and July 10, 1975; State Commission, *Minutes*, August 19 and September 3, 1975.
29. This question was raised at the state commission meeting, and Bodovitz said he doubted that public money should be spent to acquire it:

> He said courts in Wisconsin and other places have held strongly that a landowner does not have an absolute and unlimited right to take land in its natural character and convert it to something else; the buyer of marshland knows exactly what he bought, and there is no automatic legal requirement that the owner of marshland be compensated if he is not allowed to change the character of the marsh. Mr. Bodovitz said there are about 600 acres left of the productive saltmarsh around Humboldt Bay, much of it in parcels of this sort. Viewed one at a time, it is easy to have sympathy with the landowner, but looked at from the perspective of the Coastal Act requirements, it would be inconsistent with the Act to allow all that marshland to be lost, and a precedent would be set by any approval.

(State Commission, *Minutes*, September 2–3, 1975, p. 10). Walsh did file an inverse condemnation suit but lost in state district court.
30. State Commission, *Minutes*, July 6, 1976.
31. Telegram of 4 October 1976, from the applicant in the files of the state commission.
32. Interviews with Gary Monroe, California Department of Fish and Game, 20 July 1970, and Lucille Vinyard, Redwood Chapter of the Sierra Club, 21 July 1978.
33. These figures on anticipated buildout were provided by Bill Rand, manager of the Sea Ranch Association, to Michael Fischer, executive director of the North Central Regional Commission, in a series of letters in September 1973.

34. For example, Sea Ranch restrictions on abalone catch were considerably more restrictive than those of the California Department of Fish and Game. Anyone who has stayed at the Sea Ranch has to be impressed by the environmental sensitivity of the vast majority of its residents. Respect for wildlife and for the natural beauty of the Northern Coast occupied a prominent role in the developer's advertising.

35. Report on Sea Ranch by the North Central Staff, October 4, 1973.

36. In fact, as indicated in chapter 2, one of the local organizations (COAAST) that grew out of the Sea Ranch controversy was one of the principal components of the Coastal Alliance.

37. For most of the 30 miles between Jenner and Sea Ranch, Highway 1 was a narrow, winding two-lane road—much of it along spectacular cliffs emerging from the ocean. Traffic therefore tended to congregate behind the slowest vehicles. For the staff analysis, see David Dubbink (chief planner on the North Central staff), "Memo to North Central Commissioners on Sea Ranch Traffic," February 20, 1975; Dubbink's methodology and conclusions were subsequently endorsed by the state highway department (letter of 6 March 1974).

38. Letter from Joseph Picchi to State Senator Randolph Collier, 29 April 1974.

39. Recall that a local group (COAAST) had been formed in 1968 expressly to fight Sea Ranch; the group had waged a successful 3-year fight to persuade the legislature to pass the 1970 Dunlap Act requiring "reasonable" public access in all future coastal subdivisions; and its founder, Bill Kortum, was the first president of the California Coastal Alliance.

40. These options are based largely upon a letter from Michael Fischer to State Senator George Moscone, 21 August 1975.

41. While the North Coast Region preferred to deal with aggregate impacts in the planning process, the state commission essentially forced it to develop overall conditions for Irish Beach and other subdivisions (State Commission *Minutes*, May 15, 1974, pp. 3–6, June 19, 1974, pp. 6–8).

42. Of the approximately 1700 votes in the Sea Ranch Association (1 for each lot), approximately 300 were held by people with houses, 100–200 were unsold lots held by Oceanic California, and the remaining 1200 or so were held by people who had purchased lots not yet built. On the basis of past buildout of around 60 houses/year, only about 200 of the last category would be expected to build during the 1973–1976 period. Thus, the majority of votes in the association were actually held by people who expected to build sometime in the future and whose primary concern was that their development rights not be "abrogated" by the coastal commissions. It was for this reason that the question of ultimate buildout was so crucial to the negotiations between the association and the commissions.

43. The "hostage approach" to obtain access easements was first applied by the North Central Commission in the *Smith* case involving Seadrift subdivision (North Central, *Minutes*, June 7, 1973), although the basic strategy of developing overall conditions was fully articulated for the first time a few months later with respect to the Ocean Marin subdivision (North Central, *Minutes*, September 20, 1973, and March 21, 1974). Subsequently, overall conditions were developed for Seaview Highlands, Serano del Mar, Timber Cove, Sea Ranch, Bodega Harbor, and Seadrift. The figures on permit activity at Sea Ranch are based upon the Rand letter to Fischer of 18 September 1973, as well as a systematic reading of the North Central *Minutes*.

44. North Central, *Minutes*, October 19, and November 5, 1973, as well as a letter from Celia Von der Muhll of the Sierra Club to North Central Commissioners, 15 November 1973.

45. North Central, *Minutes*, November 15, 1973, p. 5; also North Central, *Minutes*, October 19, 1973.

46. North Central, *Minutes*, February 21, 1974; letters of 21 January and 31 January 1974, from Reverdy Johnson of the Sea Ranch Association (SRA) to Michael Fischer presenting a detailed proposal; letter of 19 February 1974, from Deputy Attorney General William Chamberlain to the North Central Commission outlining the attorney general's criticisms of the SRA's proposals; and memo of 20 February 1974, from Dubbink to the North Central commission on Sea Ranch traffic.

47. North Central, *Minutes*, April 4, 1974, pp. 8–18; and North Central Regional Commission, "Overall Conditions for the Review of Development Permits at Sea Ranch," April 15, 1974.

48. The state commission retained the access and tree-trimming conditions previously adopted by the North Central region, modified the septic condition to require dual leachfields, substituted an *ad hoc* review of site and bulk restrictions for the detailed specifications adopted by the regional commission, and added a condition dealing with monitoring of water supply from the Gualala River. The major change, however, was the addition of the $1500 "environmental deposit." See State Commission, *Minutes*, May 15 and June 19, 1974, as well as the state staff recommendation pursuant to the June 12 meeting.

49. Robert Jones, "Community Feud Dramatizes Shore Issues," *Los Angeles Times*, 20, November 1977.

50. See letter of 12 June 1974, from John Marchant of the Sea Ranch Association to Joe Bodovitz, executive director of the state commission; Robert Jones, "Community Feud Dramatizes Shore Issue," *Los Angeles Times*, 20, November 1977; and memo from Bob Brown (executive director) to North Central Commissioners, "Summary Report on the Sea Ranch," March 31, 1977.

51. State Commission, *Minutes*, June 12, 1974; memo from Michael Fischer to North Central Commissioners, "Sea Ranch Environmental Deposit," November 15, 1974. The Fischer memo concluded that the septic tank and water supply conditions were well within the authority of the regional water quality control board; the tree-trimming program would require new legislation and the cooperation of Caltrans; and the acquisition of the two access easements would require specific legislative authorization and the cooperation of the State Parks Department and the State Public Works Board. It also estimated the total expense of $192,000 (plus $21,000 in annual operating costs), which could be met by $1500 charges on 129 houses (plus 14 houses for annual operating costs). For the continuing skepticism of the regional commission concerning the effectiveness of the $1500 deposit, see North Central, *Minutes*, December 12, 1974, pp. 7–10.

52. See memo from Deputy Attorney General Donatas Januta to Michael Fischer, "NRDC vs. California Coastal Commission," February 23, 1976; and memo from Januta to David Dubbink, "The Sea Ranch Public Access Upheld," *Santa Rosa Press Democrat*, May 12, 1978.

53. The intense frustrations and perceived inequities of the hostage strategy are well expressed in the letter of 16 July 1977, written by James White (a Sea Ranch lot owner) to the state commission:

> We own a lot at the Sea Ranch which we purchased in 1971 as a residential lot. We do not now see any likelihood of living there and for some time have been wanting to sell it, only to find that there is little market because of the uncertainties regarding building permits.
>
> We do not appreciate how this situation can be allowed to continue! It is our understanding that the Sea Ranch subdivision met all legal requirements. . . . Lots were sold as approved buildable lots. . . . Now, suddenly, the clock is apparently supposed to be turned back and a new start made.
>
> In the case of our particular lot, moreover, we feel that we are the victims of a quarrel that really shouldn't involve us. Our lot is east of the highway, back among the trees. Building on it does not block any view and the building would be well-hidden by the trees. What earthly reason is there for casting any question on the rights of its owner to use the lot for establishing a residence, as originally approved when the Sea Ranch was set up?

See also the comments of Joe Bodovitz in the State Commission, *Minutes*, June 19, 1974, pp. 7–8; and the column by Margaret Azevedo (former chair of the North Central Commission) in the *San Rafael Independent Journal*, March 24, 1979.

54. This assumes a rather conservative price of $40,000 per lot for the 600 lots.

55. Memo from Michael Fischer to Joseph Bodovitz, "1974 Annual Report of North Central Regional Commission," pp. 5–8.
56. Security Pacific Bank, *Area Profile*, pp. 2–5. In contrast, only about 1.2 million people used the Sonoma County beaches over an entire *year* (Dubbink memo on Sea Ranch traffic).
57. California Coastal Commission, South Coast Region, "Coastal Access Study," June 1978, pp. 1–2.
58. Robert Warren, Louis Weschler, and Mark Rosentraub, "Local–Regional Interaction in the Development of Coastal Land-Use Policies: A Case Study of Metropolitan Los Angeles," *Coastal Zone Management Journal* 4 (1977), pp. 331–362.
59. State Commission, *Minutes*, October 17, 1973, and November 28, 1973.
60. State Commission, *Minutes*, December 8, 1974.
61. Conversation with Mel Carpenter, executive director of South Coast Commission, June 1979, and State Commission, *Minutes*, February 4, 1975, pp. 16, 28, and March 5, 1975, p. 20. In addition, Warren *et al.*, indicates that 246 single-family residences (many of them presumably west of the highway) were approved by the regional commission in Malibu during 1973–1974; Warren, Weschler, and Rosentraub, "Local–Regional Interaction in the Development of Coastal Land-Use Policies," p. 352.
62. California Coastal Commission, "Coastal Access Study," pp. 1–2.
63. State Commission, *Minutes*, May 16, 1973; March 5 and March 26, 1975.
64. The appeal numbers and the state commission decision dates were as follows: (1) 93-73 (July 18, 1973), (2) 183-74 (September 4, 1974), (3) 201-74 (September 18, 1974), (4) 261-74 (December 4, 1974), and (5) 20-75 (February 18, 1975). For a discussion of the changes in local ordinances, see Warren *et al.*, "Local Regional Interaction," pp. 355–356; also interview with Mel Carpenter, 10 June 1975.
65. For somewhat similar models, see Richard Hofferbert, *The Study of Public Policy* (Indianapolis: Bobbs-Merrill, 1974), chap. 4–7; Daniel Katz and Robert Kahn, *The Social Psychology of Organizations* (New York: Wiley, 1966), chap. 2; Heinz Eulau and Kenneth Prewitt, *Labyrinths of Democracy* (Indianapolis: Bobbs-Merrill, 1973), part IV; and Daniel Mazmanian and Paul Sabatier, "A Multivariate Model of Public Policy-Making," *American Journal of Political Science*, 24 (August 1980), pp 439–468.
66. The denial rate varied from 1% to 40%. In order to normalize this skewed distribution, an arcsine transformation was applied to the dependent variable [J. Cohen and P. Cohen, *Applied Multiple Regression/Correlation Analysis for the Behavioral Sciences* (New York: Wiley, 1975), p. 256]. Nevertheless, this had very little effect on either the correlation coefficients or the R^2. For a more extended discussion of this entire analysis, see Mazmanian and Sabatier, "A Multivariate Model."
67. In the case of county supervisors, "constituency" was defined as the county because it was the county board of supervisors—not the voters of a specific supervisorial district—that selected the representative to the regional commissions and because little demographic information is available by supervisorial districts. For municipal representatives, it was the city from which they were actually elected; in some cases, e.g., Los Angeles and Long Beach, it was the city council that selected its representatives, while in other places it was the association of city officials, "constituency" was defined as the entire region on the assumption that most public members saw themselves as representing the broader "regional interest" (albeit with no direct accountability).
68. Selection of these need/resource variables began with an extended list of traditional demographic indicators, several measures of recreational demand, and a variety of indices of economic structure and well-being. Unfortunately, unemployment (either aggregate or in the construction industry) was not available for muncipalities and thus could not be included. Each indicator was then correlated with Commissioners' Denial Rate. The five with the highest correlations (the

three eventually selected plus median income and per capita income) were then subjected to a stepwise regression. The three selected all added at least 5% to the R^2.

69. Here, again, a variety of measures of liberalism, environmental concern, party identification, and perceptions of coastal-related problems were subjected to a stepwise regression. The Environmental Protection Scale and the Localism/Market Solutions Scale each had a simple correlation of about .70, while the Coastal Degradation Scale was slightly lower (and still added over 5% to the variance accounted for). Since the first two were highly intercorrelated, the Environmental Protection Scale was selected because it had no missing cases. For a more extended discussion, see Mazmanian and Sabatier, "The Integration of Five Approaches to Decision-Making." The items on the Environmental Protection Scale were presented previously in chapter 4. The items on the Coastal Degradation Scale were all selected from our survey questionnaire administered to commissioners and staff in 1974–1975 in which they were asked to rate the seriousness of a variety of coastal-related issues on a 7-point Likert scale from Very Serious Problem (7) to Not at All a Problem (1). They were then factor-analyzed, with four factors emerging. Following are the items (and factor loadings) that loaded at least .40 on the Coastal Degradation Factor: The destruction of the scenic resources of the coast (.77); the shortage of public access to the beach (.68); the gradual growth of commercial strips and other facets of urbanization (.68); the destruction of intertidal and beach organisms (.66); the construction of ugly buildings and developments (.63); air pollution (.50); the loss of a sense of community associated with small towns and groups of like-minded people (.45). Other factors dealt with housing and unemployment, logging and limited industrial bases, the intrusion of outsiders, and camping. For a more extended discussion, see Mazmanian and Sabatier, "The Integration of Five Approaches to Decision-Making."

70. The difference in denial rates for each commissioner ranged from a high of 80% to a low of 7%, with an overall mean of 40%.

71. The difference in denial rate for each commissioner ranged from a high of 34% to a low of 1%, with an overall mean of 17%. Because of the skewed nature of the distribution, an arcsine transformation was applied to more closely approximately a normalized distribution.

72. Following are the results of this simplified model:

	Simple r	Beta	R^2 added	Total R^2
1970 community population	.65	.44	.42	.42
Coastal Degradation Scale	.69	.39	.33	.75
Response to outsiders	.76	.43	.13	.88

73. The path analysis was based upon a hierarchical regression in which a single indicator was selected for each variable in the model, with 1970 population entered first and the remaining variables entered together. This revealed a simple r between 1970 Community Population and Commissioner's Denial Rate of .65, with a beta of .38. Thus, .38/.65, or 58%, of Community Population or Denial Rate was direct rather than indirect.

74. The study that comes closest (although not intentially) to a test of our model is Lettie Wenner's analysis of the factors affecting the stringency of water pollution control regulations (not permit decisions) in 50 states (Lettie Wenner, "Enforcement of Water Pollution Control Law," *Law and Society Review*, May 1972, pp. 481–507). Wenner's model contained indicators of what we would term (1) community needs and resources (e.g., water quality), (2) a variety of statutory charcteristics, including appointment criteria, and (3) two general measures of policy outputs—

water quality standards and enforcement effort. She did not, however, look at agency permit decisions, nor did she examine the attitudes of agency officials or the participation of interest groups in agency policy making (although she did attempt to measure their "strength" within each state). For other quantitative studies that have dealt with several variables in the model, see Hofferbert, *The Study of Public Policy*; Eulau and Prewitt, *The Labyrinths of Democracy*; L. Harmon Zeigler and M. Kent Jennings, *Governing American Schools* (North Scituate, Mass.: Duxbury, 1974); Harrell Rodgers and Charles Bullock, *Coercion to Compliance* (Lexington, Mass.: Lexington Books, 1976); George Downs, *Bureaucracy, Innovation, and Public Policy* (Lexington, Mass.: Lexington Books, 1976); and Nelson Rosenbaum, "Statutory Stringency and Policy Implementation: The Case of Wetlands Regulation," in *Effective Policy Implementation*, ed. Daniel Mazmanian and Paul Sabatier, (Lexington, Mass.: D. C. Heath, 1981).

6

IMPACTS—REGULATION TO WHAT END?

As shown in Chapter 5, throughout their 4 years the commissions reviewed thousands of development proposals for everything from the refurbishing of existing dwellings to substantial new commercial complexes and residential subdivisions. Although the overall percent of permit denial was small, for most projects that were deemed to have a potentially adverse affect on wetlands, natural habitat, scenic vistas, or beach access the commissions used their power to impose modifications on the project (through the mechanism of specifying conditions) and in some instances resorted to outright denials. In terms of the framework of analysis presented in Chapter 1, the permit decisions constituted a major part of the "policy outputs" (stage one) of the implementation process.

The fact that the commissions made a wide range of decisions under their permit authority does not mean that they would be automatically translated into action or that they would have the intended effects. There are several factors bearing on the translation of agency decisions into actual impacts. A major one is the extent of compliance with agency decisions by target groups (permit applicants)—stage two of our framework. It may be insignificant or even irrelevant that the commissions voted to restrict or place conditions on development to protect wetlands and open space if applicants ignored the commissions' rulings. As important, it may be irrelevant or insignificant that permit applicants complied with the commissions' rulings if a large enough number of property owners and developers simply evaded the permit process altogether. A critical link between agency decision and the impacts of those decisions, therefore, is the degree of compliance by those being regulated.

Even if all those regulated fully complied, desired impact(s) (those prescribed under the act and by the commissions) still might not be realized, depending on the comprehensiveness and validity of the theory underpinning the regulatory rulings. This is an important *caveat*. For example, the Coastal Act assumed (had as an implicit theory) that greater public access to the beach could be achieved by requiring dedications of access or access easements as a condition of permit approval. In practice, this was seldom sufficient to assure access as it ignored the problems of site improvement, maintenance, clear marking of access points, and liability. In short, the police power authority provided under the act was insufficient to adequately accomplish the objective.

Finally, any attempt to assess the distinct contribution of a regulatory agency such as the coastal commissions in bringing about changes in the behavior of persons or situations over which it does not have direct control is invariably complicated by the fact that whatever changes do come about may partially be the result of other causes. To the extent this is a possibility, assessment of impacts requires estimating the distinct contribution made by the regulatory agency to the observed change. The need to sort out multiple causes is vividly illustrated by the situation that developed early in the life of commissions when representatives of the construction industry repeatedly charged that the commissions were "causing" the dramatic decline in construction throughout the state and the widespread unemployment in the industry in the 1973–1974 period. But were the commissions really at fault? While they may have contributed by adding to the time and costs of winning project approval from public authorities and by denying some projects, the major cause was probably the substantial rise in interest rates and general downturn in the economy that, unfortunately for the commission, coincided with their establishment.

With these points in mind, we shall first consider the extent of compliance with the commissions' general permit process as well as the specific conditions that were mandated. We then turn to the actual impacts of the commissions in three areas: (1) the provision of greater recreational opportunities through opening accessways and expanding coastal parks and recreation facilities; (2) the protection of wetlands; and (3) the adverse effect of the commissions on the housing market by driving up the cost of land, new construction, and the purchase price of existing units. Each of the three impacts was chosen either because it was a major objective under the act (wetlands protection, access), or because it became a major focus of the commissions during the implementation process and was highlighted in the coastal plan (acquisition), or, in the case of housing costs, because it was one of the more contentious issues throughout the life of the commissions and served to bolster the positions of those attempting to reduce the powers of, if not eliminate, the agency. As will be seen through these examples, the path from the commissions' permit decisions to their actual impacts was a long and complicated one.

I. Compliance with the Coastal Act

In general, most citizens appear to be law-abiding. However, some people fail to comply with specific laws with which they strongly disagree, and many cheat "just a little" even when in agreement with a law's basic intent. Examples of the latter are especially numerous: exceeding posted speed limits, taking more than the permitted fishing catch, and probably most common of all, cheating on income taxes.[1] Students of the criminal justice system and school desegregation have conducted extensive research on why people either do or do not comply with the law, and have concluded that several factors contribute, ranging from the clarity of a law to enforcement efforts to personal values. Only recently has research moved beyond compliance with criminal codes or nationally controversial issues like desegregation to the kinds of compliance issues raised by the myriad state and local regulatory rules being promulgated by agencies such as the coastal commissions.

Nelson Rosenbaum argues that the extent of compliance in this area is a product of four basic factors. First is the extent to which citizens know that their activities fall under a given law. In the area of criminal justice, ignorance is rarely a source of noncompliance. But in the regulatory area, particularly when dealing with new laws at the state and local level that affect a large number of persons, a significant portion of the citizenry may simply not be aware of the law.

In the case of the Coastal Act, awareness of the law, at least of its permit requirement, was not a serious problem. The act received extensive attention in the media and through the Initiative campaign. Also, the application for a permit from the commissions came at the end-point of a series of local, county, and other agency permits that virtually all development activities required. Although the other permitting agencies were not obliged to alert applicants of the additional coastal commission permit requirement, most did so.

Second on Rosenbaum's list is the "benefit-cost" analysis of noncompliance conducted by each potential violator. Most students of compliance believe, even if on the basis of somewhat crude estimates, that before failing to comply most people will consider the monetary benefits and costs of their actions.[2] In the case of land use regulation of interest to Rosenbaum, and relevant to the coastal commissions, the benefits of noncompliance would be, for example, the expenses saved by not completing the permit application, by avoiding the inevitable delays in construction caused by the regulatory process, and by modifying the project to satisfy the regulators. For commercial developments in particular, such costs if incurred might be fatal to their economic viability. The costs of noncompliance, of course, would be whatever monetary and other real penalties could be imposed if the noncompliance were detected, prosecuted, and convicted.

Under the Coastal Act the cost of noncompliance could be severe. The penalties that could be imposed were a fine up to $10,000 and a jail sentence up to 2 years, or both, for each conviction of a violation under any provision of the act. In addition to any other penalties, the actual execution of any development in violation of the act was subject to a fine of up to $500 per day for as long as the violation persisted.[3]

In analyses of the criminal justice system a good deal of attention is given to a third factor, the social stigma associated with conviction for noncompliance. Obviously a stigma is attached to a conviction for rape, murder, and similar high crimes. Rosenbaum views this factor as less important in the noncriminal area of regulatory law. In the latter case, public sympathies are less emotionally charged, enforcement efforts are usually low-keyed, penalties are usually monetary, and convictions are rarely publicized. However, in context of the Coastal Act, the fear of social stigma may have been an important factor in contributing to compliance, particularly with respect to most of the large developers and businessmen who had both "goodwill" and professional reputations to maintain. It would also be embarrassing for the numerous governmental actors who sought permits from the commissions to be found in noncompliance.

Finally, the single most important factor identified by Rosenbaum is the climate of enforcement—that is, the likelihood of detection (as perceived by those subject to regulation) as well as the certainty of prosecution and penalization if detected. In certain respects the commissions were at a distinct disadvantage in both areas. They had no enforcement branch or officers, no formal follow-up inspections, and variable cooperation from the county building inspectors who routinely conducted inspections. Moreover, the commission had to rely on the attorney general's office for prosecution of any violations that might be detected. As we shall see, however, detection of any major case of noncompliance with the permit process was quite likely, and with permit conditions it was reasonably likely, though prosecution for violation was more problematic.

A. Gauging Noncompliance

Short of an exhaustive comparison of the status of every parcel of coastal land before enactment of the Coastal Act in 1972 and at its conclusion at the end of 1976, it is obviously impossible to gauge with certainty the extent of compliance with the commissions' permit process. A similarly diligent effort would be required to inspect every approved permit to determine the extent of compliance with any conditions that may have been attached. Nevertheless, three sources of information are available that, while not nearly as exhaustive as we might like, suggest that the climate of enforcement was weighted heavily in favor of compliance. The first is material collected by Nelson Rosenbaum on the extent of noncompliance in each of the six commission regions. The data were collected through interviews with each region's executive director in the latter part of 1977.[4] The second source is our own research and interviews with many persons within and outside the commissions in the North Coast, North Central, and South Coast regions. The third is the material recently compiled by the successor commissions on the extent of noncompliance, or more accurately incomplete compliance, by permit applicants who made assurances to the commissions that public access dedications would be made as a condition of their permit approval.

The Rosenbaum data are presented in Table 6-1. In three of the four regions where annual figures are available—the North Central, Central, and South Coast—it would appear that target groups quickly learned to comply with the act. Noncompliance ran at approximately 4%. The North Coast region obviously presents an exception, where noncompliance was as high as 9%. Overall, however, the level of noncompliance with the application process across all the regions over the 4-year period was perceived to be limited (2–3%). Noncompliance with permit conditions was even less of a problem (1%). In terms of the actual number of cases, the total of both types of noncompliance over the 4 years and six commissions was approximately 900, which stands in contrast to the more than 24,000 permit applicants who petitioned the commissions and apparently complied with their decisions.[5]

TABLE 6-1

NONCOMPLIANCE WITH PERMIT PROVISIONS OF THE COASTAL ACT
(DETECTED CASES)[a]

	North Coast		North Central		Central		South Central[b]		South Coast		San Diego[c]	
	%	N	%	N	%	N	%	N	%	N	%	N
A. Noncompliance with permit application process, by region and year												
1973	1	(6)	3	(10)	3	(25)	—		7	(154)	—	
1974	7	(24)	6	(8)	3	(19)	—		1	(28)	—	
1975	9	(36)	4	(6)	2	(15)	—		3	(52)	—	
1976		(24)		(5)		(20)				(39)		(25)
Total		(90)		(29)		(79)		(100)		(273)		(25–125)
Grand total N = 596–696												
% = 2–3[d]												
B. Noncompliance with development conditions, by region and year												
1973	0	(0)	0	(1)	1	(10)	—		0	(0)	—	
1974	0	(1)	0	(0)	3	(15)	—		0	(4)	—	
1975	1	(3)	1	(2)	2	(12)	—		0	(8)	—	
1976		(5)		(1)		(18)			0	(2)		(10)
Total		(9)		(4)		(55)		(100)	0	(14)		(10–50)
Grand total N = 192–232												
% = 1												

[a]Source: Nelson Rosenbaum, *Environmental Law Enforcement: A Case of Wetland Regulation* (Washington, D.C.: Urban Institute, in preparation).
[b]Annual records were not kept, the total shown is the executive directors' best estimate.
[c]Annual data missing except for 1976.
[d]Percent computed on basis of 25,000 permits for the 4-year period.

Reports by the regional executive directors naturally may be suspect. It might not look very good if, on the basis of their own records, a significant amount of noncompliance could be shown to exist. On the other hand, exaggerating the extent of noncompliance might be a way of generating new money and staff. We have no reason, however, to doubt the reporting of the executive directors. Furthermore, the results of Rosenbaum's survey are corroborated by our independent discussions with persons equally familiar with the coastline from other state agencies and in local government. The general consensus is that for virtually all major projects there was compliance with the permit process and probably with the conditions imposed as well. Where noncompliance did exist, it was usually on smaller developments that most often involved remodeling or additions that were difficult to monitor and trivial in terms of their impacts on coastal resources: the very kinds of evasions typical of normal city and county building permit requirements.

The three cases of more substantial violations that we uncovered appear to be exceptions that prove the rule. One involved residential construction at Whale Gulch, in northern Mendocino County, where there was extensive violation of both coastal commission and county regulations.[6] The violations were known to exist but, due to the sensitive nature of the community involved (an alternative life-style commune) and proposed changes in state law to deal with such communities, the matter was never pursued by either county or coastal commission authorities. A second violation was at Bolinas, in Marin County, where a significant amount of illicit building and conversion had taken place prior to the establishment of the commissions and continued unabated throughout the life of the commissions, resulting in a substantial increase in the community's 1000-member population. Yet virtually no permits were applied for. This became known as the "10 by 10" problem. It involved widespread conversion of garages, tool sheds, and other small structures into residential units. The violations were difficult to deal with because the county did not require building permits for structures less than 10 feet by 10 feet, the local population was hostile to outside interference, the conversions were often hard to detect, and the direct impact on coastal resources was believed to be minimal.[7]

The third case involved the refusal by the Army Corps of Engineers' Los Angeles District to recognize the permit authority of the commission when it came to its harbor dredging activities. The corps asserted its authority under federal statute to dredge and maintain all ports and harbors, new and existing. The only instance where this prerogative was exercised, however, was the dredging of the San Diego harbor.[8] The commissions did not formally challenge the corps, and the issue never surfaced in the media or even widely among commission circles.

The major area of noncompliance under the 1972 act was overlooked in both Rosenbaum's and the majority of our interviews because it did not surface until 1978. It concerned the failure of many permit holders to follow through with the formal recordings of an access dedication with the proper county record-er and notification of the commission of such. The full extent of the problem was not investigated by the commission staff until 1978. Due to the time that has elapsed since issuance of the permits under the 1972 act and the questionable wording (legally speaking) of many, enforceability of many unrecorded dedica-tions under the 1972 act was dubious.[9] The problem arose, at least in part, because under the 1972 act (and, until recently, under the 1976 act) permits were routinely issued following the vote of the commission. The permit holder could therefore proceed with construction without recording his or her dedica-tion with the county recorder, and nobody would know.[10] The issue of non-compliance with access conditions did not surface until 1978 because it was only then that the state commission staff began an inventory of all access conditions. In the process it was discovered that few systematic records had been kept by either the regional or the state commissions on access, and little monitoring of compliance with the dedications had been done.

B. Accounting for Compliance

Why was the rate of compliance so high overall, on the one hand, yet so problematic when it came to the access dedication condition? As for the gener-ally high level of compliance with the permit process, we have already men-tioned the extensive publicity given the new law, the stiff penalties that could be imposed, and the stigma attached to noncompliance. But the crucial factor seems to have been the climate of enforcement. All new major building or dredging or even small nonstructural development (such as small farm dikes) were quite visible: visible to concerned neighbors, to environmental groups, to personnel from the Fish and Game Department, and to the commissions' permit staff during their frequent field visits.

Second, almost all of the early violators detected were required to obtain a permit, although many had almost completed their development by the time the commission staff could approach them. Most pleaded ignorance of the law and were not fined. In several cases of known willful violation, however, the attorney general brought suit. Although out-of-court settlements of only a few hundred dollars were normally reached,[11] the deterrent effect of a possible court action seemed to be a credible one.

In general, the relatively high degree of compliance with the Coastal Act appears to be attributable in part to the low direct cost to the commissions for their own detection efforts (the permit staff were already in the field). Probably more important was the active monitoring by the commissions' supportive con-stituency, especially the large number of environmental groups.

The case of incomplete compliance with respect to access dedication provisions highlights the importance of visibility (or, as in this instance, the lack of visibility) of noncompliance. Keeping track of the documents that were required to make the promised dedications enforceable was not a task that could be carried out as part of the routine activities of the commissions' permit staff, nor would it likely be undertaken by citizens.

With the exception of access conditions, noncompliance was apparently kept to a minimum. One might thus infer that the permit decisions discussed in Chapter 5 had their intended effects on coastal resources. Yet the issue is not quite that clear-cut. Thus, we turn to a closer inspection of the commissions' impacts in three specific areas.

II. ACCESS AND ACQUISITIONS

Article X, Section 4 of the California Constitution states unequivocally that access to the state's shoreline and adjacent waters shall be provided the public:

> No individual, partnership, or corporation, claiming or possessing the frontage or tidal lands of a harbor, bay, inlet, estuary, or other navigable water in this state, shall be permitted to exclude the right-of-way to such water whenever it is required for public purpose . . . and the Legislature shall enact such laws as will give the most liberal construction to this provision so that access to the navigable waters of this state shall always be attainable for the people thereof.

Despite this provision and the many public parks along the coast, the need for even greater access, especially in urban areas, was a major impetus to the passage of the Coastal Initiative. As noted in Chapter 2, the combined pressures of an ever-increasing population, the mismatch between substantial access in the northern counties and the population in the south, and the walling off of large stretches of the coast made the provision of access a central concern of the commissions.[12]

This is reflected in the permit process, as shown in Table 5-4, where one-sixth of the permit cases that went to public hearing in the North Central and South Coast commissions concerned public access. Moreover, ensuring greater access was ranked as third most important among 19 objectives of the Coastal Act by a cross section of nearly 500 public officials, interest groups, permit applicants, and citizens that we surveyed.[13] Yet, how much new access actually was provided? Equally important, how much could have been provided given the commissions' authority under the Coastal Act?

To begin with, the commissions exercised their permit authority to win numerous dedications of lateral accessway across beachfronts and a more limited number of vertical accessways from the nearest public road to the beach. Table 6-2 shows the number of dedications that were incorporated into approved permits in the North Coast, North Central, and South Coast regions. Obviously,

TABLE 6-2
ACCESS DEDICATIONS/EASEMENTS/AGREEMENTS

	North Coast[a]	North Central[a]	South Coast[b]
Lateral	—	1	111 (1973–1976)
Vertical	2	a. 8 on individual permits	12 (1973–mid-1978)
		b. 5 subdivision permits	

[a]From our file of all permits, 1973–1976.
[b]From "Coastal Access Study," California Coastal Commission, South Coast Region, June 1978.

the South Coast Commission made the greatest number of demands on permit applicants to provide beach access, particularly along the 27-mile Malibu coastline, followed by the North Central Commission and then the North Coast. The varying interest by commissions appears to have paralleled need. In Los Angeles and Orange counties much of the coastline is developed, and one of the few viable ways of creating more access to the beach short of acquiring the remaining undeveloped parcels is through permit conditions. In contrast, in the rural counties of the North Coast, where access is far less of a problem, the issue was seldom raised.[14]

Beyond the point of requiring access dedications the theory underlying the act fell short. A dedication of access could be required, but the commissions could not guarantee that the accessway would be opened. For lateral accessways this was not too serious a problem, at least under the 1972 act. The property owner, in effect, promised not to cordon off or develop the strip of land covered by the dedication (usually from 3 to 25 feet from the mean high tide) but little more. The accessway was simply left in its natural state, and questions of improvements, maintenance and liability costs were rarely at issue.[15] For the all-important vertical accessways, however, paths and sometimes roadways from Coastal Highway 1 (or nearst public road) to the beach were involved. These typically required improvements, and always maintenance and liability costs. Therefore, most vertical access dedications stipulated that the accessway would be made available for public use contingent upon a public or private agency assuming the expenses and legal responsibilities involved.

The coastal commissions could not assume such responsibilities or costs themselves, however, nor could they require others to do so. When they turned elsewhere their pleas generally fell on deaf ears. It would not be until an access-way bill was passed in 1979 that the commissions' powers and capabilities in this area were enhanced. And with only a couple of exceptions neither other state agencies, such as the California Department of Parks and Recreation, nor local governments were willing to accept responsibility for opening and maintaining commission-won accessways. The end result, of course, was that even after the heated battles over the commissions' insistence that access be provided, little materialized. For example, in the very congested Malibu area, only three vertical accessways were actually dedicated between 1973 and 1976, and of these, only one had been improved, clearly marked, and opened for use.[16]

The only other instances where vertical accessways were opened as a result of permit conditions imposed by the commissions were those incorporated into larger commercial or residential developments not requiring action by a local or state agency. For instance, the Hotel del Coronado (along the San Diego coastline) opened a well-marked accessway across the hotel property in an area of the coast where the public would otherwise have been denied entrance to the beach.

It may seem a bit unfair to fault the Coastal Act for not containing adequate mechanisms for connecting permits to the actual opening of accessways. Preserving options by denying some permits outright and placing access conditions on others obviously preserved options for future action. Yet the proponents of the act promised access, the voters of the state surely expected access, and the commissions demonstrated their awareness of this by trying hard within their limited capacity to encourage property owners, other government agencies, or anyone else to assume the dedications that they had won. This suggests that both the public and the commission viewed their responsibility as more than simply wringing easements and promises of dedication out of coastal property owners. The fact that so few of the critical vertical accessways were opened through 1976 points to a serious weakness in the assumption in the act that it was a simple step between permit conditions and actual impacts.[17]

A related problem surfaced with respect to the access to be provided through acquisitions for coastal parks, recreation, and wetlands protection. The commissions could deny development as a means of preserving options and did so on numerous occasions. But they could not by themselves act affirmatively. However, in this instance the issue was resolved in a more positive manner.

Caught in the dilemma of being expected to provide more acquisitions (somehow) but lacking the formal authority to purchase or take title to property, the commissions decided upon another course of action. As part of their planning functions (see Chapter 7), they generated considerable publicity about the need for greater acquisitions and developed criteria for prioritizing and ranking 154 prime acquisition sites.[18]

The coastal plan listed priorities as (1) lands best suited to serve the recreational needs of urban populations, (2) lands of significant environmental importance, such as habitat protection, and, of highest priority, (3) lands in either of the above categories proposed for development or use incompatible with their basic resource and recreational value.[19] The development of specific criteria and the emphasis on urban sites was unusual in the context of the traditional acquisitions approach used by the state legislature and Department of Parks and Recreation. Up to that time, Parks and Recreation did only limited prioritization and long-range planning for acquisitions, though coastal acquisitions in general had been evolving since the late 1960s as a high priority within the department.[20] The legislature, not surprisingly, traditionally treated park acquisitions as another pork barrel issue, with selections based as much, if not more, on the power and needs of legislators than Parks and Recreation recommendations.

It followed from the commissions' focus on urban area acquisition and sites of eminent development that the region of greatest need was the South Coast. Repeatedly throughout the hearings and discussions of 1975 and early 1976 it was asserted by the commissions that approximately one-half of whatever funds were authorized by a 1976 bond act should go to the urgent needs of Los Angeles and Orange counties.[21] For example, in the Malibu area of Los Angeles County a substantial number of smaller acquisitions were targeted to complement existing local, county, and state facilities as well as several large purchases, such as the 130 acres at Malibu bluffs and 72 acres at La Costa Beach West. The Malibu acquisitions, in total, amounted to approximately 420 acres of prime beach-front bluff and wetlands properties, at an estimated cost of over $10,000,000. Likewise, in Orange County, an incredibly expensive 1050 acres of the Irvine coast (estimated cost, $20 million) was targeted.[22]

Overall, the cost of 131 of the 154 sites on the commissions' acquisition list, based on available estimates, was $200 million. Due to the conservative nature of the original estimates, inflation, and the time lag between when the list was developed and when acquisition could actually take place, the cost of the 131 sites (as well as the 23 sites without cost estimates) would likely be much higher than suggested by the commissions.

While there was sentiment in the legislature for park acquisitions, it was not exclusively for coastal parks and wetlands, or for the specific sites identified by the commissions. Consequently, after extensive negotiations involving key legislators, the commissions, the governor, local governments, and environmental and development groups (in particular CCEEB), the Nejedly-Hart State, Urban and Coastal Bond Act of 1976 was passed and placed on the November ballot.[23] The bond act was for a total of $280 million but included only $120 million for coastal acquisitions, as part of a package involving $85 million to cities and counties for local recreation projects, $34 million for improvement of existing state parks, $15 million for wildlife management, and $26 million for inland water sports and camping areas.[24]

At a time when most government spending measures were going down to defeat at the polls in California and across the nation, the bond act passed by 51.1%. Interestingly, some of the areas it was intended to benefit most were not overly enthusiastic. It received only a 47% vote in both Los Angeles and Orange Counties. It was carried, essentially, by the traditional constituencies supportive of state parks and wetlands acquisitions in the San Francisco Bay area.

The bond act authorized the state to expend funds but did not include a list of specific acquisitions. Assembly Bill 924 of the following year spelled out the acquisitions for 1977–1978. The commissions' effort at prioritizing sites was not lost on the legislature. Of the 25 sites in the bill, 16 were from the commissions' March 1976 list of 154 sites. Of the 16, 5 were from the highest priority category I-A, 4 from category I-B, and 5 from II-A.[25] Clearly, the commissions' grand acquisition design was not being carried out in a comprehensive manner (and probably would not be in the foreseeable future). Yet an important step had been taken. Thus, for example, of the 420 acres designed by the commissions for purchase in the Malibu area, 104 acres, at Malibu bluff, were part of the AB 924 acquisitions.[26]

This brings us to the question of whether the acquisitions provided for in AB 924 (and others that would eventually flow from the 1976 bond act) should be viewed as an impact of the commissions. In a direct causal sense, probably not. The commissions began the acquisition process but had no authority to see it through to completion. Once the commissions had established criteria and identified prime acquisition targets, the initiative was passed to the legislature, the voters of the state, and then back to the legislature. While the commissions and their supportive constituency were important actors throughout, they could not be viewed as determinative. The question really comes down to how important were the commissions in bringing about the observed impact? The answer, we think, is that it is difficult to imagine the flurry of acquisition activities or the proposal and passage of the 1976 bond act in the absence of the commissions. Despite their lack of authority to implement their program of action, they went a long way toward bringing about its end. This is especially true with respect to the emphasis on urban areas sites and those prime for development.

III. WETLANDS

There are few coastal resources in greater need of protection from man's encroachment than the state's remaining wetlands. Remove the wetlands and the marine food chain may be irreparably damaged. Yet coastal estuaries and wetlands have been dredged for ports and marinas, subjected to sedimentation from upland erosion, filled in for development, used as sumps for urban sewage and industrial waste, and denied needed freshwater inflow through water diversions. The result has been a two-thirds reduction in the coastal wetland acreage since 1900, with two-thirds of that remaining in the San Francisco Bay. In the southern half of the state (from Morro Bay to the Mexican border) the reduction has been even more dramatic, to one-fourth the 1900 level, to a present estimated 31,700 acres, of which 18,600 are open water.[27]

Marine biologists and ecologists generally have been aware of the implications of encroachment of man on the wetlands. So was the state's Department of Fish and Game.[28] But these groups were never an especially well-organized or potent political force. The campaign for the Coastal Initiative brought the wetlands protection forces together, however, and gave the wetlands issue much needed visibility. Though many people saw the Initiative as a means of providing more access to the beach and shoreline, the respondents to our 1978 elite survey viewed as the number one objective of the act "preserving further reduction of coastal natural habitats such as wetlands, marshlands, and estuaries."[29] To prevent their further destruction, wetlands were listed as one of the six categories for which no permit could be approved without a two-thirds affirmative vote of the total commission membership.[30]

What did the commissions accomplish in their efforts to redress the encroachment on wetlands? To begin with, through the permit powers they imposed a virtual moratorium on all building into the wetlands and conditioned development along feeder streams and floodways. Though the actual number of permit applications involved were relatively few in number (see Table 5-4), the consistent denials or strict conditioning by the commissions clearly signaled their position. This was illustrated by the case study of Humboldt Bay discussed in Chapter 5.

A similar example is presented in the case of the Los Penasquitos Lagoon in San Diego County. The salt marsh area had already been under the management of the Department of Parks and Recreation for some time but the upland freshwater marsh and surrounding buffer zone was not. Under the 1972 act the commissions allowed no new development into or around the lagoon, within their 1000-yard permit authority. But they had no authority over proposed development (e.g., an industrial park) within the watershed further inland. The commissions' expanded authority under the 1976 Coastal Act enabled them to condition most new developments in the industrial park to protect the creek's floodway, with one outright denial.[31] Also, of great significance, purchase of the inland freshwater marsh portion of the lagoon along with a buffer zone was provided for in AB 924. The combination of the acquisitions program in and about the lagoon with restricting development through the permit process along the feeder streams was a major step in assuring the long-term protection of the lagoon.

The coordination between Fish and Game and the commissions on wetlands protection provides possibly the best example of continuing interagency cooperation by the commissions. What evolved, in effect, is that the commissions provided Fish and Game with a surrogate regulatory authority, and Fish and Game provided the commissions with expertise and a monitoring capability.[32]

The importance accorded wetland protection by the commissions was also reflected in other acquisition efforts. Of the commissions' 154 recommended purchases in 1976, 46 were for habitat wetlands. And of the 16 acquisitions on the commissions' list included in AB 924, 5 were from the wetlands category. In sum, the commissions effectively used their powers to halt the intrusion of man into the wetlands, cooperated with Fish and Game in doing so, and made every effort to bring about state purchase of major sites. As in the case of parks acquisition, it is difficult to imagine the attention given the issue in the absence of the commissions.

IV. HOUSING

One of the most controversial impacts of the commissions under the 1972 act was their effect on the cost of housing. Commission opponents contended that prices rose dramatically for coastal residential sites due to the commissions' restrictive policies. Moreover, they argued that the commissions had little regard for the desires of many moderate- or lower-income families to live along the coast, thus giving rise to charges of "environmentalists as elitist."

The rebuttal by commission defenders was, first, that the commissions' mandate called only for environmental protection and providing greater coastal access, without consideration of economic implications; second, that the commissions were instrumental in protecting existing neighborhoods of moderate-priced homes and preventing the conversion of rental apartments to condominiums (e.g., Venice and Santa Monica, respectively) in instances where local governments had failed; and third, that the commissions generally required provision of at least some moderate-priced homes and rentals in approving large multiple-family developments or subdivisions.

For all the hyperbole about housing costs, there has been an amazing lack of good empirical evidence.[33] Nevertheless, the logic behind the critique of the commissions is a fairly straightforward application of economic theory. If the commissions imposed new restrictions on and/or prevented development of sites in the coastal zone while demand for coastal sites continued to rise, several results should follow.

First, the value of undeveloped land, especially larger holdings, that came under the regulation of the commissions should have decreased in market value. Indeed, some evidence indicates that the assessed value of undeveloped parcels in the coastal permit zone were revised downward as a result of the uncertainties in their developable prospects—thus market value—under the commissions. In

Los Angeles County, for example, owners of undeveloped land in the permit zone were granted a 10% reduction in assessed valuation.[34] A comparison of the rise in value of undeveloped land in the permit area in the City of Ventura and inland several miles between 1971 and 1975 showed the rise inland to be 15% higher than along the coast. This suggests that, all else being equal, the commissions' presence "depressed" the market value, as reflected in assessed value, of undeveloped land in the permit area.[35]

Second, the price of developed sites, of which housing makes up the majority, should have risen in value. As demand continued to grow (presumably people were still eager to live along the coast) yet the supply of new houses was artificially depressed by the commissions' regulatory policies, the resale price of existing units would be bid up by buyers competing for the limited number of resale units. Stories of drastic escalations in resale value of homes along the coast abounded. But as shown in Chapter 3, during the commissions' existence a substantial (though lesser) rise also occurred inland, thus leading to the suspicion that something other than the commissions was behind the rise in prices on (and off) the coast. The appropriate question, then, is what proportion of the rise was due to the commissions and what proportion to other factors?

An initial attempt to determine the answer empirically was undertaken by economists examining undeveloped land and developed residential sites in the permit zone and inland in Ventura County.[36] They found, surprisingly, that the value of developed sites appreciated *less* rapidly along the coast than inland. Cautious use of these results is in order, however. The analysis was based on assessed value, not market or actual sale value. And the comparison was of assessed value of the land less any development, not the total assessed value of the land and improvements.

A more complete study of the effects of the commissions on housing under the 1972 act was conducted by Robert Kneisel.[37] He analyzed all resale housing prices for single-family dwellings in the 1000-yard coastal permit zone, a border zone approximately 800–900 yards wide just landward of the permit zone, and a sample of resales several miles inland from the coast in Los Angeles County. The trend in housing prices, as measured in quarterly mean prices, from June–August 1967 through September–November 1975, are plotted in Figure 6-1. The figure shows that throughout the entire period the average sale price of housing in the coastal permit zone was higher by several thousand dollars than in either the border area or inland. More important, in light of the debate over the commissions' effect on housing, although there was a precipitous increase in prices that coincided with the starting of the commissions in January of 1973 (the 23rd quarter in the series), prices were rising in all three areas, presumably due to more systematic factors such as inflation.

Nevertheless, it is important to gauge the precise contribution of the commissions to the rise in prices in the permit zone. To estimate the size of this

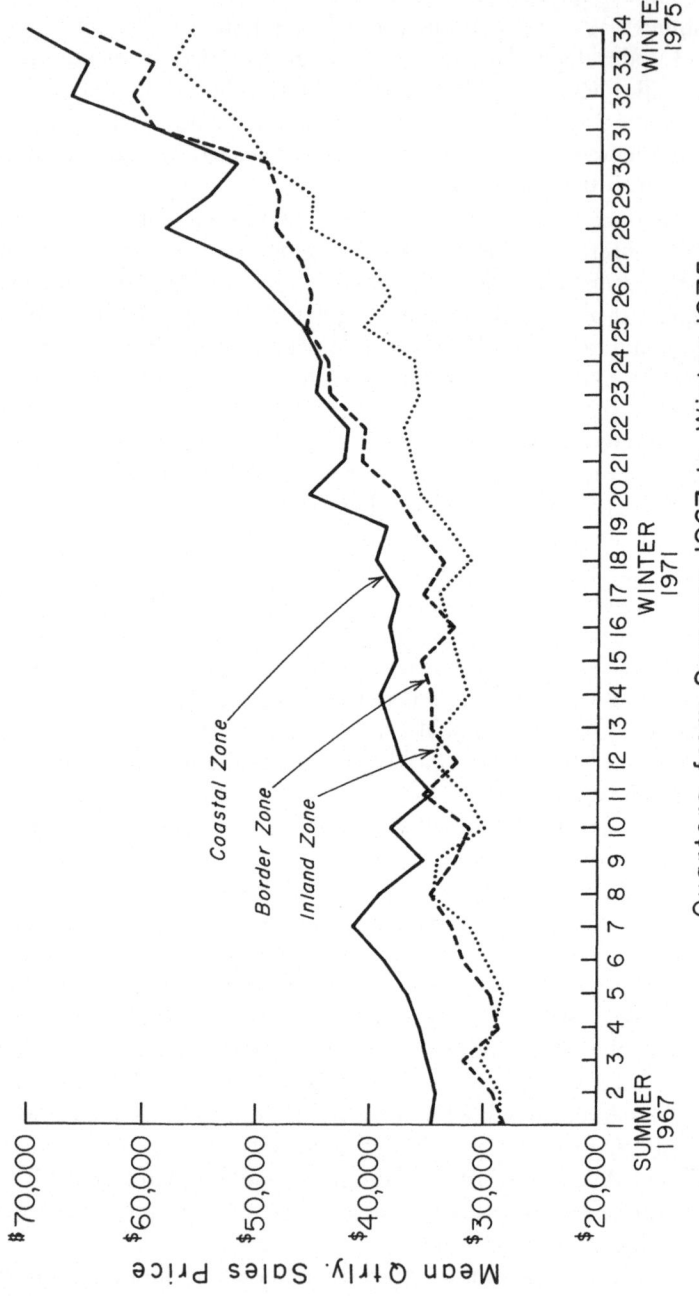

FIGURE 6-1.   Sale price of houses in the coastal permit zone, an adjacent border area, and inland, 1967–1975.

impact, Kneisel first adjusted all prices for a variety of housing characteristics that would otherwise cause differences in the prices of houses in the three areas (e.g., number of rooms, floor space). Using several different regression models, he then examined the contribution of construction costs, mortgage rates, unemployment, population growth, density, and the presence of the coastal commissions on the adjusted (hedonic) resale prices in each area.

The analysis showed that after partitioning out the impact of construction costs, etc., on prices, a net increase in the permit zone of approximately $4000 per house remains to be accounted for. The $4000 is the best estimate available, then, of the *maximum* dollar impact that could have resulted from the various restrictions on development (real and perceived) of the commissions on the sale price of homes in the permit zone of Los Angeles County. It represents 7% of the mean sale price of houses in the permit zone from 1973 through 1975. An important factor not captured by the Kneisel analysis—basically because no systematic data on the factor exist—is the extent to which down-zoning and new building requirements by local coastal governments during 1973–1976 contributed to the price rise in the permit zone. With this omission in mind, it nevertheless seems reasonable that the 7% figure is fairly accurate.

The third result that should have followed from placing restrictions on developments along the coast was an increase of value for unregulated sites immediately adjacent to the permit zone. This should have occurred because people wishing to reside close to the coast but deterred from buying at the new level of home prices in the permit zone could be expected to attempt to move to the next closest location. This increased demand would cause a rise in the price of houses in the adjacent area, but not as great a rise as in the regulated area. Kneisel's results suggest that this is what happened. Repeating the regression analysis, this time for the border zone, resulted in an estimated maximum impact of $1000 on the average sale prices of homes. In this instance, as the theory suggested, buyers moving to the area adjacent to that regulated bid up prices beyond what would otherwise have been the case.

These results provide the best empirical evidence available, in Los Angeles County, of the impact the commissions had on housing[38]—an impact that was inevitable given the imposition by the commissions of land use controls to limit development in the highly prized coastal zone. More noteworthy than the fact that an impact occurred is that it was not nearly as enormous as opponents had often suggested. Yet it cannot be denied that it was a $4000 "windfall" profit, on the average, to every pre-1973 owner of a coastal house who sold during the observed period, which resulted directly from an agency using the police power of the state to restrict development. It was not exactly the massive transfer of wealth from the "poor to the rich" or the kind of added cost that generated "class conflict" that the commissions had been accused of.[39] More accurately, the transfer was between the moderate-to-wealthy residents of the coast before 1973 and the generally wealthier ones who gradually replaced them.

V. Conclusions

Given the extraordinary difficulty most public agencies face in achieving a significant positive impact, the coastal commissions must be judged fairly successful in achieving their main objectives. To begin with, noncompliance by target groups was kept to a minimum. This resulted less from a well-structured enforcement program by the commissions, however, than from the visibility of noncompliance in conjunction with the active monitoring by the supportive constituency, and the possibility of fairly severe penalties being imposed if detected. The one exception was the failure of the commission to ensure that all legal documents required pursuant to access dedications were actually filed.

In moving from compliance with permit conditions to the actual opening of accessways, the implementation process was flawed not only by the technicality of recording legal documents. The act did not provide the commissions with the capability to physically open and maintain the accessways, nor did it assign this responsibility to other state or local agencies. At a minimum, this suggests a shortcoming in the formal authority granted the commissions and probably a flaw in the causal theory underlying the act. Formal authority was also lacking for obtaining acquisition of whole parcels of beach-front property for the purpose of access and recreation. The commissions did achieve some success in this area, however, through stimulating interest in and developing a program of acquisitions that was partially realized through the Nejedly-Hart Bond Act.

The commissions were probably most successful under the 1972 act with regard to their impact on wetlands protection. In this instance, the policy directives of the act were quite clear, the commissions' permit authority could be used effectively to preserve remaining wetlands (at least those portions of wetlands within the 1000-yard permit boundary), the agency leaders were committed to the goal of wetland protection, and, finally, changes in public opinion, socioeconomic conditions, and other factors during the 1973–1976 period did not undermine the pursuit of this objective. The high priority given wetlands in the commissions' 1976 acquisitions list, and the Nedjedly-Hart Bond Act, provided permanent wetland protection in many cases.

The adverse impact that the commissions had on the cost of housing was not intended under the Coastal Initiative, though it may have been inevitable. It is hard to imagine that any statute that restricts certain types of development, lowers the density of development, and requires additional checkpoints in the development process would not contribute to the rising cost of both developed and the remaining developable sites.

Notes

1. Harrell Rodgers, Jr., and Charles Bullock, III, *Coercion to Compliance* (Lexington Books, 1976), pp. 2–3.
2. An interesting formal cost–benefits theory is developed by Rodgers and Bullock, *Coercion to Compliance*.
3. Sections 27500 and 27501 of the 1972 act. Under the 1976 act a prison term was eliminated as a possible penalty. The maximum per-day fine for willful development in violation of the act was raised to $5000, however (Sections 30820 and 30821).
4. Nelson Rosenbaum, *Environmental Law Enforcement: A Case of Wetland Regulation* (Washington, D.C.: Urban Institute, 1980).
5. The average annual permit load ran approximately 400 in the North Coast, 200 in the North Central, 750 in the Central Coast, 900 in the South Central, and 3000 in the South Coast regions, and 1000 in San Diego.
6. Interview with Jack Lahr, executive director, North Coast Regional Commission, 1973–April 1976, 27 July 1978.
7. Interviews with Bob Brown, executive director, North Central Regional Commission, since August 1976, 18 July 1978; and Kathleen Ohlson, Marin County Planning Department, 1 August 1978.
8. Statements by Hugh G. Robinson, Los Angeles, as district engineer, 14 November 1977. Maintenance dredging of existing navigation channels was already exempt from the Coastal Act, Section 27405.
9. Interview with Bonnie Woostencraft, who began the access study in 1978 while serving on the staff of the state commission, 13 February 1979.
10. Today, the commissions issue an "intent to issue permit," which specifies conditions. Only upon receipt of all legal documents that the conditions have been complied with, e.g., notice from the county recorder that a dedication has been filed, is the permit issued.
11. Interviews with Rick Rayburn, executive director, North Coast Region, since August 1977, 20 July 1978; Don Peterson, North Coast Region commissioner, 1975–1976, 21 July 1977; and Jack Lahr.
12. For example, two of the six categories of development explicitly listed in the act requiring a ⅔ affirmative vote of all commission members pertained to access: (1) any development that would reduce the size of any beach or other areas usable for public recreation; (2) any development that would reduce or impose restrictions upon public access to tidal and submerged lands, beaches, and the mean high tide line where there is no beach.
13. "California Coastal Commissions Evaluation Survey," Winter 1977–1978.
14. Only under the 1976 act, which was much more explicit about requiring access across all shoreline developments, did the number of access dedications rise dramatically in the North Coast and other regions that had previously paid less attention to the issue.
15. Under the 1976 act even lateral accessways became a problem because of the specific provision in the act that maintenance and liability must be provided before the accessway could be opened: "Dedicated accessways shall not be required to be opened to public use until a public agency or private association agrees to accept responsibility for maintenance and liability of the accessway" (Sec. 30212).
16. Joan Sweeney, "Beach Access: Slow Gains," *Los Angeles Times*, 16 May 1977.
17. Since passage of the 1976 act intermittent negotiations have taken place between the commissions and legislators, and the commissions and other state agencies and local governments, but no resolution to the issues of funding and assigning of legal responsibilities have been achieved. Under the new act such resolutions are made all the more pressing in that even lateral access

dedications may not be opened until assumption of maintenance and liability by a public or private agency.

18. *Coastal Plan*, Policy 155.
19. "Recommended Coastal Properties for Public Acquisition," California Coastal Zone Conservation Commissions, March 1976, pp. 1–2.
20. Interview with Ross Henry, California Department of Parks and Recreation, 1 April 1979.
21. Statements made by Joseph Bodovitz, executive director of the state commission, *Los Angeles Times*, 20 February 1976 and 28 August 1979.
22. Figures based on information provided in "Recommended Coastal Properties for Public Acquisition," pp. 35–36.
23. The bond act was only one of the three pieces of coastal legislation before the 1976 legislature. For a more complete discussion, see chapter 9.
24. November 1, 1976.
25. Not all the monies authorized by the legislature in AB 924 came from the 1976 bond act. Of the nearly $53 million authorized and approved by Governor Brown, almost $7 million came from general funds. The governor deleted approximately $12 million from AB 924 before signing it. The comparison of acquisitions, real and suggested, yields the following general conclusions: (1) 154 sites were suggested for acquisition by the coastal commissions; of these, 16 were selected to be funded by AB No. 924; (2) in other words, of the 25 selected sites for acquisition, 16 had been suggested by the coastal commissions.
26. Designation for purchase in an authorization bill such as AB 924 and actual purchase by the state, however, is not necessarily a smooth or a quick process. Authorized acquisitions are turned over to the Public Works Board and the Director of Parks and Recreation for action. Before they can act they must receive an appraisal of the site's value from the General Services Administration, with value set at the time of purchase, *not* the time of the legislative authorization of purchase. Just getting this appraisal may take up to a year. Once the appraisal is made, property owners may resist the state's attempt to purchase or disagree on the selling price and, if private negotiations fail, turn to the courts, which can extend the process up to several years. Ultimately, due to differences in authorized expenditures and the ultimate selling price or modifications of the area designated for purchase resulting from negotiations involving the Public Works Board, the Parks Department, and the seller, the actual acquisition may be a good deal different from that originally designated by the legislature: *Minutes* of the State Commission Meeting, September 22–23, 1976; interview with Peter Douglas, state commission staff, 15 February 1978.
27. Composite from *Coastal Plan*, and Larry Oglesby, "Southern California Estuaries and Bays" (Unpublished paper, 1976).
28. Fish and Game had been attempting to generate awareness of the plight of wetlands through a series of publications beginning in the 1970s focusing on major remaining wetlands. It also issued a report in 1974 identifying 25 priority wetlands recommended for state acquisition, 19 of which were along the coast. The 19 were incorporated directly into the commissions' *Coastal Plan*. Interview with Bruce Browning, Department of Fish and Game, 1 January 1979; *Coastal Plan*.
29. "California Coastal Commissions Evaluation Survey," Winter 1977–1978.
30. A ⅔ vote was required for any request for "Dredging, filling, or otherwise altering any bay, estuary, salt marsh, river mouth, slough or lagoon" (Section 27401).
31. The denial is on appeal to the state commissions. Interview with James McGrath, state commission staff, 19 April 1979.
32. Interviews with Gary Monroe, California Department of Fish and Game, 10 July 1978; and Bruce Browning, California Department of Fish and Game, August 1978.
33. Robert Kneisel, *Economic Impacts of Land Use Control: The California Coastal Zone Conserva-*

tion Commission (Davis, California: Institute of Governmental Affairs and Institute of Ecology, February 1979).

34. Robert Healy, "Saving California's Coast: The Coastal Zone Initiative and Its Aftermath," *Coastal Zone Management Journal*, 1(4, 1974), p. 386.
35. H. E. Frech, III, and Ronald N. Lafferty, "The Economic Impact: Land Use and Land Values," in *The California Coastal Plan: A Critique* (San Francisco: Institute for Contemporary Studies, 1976) pp. 69–91.
36. Frech and Lafferty, "The Economic Impact."
37. Robert Kneisel, "The Impact of the California Coastal Commission on the Local Housing Market" (Unpublished Ph.D. dissertation, University of California, Riverside, 1979).
38. Kneisel repeated the analysis for housing prices on and off the coast in Ventura County. The results show a far less discernible impact of the commissions. However, because of the limited number of permits involved, the reliability of the findings is questionable.
39. Vitriolic statements are standard fare for such authors as Tom Hazlett in attacking the commissions. See "A Case of Cost Coastal Piracy," *Inquiry*, 19 March 1979.

THE CALIFORNIA COASTAL PLAN

Its Development and Consistency with Statutory Objectives

Thus far we have focused on the classic adjudicatory function of the coastal commissions, namely, the review of applications for development within the 1000-yard permit area. We have examined the general stringency of permit decisions, their consistency with statutory objectives, the major variables affecting those decisions, and the impacts of the permit process on public access, wetlands, and housing costs.

We now turn to the other principal task of the commissions—the preparation of a coastal plan to be presented to the 1976 legislature. The two processes were, of course, interrelated. Permit decisions were to protect planning options, and, as a practical matter, the permit review experience would provide guidance in developing plan policies, particularly in areas such as the location and intensity of permissible development.[1] Likewise, as the plan policies evolved, they served, at least implicitly, as guides in permit review, particularly starting in late 1975 after the adoption of the coastal plan by the commissions.[2] In the latter instance, the development of plan policies was analogous to the classic rule-making responsibility of regulatory agencies.

Nevertheless, the principal function of the coastal plan was not to guide the commissions' adjudicatory decisions during the 1973–1976 interim period but rather to serve as a proposal to the 1976 legislature concerning the detailed policies to guide the long-term utilization of coastal resources and the institutions necessary to administer those policies, consistent with the objectives of Proposition 20. The planning process was to be an open one, with extensive public participation on the first round of planning to take place separately in each of the six regions, and then again when the state commission synthesized proposals from the six regions.

The measure of success of the planning process is therefore threefold. First, to what extent was it an open process, with the differing public views expressed during the regional planning hearings and incorporated into the regional proposals to the state actually acted upon by the state staff and ultimately adopted by the state commission? Second, was the plan adopted consistent with the objectives of the 1972 Coastal Initiative? Third, did the planning effort generate sufficient political support to persuade the 1976 legislature either to adopt it outright or to incorporate its principal components into law?

This chapter will first sketch out the development of the planning process and the strategy pursued by the commissions. A close analysis of three of the nine plan elements is then presented in an attempt to gauge the relative openness of the planning process between regions and the state. Finally, the content of the plan is examined in light of the objectives of the 1972 Coastal Act. The following chapter will examine the final stage of this overall process in terms of the 1976 legislature's reaction to the plan's policies and institutions proposed for their implementation.

I. The Strategy of Plan Development

The 1972 Coastal Initiative did far less to structure the planning process than it did interim permit review.[3] It required the plan to be consistent with the following very general set of objectives, all of which gave a strong environmental "tilt" but little meaningful guidance: (a) the maintenance, restoration, and enhancement of the overall quality of the coastal zone environment, including, but not limited to, its amenities and aesthetic values; (b) the continued existence of optimum populations of all species of living organisms; (c) the orderly, balanced utilization and preservation, consistent with sound conservation principles, of all living and nonliving coastal zone resources; (d) avoidance of irreversible and irretrievable commitments of coastal zone resources. It also required the plan to contain a number of components, only a few of which provided much substantive direction: The public access element was to contain provisions "for maximum visual and physical use and enjoyment of the coastal zone by the public," and the public services and facilities element was to provide for "the general location, scale, and provision in the least environmentally destructive manner of public services and facilities [including power plants] in the coastal zone." The statute likewise said very little about the manner in which the plan was to be developed—only that the regional commissions were to present their recommendations, after public hearings in each county, to the state commission by April 1, 1975, and that the state commission was to adopt the plan by December 1, 1975. In short, the only explicit requirements in the Initiative for the plan and planning process were that (1) public access and environmental protection be accorded high priority, (2) the plan be completed within definite deadlines, and (3) the state commission be ultimately responsible for the content of the plan.

Given this minimal guidance and the enormous conflicts involved in proposing policies for the very large and diverse expanse of the coastal planning
zone—an area 1100 miles long and approximately 5 miles wide[4]—it was no
mean achievement that the commissions managed to prepare on time a coastal
plan generally consistent with statutory objectives and, as we shall see in the next
chapter, to build sufficient political support to shepherd a reasonable facsimile
through the 1976 legislature. This was achieved through an enormous amount
of work by staff and commissioners and by a general strategy that emphasized
building public understanding and support and developing a cooperative relationship between the state and regional commissions.

Instead of attempting to anticipate preferences of the state's legislators, the
basic approach of state commissioners and staff was to develop a plan that would
be consistent with their preferences and the act's objectives, in the process
developing sufficient public support to persuade the legislature to go along.[5] This
was based partially on perceptions about the legislature, including uncertainty
about who the key members would be in 1976, doubts that even key legislative
committees were sufficiently interested to devote large periods of time to plan
formulation, and past experience that had indicated that development interests
and local governments were sufficiently strong that a plan consistent with the
act's objectives would not emerge without a long campaign to mobilize public
support. The strategy was also based on the belief that the success of the Coastal
Initiative had provided them with a strong popular mandate that was at least
implicitly critical of the previous performance of the legislature and state and
local agencies. Finally, it was based on the knowledge that a similar strategy of
using the preparation of a plan as a means of mobilizing political support for a
stronger legislative mandate had been successful with protection of San Francisco Bay[6] and was worth trying to repeat even in the far more complex and
politically controversial case of protecting of the state's entire 1100-mile
coastline.

The strong emphasis on "participatory planning" involved a number of
tactics.[7] First, in order to facilitate public participation and understanding, the
plan was broken into nine elements—marine resources, geology, coastal land
resources, energy, recreation, appearance and design, transportation, intensity of
development, and government powers and funding—and attention was focused
on the development of fairly general policies by the regional and state commissions after extensive public hearings. The separate elements were then to be
integrated into a preliminary plan, which would be subjected to additional
hearings before being revised and adopted as the final plan by the state commission. In short, this was to be a plan constructed from "the bottom up" with
extensive public participation rather than a comprehensive plan developed by
staff and then presented to the public and commissioners. Second, partially to
facilitate public discussion but also because of time and resource constraints, the

emphasis was on the resolution of conflicts rather than on the development of extensive data bases and sophisticated technical analyses. Third, the consideration of each element by the regional and state commissions was to be accompanied not only by public hearings in each county but also by informal meetings, opinion surveys, the solicitation of comments on draft documents, and a variety of other measures to stimulate informed participation by as wide a variety of interest groups and citizens as possible. A final aspect of the process, which emerged from demands by the regions and much of the public, was the application of policies to general locales within each region. This occurred, however, only after the development of general policies so that people would not initially become preoccupied with local issues; moreover, ·the mapping remained fairly general and never became so detailed as to apply to specific parcels or to involve precise zoning classifications—both of which were likely to increase opposition.

The second basic component of the planning process was the development of a cooperative arrangement between the state and regional commissions.[8] As previously indicated, the act was essentially silent on this score. Thus, it would have been possible for each of the regional commissions to develop their proposals independent of each other and of the state commission. At early meetings of the state and regional staff, however, it was decided that the absence of effective coordination throughout the process would result in a hopeless task for the state staff and commission in April 1975, thereby largely vitiating the efforts of all involved at the regional level and resulting in almost certain defeat of a hastily prepared state plan. Moreover, the state staff clearly recognized that the participatory strategy would have to be implemented primarily by the regional commissions, because they were more accessible to people and because coastal issues were more easily comprehended at the local/regional level.

As a result, a basic process emerged for each element whereby the state staff first developed a background technical document identifying the basic issues and compiling relevant information from state agencies and other sources concerning baseline data and important trends. After review by outside experts and regional staff, this document was distributed to the regions with a set of policies proposed by the state staff. The regional staff then revised these proposals and submitted them to public hearings and otherwise solicited comments from local agencies and interest groups. After at least one public hearing in each county, these were further revised and approved by the regional commission. The state staff then integrated the often conflicting proposals from the six regions and submitted them to further hearings at the state level. These were approved (after appropriate revisions) by the state commission. The adoption of specific elements took roughly 18 months from the fall of 1973 to March 1975. The elements were then combined by the state staff into a preliminary plan, which was subjected to a series of 20 joint hearings by state and regional commissioners throughout the state in the spring of 1975. At the same time, the regions were applying the policies to geographical areas within their boundaries and developing lists of suggested areas for public acquisition (usually either to preserve wildlife habitats or to provide public access). In the summer of 1975 the state commission at a series of meetings revised the preliminary plan policies, the proposed maps, and the acquisition lists, and finally in September adopted the coastal plan.

The potential clearly existed for the state staff to be the dominant actors in this entire process. In preparing the background technical document and the original policy proposals for each element, they controlled the initial agenda. Moreover, through their synthesis of recommendations from the six regions (and from the preliminary plan hearings) for presentation to the state commission, they played a critical "gatekeeping" role for part-time state commissioners under substantial time and resource constraints. Not surprisingly, there were periodic complaints from regional staff and commissioners and representatives of development groups that their views were being filtered out by state staff.[9] If true, this would suggest that the entire participatory process was somewhat of a sham— that, for all the public and regional participation, the important decisions were still being made by professional planners and that participation was simply a means of developing political support for those decisions. To the extent that the perception of state staff domination became widespread, it would probably affect the amount of public and regional commission support for the plan, and ulti- mately its fate before the 1976 legislature.

It is to examine the extent to which the basic strategy of participatory planning and state–regional cooperation was actually followed in practice that we now turn to an analysis of the relative influence of various actors in the development of the coastal plan in an illustrative sample of cases.

II. The Planning Process in Operation

A comprehensive analysis of the major actors responsible for the develop- ment of the nine elements and their final integration into the coastal plan is beyond the scope of a single chapter. Instead, we shall summarize the results of a detailed study of the development of three of the most important elements— coastal land resources, intensity of development, and government powers and funding—as they were addressed by the North Coast, North Central, and State commissions.[10] We shall first briefly discuss the subject of each of the three elements, the original state staff recommendations, and the locus and direction of subsequent policy changes. Through tracing the ebb and flow of the process and by comparing the original state staff proposals with the policies ultimately adopted, we are able to gain some sense of the major factors affecting plan development and the extent to which the actual process conformed to the pre- cepts of participatory planning and state–regional cooperation.

A. *The Coastal Land Element*[11]

The coastal land element was one of the three dealing with the natural resources of the coast (marine environment and geology were the other two). It dealt with such topics as terrestrial and freshwater habitat, agricultural land, timber production, flood control and flood plain development, nonpetroleum mineral resources, and air quality.

In March 1974 the state staff proposed a number of policies covering protection of anadromous fisheries (e.g., salmon), restriction on development in floodplains, protection of natural habitats and rare and endangered species, protection of agricultural land from urban encroachment, development of soil maps, regulation of nonpetroleum mineral extraction (principally sand and gravel operations), and regulation of air pollution sources.

Of the 18 policies originally proposed by the state staff on the coastal land element, 8 were essentially unchanged by the North Central region. But 8 others were substantially modified by the regional staff and eventually accepted with only minor revisions by the regional commission. These included the addition of timber policies and substantial modifications in the policies dealing with agricultural lands, sand and gravel deposits, rare and endangered species, and air quality. With the exception of the latter 2, the regional modifications proposed a broader definition of coastal-related resources and required stricter protection of the environment than those submitted to them by the state staff.[12]

A somewhat different pattern prevailed on this element in the North Coast region. The regional staff also took a relatively free hand in revising state staff proposals, substantially modifying 4 of the 18 and proposing several new timber policies as well. There were also a few changes introduced as a result of calls in the hearings for more sensitivity to economic dislocations, primarily in the timber industry. In general, the regional staff proposals attempted to find a middle ground between the environmental orientation of the state staff proposals and the *laissez faire* attitudes of local residents and commissioners. Unlike its counterpart to the immediate south, the North Coast Commission substantially revised the staff recommendations. Of the 25 policies eventually recommended by the regional staff after the hearings, the commission made substantial modifications in at least 9 to provide for less environmental protection, more economic development, and less public control of resources. In addition, it made some changes in the same direction in another half dozen policies. The ax fell heaviest on policies dealing with flood control, anadromous fish spawning areas, riparian habitats, forest resources, and air quality.

Overall, reflecting local sentiment, the North Central region substantially modified about half of the original state staff proposals. The North Coast region went even further, changing 15 of the 18 state recommendations, about half of them outright deletions or other major modifications. Moreover, both regions recommended the addition of timber management policies, although they differed substantially on the respective emphasis of environmental protection and economic development in this area. In general, however, both staff and commissions on this element appeared to be primarily concerned with developing policies applicable to their own region rather than with influencing subsequent events at the state level.

Given this active, albeit conflicting, response from two of the regions—plus additional suggestions from the other four—what was the response of the state staff and commission? Of the 18 policies originally recommended by the state staff, only 1 remained unchanged in the policies adopted by the state commission for this element in the fall of 1974. Although none of the original proposals was dropped outright, many were extensively revised, and entirely new policies on sand replacement and timber management recommended by the regions were added. Two-thirds of the policies were modified by the state staff prior to the public hearings, and further extensive revisions occurred as a result of the hearings at the state level. The modifications generally provided greater environmental protection, more public control of resources, and slightly less concern for community economic welfare than those originally proposed by the state staff.

Further changes occurred as a result of the hearings on the preliminary plan in the spring of 1975. These generally provided for slightly greater protection of wildlife habitat, significantly greater restrictions on the conversion of agricultural land, and relaxation of the very strict air quality policies proposed the previous fall. In addition, in response to pressure from other agencies, the state staff and commission recommended that other state agencies be granted some of the power previously proposed to be vested in the successor coastal agency. Overall, of 19 policies relating to the coastal land resources in the preliminary plan, all but 3 were altered prior to publication of the final coastal plan.

In short, the coastal land element revealed the state staff and commission to be quite responsive to suggestions by the regional commissions and other actors (particularly environmental groups), with respect both to the topics to be addressed and to the substantive direction of proposed policies.

B. The Intensity of Development Element[13]

Intensity of development was the last plan element to be addressed of the five dealing with the type and scope of permissible development (the others being appearance and design, recreation, transportation, and energy) and, in many ways, represented the culmination of the commissions' efforts to determine the substantive orientation of future coastal development. In addition, the formulation of the original state staff proposals on this element differed from the previous seven in that principal responsibility was ultimately assigned to staff with extensive experience in the permit process. As a result, previous state commission decisions on permit appeals played a significant role in the development of the staff recommendations.[14]

In September 1974 the state staff recommended a set of 26 policies that sought to (1) severely restrict development in potential agricultural or lumbering areas, (2) maintain scale of development in accord with the existing character of a community, (3) assign development priority to coastal-dependent uses (e.g., ports), while mitigating the environmental effects of such uses, (4) maximize public access to the coast for all people, (5) concentrate development in existing urban areas, and (6) make some effort to restore degraded coastal resources, e.g., by reducing the number of existing subdivided but unbuilt lots.

In the North Central region, consideration of this element had several noteworthy aspects. First, the subcommittee of commissioners selected to work on this element was particularly active in preparing the original set of (ostensibly staff) recommendations for public hearings,[15] in the process cutting the number of proposed policies in half by eliminating those considered redundant with other elements. Second, the recommendations for expansion of visitor-serving facilities along the coast aroused considerable opposition within the small villages in Marin County because of the allegedly detrimental effects of such facilities on natural resources and on the character of those communities. As a concession to these interests, the recommendations were altered to reduce the size of visitor-serving facilities, to move overnight facilities inland, and in general to alter the policy from maximizing public access to balancing access with the protection of natural resources and the preservation of the rural character of the coast. Third, the regional staff and commission established very restrictive buildout levels for coastal communities. While this policy was supported by Marin residents, the commission's proposed reduction in Bodega Bay from the 2750 units permitted by Sonoma County to 1270 units was vigorously opposed by local residents who wanted the additional population base to help defray the tax assessments from a new sewage treatment plant; nevertheless, the commission refused to raise the buildout figure. In general, the policy changes within this region were consistently in favor of greater environmental protection and concentration of development in existing areas than those proposed by the state staff, and somewhat less vigorous in the pursuit of public access.

Turning to the North Coast region, the general pattern of substantial commission changes in staff recommendations was repeated. In this case, however, the original policy proposals were made not by the regional staff *per se* but rather by a consulting firm that had been hired because of the lack of staff trained in urban planning. Nevertheless, of the 39 policies proposed by the firm with staff concurrence, 23 were either deleted or substantially revised by the regional commission. Some of the deletions involved perceived redundancies with other elements. But there was also a consistent commission effort to (1) relax proposed restrictions on lot splits and expansion outside existing urban areas, (2) defer decisions concerning ultimate population levels pending further study, and (3) place all future authority for development review in the hands of local governments and state agencies antedating the Coastal Initiative, i.e., to reject any role for a future coastal agency.

Consideration of the intensity element by the state staff and commission was collapsed with the hearings on the preliminary plan. In synthesizing the regional proposals for presentation as tentative policies in the preliminary plan, the state staff and commission incorporated several entirely new policies suggested by various regions that dealt with the constraints of water supply on development and the restoration of degraded urban areas by redevelopment agencies. They also developed a more detailed policy on land divisions and somewhat relaxed the restrictions on coastal-dependent development. As a result of subsequent hearings, a number of refinements were made, although seldom involving changes in overall policy orientation.

In general, however, if one compares the original (September 1974) state staff recommendations with the policies ultimately incorporated into the coastal plan, the intensity element underwent less change than either of the other two we shall examine. Of the 21 policies coded for this element, 7 were ultimately adopted by the state commission in the same form as had been originally recommended by the state staff. The changes, many of which were the result of similar recommendations from several regions, generally involved greater protection of natural resources, less public control over private actors, somewhat greater public access to the coast, and considerably stronger restrictions on development outside of existing urban areas than had been originally recommended. The relative stability of this element was probably a function of its placement near the end of the process, thereby benefiting from the state commission's extensive permit experience and its planning decisions on the previous seven elements.

C. *The Government Powers and Funding Element*[16]

The final element addressed in the planning process, government powers and funding, was probably the most important and controversial. While the first eight elements sought to develop policies to guide the future utilization of coastal resources, this one dealt with the institutions and the procedures by which those policies were to be carried out. It concerned such critical issues as (1) the distribution of authority for future planning and permit review among a successor coastal agency, other state agencies, and local governments; (2) the structure and composition of the future coastal agency; and (3) the need for institutions to deal with such problems as the management of accessways, the restoration of degraded urban areas, the consolidation of unbuilt lots in existing subdivisions, and the maintenance of marginally productive areas in agricultural use.

The state staff began work on this element in the spring of 1974 with the commissioning of background studies, generally on legal issues. In October it proposed a set of policies to (1) retain the state and regional commissions in essentially their same form, but with additional authority over public works projects within the entire coastal zone and in watersheds outside the coastal zone that affected coastal resources; (2) delegate ongoing permit authority to local governments provided that the coastal commissions retained authority over the nearshore area and appellate review over all local decisions; and (3) establish a coastal restoration agency with authority to acquire land and operate public facilities, issue bonds, and receive funds from tidelands oil production. The state staff did not, however, deal with the critical issue of the geographic scope of authority of the successor coastal agency, preferring instead to wait for regional suggestions and for better information concerning the boundaries of coastal-relevant resources. On the whole, however, the state staff recommended a substantial increase in the permit review authority of the coastal agency from the 1000-yard area during the 1973–1976 period to include virtually all projects within the 5-mile-wide coastal planning boundary and even dams outside the zone.

As was their custom, the North Central staff liberally rewrote the state staff's recommendations, in part to reduce the number of policies involved. While agreeing that both the state and regional commissions should be continued, the regional staff suggested numerous changes aimed at giving more authority to local governments and to other state agencies. Probably its major proposal was that the commissions continue to have permit authority in the nearshore area (involving critical wetlands, recreation areas, and viewsheds) but that review within the broader coastal management zone be turned over to local governments once their plans and ordinances had been brought into conformance with coastal plan policies (with appeals limited to variances and other nonconforming uses). After a series of informal meetings and public hearings, these recommendations were accepted by the North Central Commission in March 1975 with only minor modifications, chiefly to clarify that the commissions would have permit authority throughout the coastal management zone in the interim period pending the approval of local plans. On the whole, however, this element aroused very little controversy in this region.

Such was not the case in the North Coast region. The regional staff hired Bruce Haston, a political science professor from Humboldt State University, to act as a consultant on this element. After interviewing approximately 70 opinion-leaders throughout the region, Haston (with staff concurrence) substantially revised the state staff recommendations to provide for (1) continuation of the state commission but abolition of the regional commissions after an interim 2-year period; (2) restriction of the commissions' permit authority to a 100-yard near-shore area, with local governments assuming jurisdiction in the remainder of the previous (1973–1976) permit zone after their local plans had been brought into conformance with coastal plan policies; (3) assignment of the functions of the restoration agency to existing state agencies; and (4) a number of measures to reduce the burden on permit applicants and to protect the economically depressed counties of the region. The Haston proposals, as approved by the regional staff, were greeted with a firestorm of protest by local residents angry over the perceived adverse effects of the commissions and other state agencies on the local economy and on their property rights.[17] Their essential message was, "We didn't vote for you in 1972. You arrogant bureaucrats have been an unmitigated disaster. Get out of our lives. Local governments can do a much better job and, if they don't, we'll throw the rascals out." Despite efforts by the regional staff to respond to these criticisms by removing coastal commission authority in even the 100-yard zone (except on appeal from local decisions) and other minor adjustments, the North Coast Regional Commission in April 1975 voted simply to abolish the coastal commissions and leave implementation of coastal plan policies entirely in the hands of local governments.

By the late spring of 1975, it was clear that the original proposals of the state staff for the continuation of state and regional commissions with considerably expanded permit authority and only a minor role for local governments was no longer viable. In fact, a rough consensus was forming around a proposal made by the County Supervisors Association of California (CSAC) that primary jurisdiction for permit review be delegated to local governments after their plans had been brought into conformance with the coastal plan policies approved by the legislature. Thereafter, the regional commissions would be abolished, with the state commission restricted to proposing plan amendments and hearing appeals from local decisions within the nearshore area. Moreover, other state agencies would be obligated to comply with the state-approved local plans. This general outline was consistent with the recommendations of four of the regional commissions (all but the North Coast and North Central) and drew considerable interest from both environmental groups and state staff because it represented a breakthrough in terms of local government recognition of the legitimacy of state review of local plans. Moreover, the required conformance of other state agencies with these state-approved local plans provided the commissions an important potential ally (i.e., local governments) in their effort to force other agencies to conform to the coastal plan.[18]

After a series of informal meetings with state commissioners, the state staff in June 1975 proposed that the commissions continue to have primary jurisdiction over development *seaward* of the mean high tide but that in the nearshore area (roughly coterminous with the 1000-yard permit zone)—as well as in floodplains and on agricultural land throughout the more extensive coastal zone—primary authority would be delegated to local governments after their plans and implementing (e.g., zoning) ordinances had been approved by the state commission as conforming to coastal plan policies. The regional commissions would then be abolished, with the state commission empowered to hear appeals from local government decisions in the nearshore area and to act on amendments to local plans.[19] The state staff also renewed its proposal for a Coastal Conservation Trust with authority to consolidate subdivided lots, protect agricultural lands through purchase and leaseback, and acquire access and scenic easements.

These proposals were then aired at a public hearing, where development groups objected to the need for any successor agency, many state agencies and special districts (e.g., ports) objected to commission review of their projects, local governments and CCEEB sought to restrict the range of postcertification appeals, and environmental groups attempted to increase the scope of the appeals process. The proposals were modified somewhat by the state commission following the hearings—chiefly to clarify the geographic jurisdiction of the commissions and the scope of appeals—and then were approved formally (with the remainder of the coastal plan) at its meeting of September 18, 1975.

The result of this entire process was the California Coastal Plan, a rather massive document involving 150 pages of substantive policies from the first eight elements, an 18-page discussion of institutional arrangements necessary to carry out those policies, and over 200 pages of maps and notes applying the policies in a rather general fashion within each of the six regions and suggesting numerous acquisitions.

D. Openness of the Planning Process

On the basis of our examination of the three elements of the coastal plan and our general knowledge of the planning process in the other four coastal regions, several conclusions seem warranted. First, contrary to the often assumed domination of the process by the state staff, we found that while the staff played an important and not-to-be-underestimated role in setting the original agenda and in synthesizing the regional recommendations for the state commission, it certainly did not come close to monopolizing the planning process. On the coastal land element, for instance, almost all of the original state staff recommendations were altered by the regions and eventually by the state commission, and several entirely new topics on timber management and sand replacement recommended by one or more regions eventually made their way into the coastal plan. Even on intensity of development, where by far the least change occurred between original staff recommendations and final commission decisions, several new topics (e.g., the constraints of water supply on development) suggested by one or more regions were incorporated into the coastal plan. Moreover, much of the stability in this element could be attributed to the extensive guidance given the staff by the state commission in its decisions on previous elements and on permit appeals. Most important of all, on government powers and funding there was a dramatic shift in basic strategy from the autonomous coastal agency with expanded authority originally recommended by the state staff to the emphasis on local delegation recommended by virtually all of the regions and subsequently accepted by the state commission. In short, even though individual regions may periodically have felt their recommendations were being ignored or overruled by the state staff, there can be little doubt that regional recommendations had a major impact on the content of the coastal plan.

Second, as was the case in permit review, the differences in policy decisions between the North Coast and North Central commissions could ultimately be attributed to the basic differences in the social and economic character of the two regions. Just as North Central commissioners represented the support for environmental planning of the urban, well-educated, and relatively prosperous residents of their region, so the majority on the North Coast Commission accurately reflected their constituents' belief in individual self-reliance and distrust of centralized governmental authority—attitudes common in rural areas heavily dependent upon resource extraction with bountiful natural beauty. Thus, it was not surprising that, in contrast to the North Central Commission, the North Coast was less protective of natural resources, less willing to place constraints on development outside urban centers, and dramatically less willing to accept intervention by nonlocal authorities.

Third, even more clearly than in permit review, commission reaction to staff recommendations reflected the degree of congruence in basic policy orientation. At the state level and in the North Central region, for example, the relatively homogeneous attitudes of staff and commissioners generally produced only minor—and often essentially clarifying—changes in staff recommendations by the respective commissions. Of course, this was also a function of commissioner involvement in early stages of the planning process via the designation of commissioner-consultants in the North Central (at least after the coastal land element) and the informal workshops the state staff held with commissioners on the crucial elements dealing with the intensity of development and government powers and funding. In the North Coast region, by contrast, the commissioners provided very little guidance to staff. Given staff efforts to propose policies reflecting their own preferences and the directives of the Coastal Initiative—with some compromise toward the localist/*laissez faire* preferences of the commission majority and local residents—it is not surprising that the commission made major revisions in staff recommendations on all three elements. As a result, the policies ultimately proposed from this region reflected the policy predispositions of the commission majority much more than those of the regional staff.

Finally, there is little doubt that the commissions made a concerted effort to encourage participation by a wide variety of groups and interested citizens through not only the multitude of public hearings but also a variety of informal meetings, advisory committees, systematic interviews with opinion-leaders, and extensive opportunities for written comments throughout the long process of plan development. [20] This is reflected in Nelson Rosenbaum's survey of participants in four state land use planning efforts. By roughly a 2 to 1 margin, participants in the California coastal planning process felt that information was made easily available to all affected citizens and that affected citizens were provided with sufficient opportunities to express their preferences to decision-makers. These figures compared very favorably with the level of satisfaction in Oregon, were somewhat higher than those in Hawaii, and were considerably greater than those in North Carolina. [21] Similarly, our own survey of 480 interested observers of the commissions revealed that 55% felt the commissions had provided affected citizens adequate opportunities to express their views, while 32% disagreed (see Table 7-1, part A). [22] There were, however, wide disparities among different groups, with commission officials, environmentalists, and civic group leaders rating the commissions' performance highly, while most business leaders and coastal property owners were quite dissatisfied. [23] If attention is restricted to actors (exclusive of commission officials) who were active in the planning process, virtually the same figures (57% vs. 33%) are obtained (see Table 7-1, part B).

It is not at all clear, however, that the extensive opportunity to participate favorably affected people's ultimate evaluation of the coastal plan. On the one hand, environmental groups and other supporters of the 1972 Initiative remained active throughout the planning process and generally supported the results. [24] On the other hand, many of the opponents of the Initiative among oil companies, utilities, construction firms and unions, and realtors remained implacably opposed to the coastal plan despite their participation at hearings. [25] In fact, toward the end of the process many of them made it clear that they had given up on what they believed to be the hopelessly biased commissions and would take their case to the legislature. Moreover, there is little doubt that the economic recession of 1974–1975—which hit the construction and timber industries particularly hard (see Table 3-1 in Chapter 3)—hampered the commissions' task. Perhaps the most that can be said is that the open planning process— plus the popular support manifested during the Initiative campaign and throughout the planning process—encouraged compromise on the part of local government associations (e.g., CSAC and the California League of Cities) and some members of the development community, represented most notably by CCEEB. While these groups continued to oppose much of what the commissions proposed, a tentative dialogue had been opened and a convergence of views involving themselves, the commissions, and some environmental groups had begun. [26] In general, however, the extensive opportunities for participation did not appear to substantially affect different groups' evaluation of the coastal plan: Those who had supported the commissions' creation in 1972 were generally pleased with

TABLE 7-1
ATTITUDES OF INTERESTED ACTORS TOWARD THE ADEQUACY OF PARTICIPATION OPPORTUNITIES AND THE CONTENT OF THE COASTAL PLAN

	Commissions did *not* give affected citizens adequate opportunities to express their views on planning policies			Orientation toward coastal plan		
	% Disagree	% Ambivalent	% Agree	% Support	% Neutral	% Oppose
Part A. Attitudes by various groups in total survey of observers of commissions[a]						
Commissioners (N = 42)	78%	5%	17%	71%	17%	12%
Commission staff (N = 42)	88%	5%	7%	97%	3%	0%
Local govt. official (N = 44)	41%	18%	41%	37%	39%	24%
State/fed. agencies (N = 20)	40%	25%	35%	26%	58%	16%
Business leaders (N = 33)	36%	6%	58%	18%.	21%	62%
Environmental group leaders (N = 30)	94%	3%	3%	97%	3%	0%
Civic groups (N = 19)	68%	11%	21%	68%	0%	32%
Journalist/observers (N = 16)	62%	6%	31%	53%	0%	47%
Applicant/coastal property owners (N = 71)	30%	16%	54%	34%	10%	56%
Interested citizens (N = 51)	51%	18%	31%	63%	14%	23%
All actors	55%	13%	32%	55%	16%	29%

[a]The N for each group includes all people in group responding to both questions. The total survey is based upon random samples taken from (a) the master mailing list of the commissions, (b) the state commission's list of people "very concerned" with one or more elements, (c) permit applicants involved in appeals to the state commission, and (d) commissioners and staff, as well as legislators on the relevant committees.

(*continued*)

TABLE 7-1 (Continued)

| | Commissions did *not* give affected citizens adequate opportunities to express their views on planning policies | | | Orientation toward coastal plan | | |
	% Disagree	% Ambivalent	% Agree	% Support	% Neutral	% Oppose
Part B. Attitudes of only those who were active in planning process (Exclusive of commission officials)[b] (N = 117)	57%	10%	33%	60%	9%	31%
Part C. Controlling for (reported) vote on Prop. 20[c]						
1. Views of those who voted *for* Prop. 20 (N = 77)	69%	10%	21%	79%	9%	12%
2. Views of those who voted *against* Prop. 20 (N = 31)	29%	6%	65%	13%	6%	81%

[b]This is based solely on the second of the four populations, i.e., those listed as "very concerned" with one or more elements of the plan.

[c]These figures involve the activists in the planning process because they involve a random sample, while the total evaluation survey contained several subsamples with no clear criteria for weighting each.

their performance, while those who had opposed the Initiative had their suspicions confirmed (see Table 7-1, part C).[27]

III. THE 1975 COASTAL PLAN: CONSISTENCY WITH STATUTORY OBJECTIVES

There is little doubt that the coastal commissions conformed to most of the statute's procedural requirements for plan preparation.[28] They met the December 1975 deadline for presentation of the plan to the legislature. They held many more hearings and other forums for public participation than were statutorily prescribed. And, with one possible exception, the coastal plan contained all of the components prescribed by Section 27304 of the Initiative. The exception involved the requirement for "a population element for the establishment of maximum desirable population densities." If this were interpreted to mean *human* populations, the commissions had a mixed record, with regions such as North Central establishing maximum population levels for most communities, and with other regions (such as the North Coast) postponing such decisions pending further study. On the other hand, if the provision were interpreted to mean "optimum populations of all species of living organisms" (as would be suggested by Section 27302b), then the statute prescribed a clearly impossible task given the absence of incredibly detailed knowledge of coastal resources, the rather rudimentary state of the science of community ecology, and the ambiguities involved in the critical term *optimum*.[29]

We now turn to the important topic of the conformity of the coastal plan with the substantive goals of the 1972 Coastal Initiative. This task is, however, complicated by the very size of the coastal plan and the ambiguity of planning objectives specified in Section 27302 of Proposition 20: (a) the maintenance, restoration, and enhancement of the overall quality of the coastal zone environment, including, but not limited to, its amenities and aesthetic values; (b) the continued existence of optimum populations of all species of living organisms; (c) the orderly, balanced utilization and preservation, consistent with sound conservation principles, of all living and nonliving coastal resources; (d) avoidance of irreversible and irretrievable commitments of coastal zone resources. All that can possibly be said is that the first and third objectives provide a general environmental "tilt," with some balancing language legitimizing orderly development. As just indicated, the second prescribes an essentially impossible task. And the fourth provides little meaningful guidance, as *any* structural development implies an "irreversible and irretrievable commitment of coastal resources," at least for several decades; for that matter, so does the public purchase of a beach.

Given these difficulties, we have chosen to examine the provisions of the coastal plan relating to (1) physical and scenic access to the shoreline, (2) the protection of wetlands, (3) attempts to allow development, including energy facilities, within environmental constraints, and (4) the allocation of responsibility among governmental and private institutions for implementation of plan policies.[30] Our previous analysis of the history of the Coastal Initiative (Chapter 2) and the criteria for permit review (Chapter 5) clearly indicated that access and wetlands were two of the more important objectives of the statute as a whole. Some of the commissions' most politically important decisions concerned what constituted "orderly, balanced development," particularly with respect to energy facilities. And this entire study is premised on the assumption that policy directives are unlikely to be realized unless institutional mechanisms are established for their effective implementation.

A. Providing Maximum Visual and Physical Access to the Ocean

As indicated previously, one of the few places in which the Coastal Initiative clearly specified a planning objective was its requirement that the plan provide "maximum visual and physical use and enjoyment of the coastal zone by the public" [Sec. 27304(c)(4)]. At least 30 policies in the coastal plan dealt directly or indirectly with this topic. We shall first summarize the major provisions relating to (1) protecting viewsheds, (2) guaranteeing public access to the coastline from coastal roads, (3) developing recreational facilities and a coastal trails system consistent with resource protection, (4) providing transportation to the coast, and (5) attempting to safeguard housing and visitor facilities for low- and moderate-income groups. Then examined are how these policies were provisionally applied to the Sea Ranch and to Malibu. Finally, some possible inadequacies in the policies will be explored.

In order to protect coastal viewsheds—defined as the areas visible from the coastal highway and vista points—the plan contained 14 policies (1) prohibiting development on beaches, sand dunes, and coastal bluffs; (2) requiring that development in scenic areas be unobtrusive and subordinate to natural landforms and that utility lines in such areas be underground; (3) prohibiting the construction of non-water-related commercial and industrial facilities along the oceanfront (unless no less environmentally damaging alternative were available); (4) prohibiting the erection of signs and billboards blocking significant coastal views; (5) requiring that views of the ocean and other natural features from the nearest coastal road be protected in new developments; and (6) mandating that visually degraded coastal areas be restored to high quality wherever feasible.[31] In addition, a number of suggestions were made, including the purchase of scenic easements under existing highway law, to maintain and expand the scenic highway system and to protect views from coastal roads to the ocean.[32]

A second group of policies involved the protection and expansion of physical access from coastal thoroughfares to the state-owned tidelands.[33] These included a number of provisions to (1) fully explore the potential of so-called implied dedications of accessways under *Gion* vs. *Santa Cruz* (*supra*,Chapter 2), (2) require that new developments between the highway and the ocean provide access, and (3) explore the provision of access in energy and public service facilities (e.g., ports) and military bases along the coast. Recognizing the limitations of the police power, however, the plan also recommended that (4) a new coastal conservancy be created to manage and acquire accessways and that (5) funds from a variety of sources, including offshore petroleum royalties, be used to defray the costs of acquisition.

But simply providing paths from coastal roads to the ocean would clearly not suffice to meet the statutory objective of maximum access. Thus, the plan also recommended to the legislature that (1) oceanfront lands suitable for recreational use near urban centers be accorded the highest priority in an acquisition program; (2) visitor-serving commercial recreational facilities (e.g., motels, restaurants, water-related shops) be accorded priority over private residential, industrial, and general commercial development along the coast; (3) oceanfront areas be reserved for water-related recreation with support facilities located upland; (4) large residential and commercial developments near the coast provide internal open space to relieve the recreational demands on public beaches; and (5) a coastal trails system be established by the state parks department linking existing trails with those acquired through public acquisition.[34] The recreation policies also recognized, however, that accessways and publicly owned trails and beachfront areas would require management (e.g., liability insurance, policing, refuse collection) by a public agency prior to opening. Moreover, it was made clear that recreational use of the coast would be controlled where necessary to prevent damage to natural resources. For example, the use of off-road vehicles was to be very strictly circumscribed, and the dredging or filling of wetlands to accommodate new or expanded recreational boating facilities was expressly prohibited.

The coastal plan also attempted to link transportation policies to access by (1) seeking to maximize the scenic and recreational qualities of Highway 1 and other coastal routes, e.g., by locating major transportation corridors further inland; (2) encouraging bus and other forms of public transit along coastal routes, in part for recreational users; (3) requiring adequate parking in new residential and commercial developments along the coast; and, perhaps most important, (4) recommending that the future coastal agency have review and veto authority over the plans of transportation agencies within the (5-mile-wide) coastal zone.[35] While the plan recommended that Highway 1 in rural areas be retained as a "scenic, two-land road," thereby effectively limiting its use as a recreational corridor, it also recommended that "capacity budgeting" (such as that used at Sea Ranch) be employed as a planning tool to protect recreational users and that the expansion of coastal roads (e.g., lateral roads from inland corridors to the coast) be permitted to increase public access.[36]

Finally, in what was a significant innovation in terms of statutory objectives, the coastal plan followed several early permit decisions in seeking to provide access to the coast for people of low and moderate incomes.[37] For example, it gave priority to lower-cost tourist facilities (e.g., campgrounds and youth hostels). And in what was an even more dramatic departure, it sought to compensate in part for the rapidly rising cost of housing along the coast[38] by giving priority to new residential developments with some moderately priced units and by trying to protect existing moderate-income neighborhoods (e.g., Venice) through resisting pressures by owners who wanted to convert apartments to condominiums and to replace older single-family residences with multiple-family and more expensive units.

In order to gain some sense of how this multitude of policies interacted in specific settings, we can turn to their proposed application in two important cases, Sea Ranch and Malibu.[39] Not surprisingly, the proposals for the Sea Ranch closely followed the overall subdivision conditions discussed in Chapter 5. The plan endorsed the provisions concerning tree trimming and bulk and site restrictions in order to protect views from the highway along this 10-mile stretch. With regard to access, however, it recommended that land be purchased in four areas to provide accessways, parking facilities, and a campground. Transportation continued to be the crucial constraint on ultimate buildout. Arguing that sufficient data on highway capacity were not yet available to permit any definitive decision, the commissions nevertheless suggested that development of the approximately 2200 lots already sold could perhaps be accommodated either through revised projections of the capacity of Highway 1 or through substantial improvements to one or both very primitive roads (one of them privately owned) leading from the coast nearly 30 miles inland to Highway 101. In short, the coastal plan provisions proposed somewhat greater access and visitor facilities than the overall subdivision conditions adopted in June 1974, while at the same time substituting acquisition for the enormous difficulties encountered in attempting to use the police power to wring concessions from the Sea Ranch Association.[40]

Turning to Malibu, the coastal plan clearly envisaged this 27-mile stretch of coast in northern Los Angeles County as one of the principal areas to meet the enormous recreational demands of the county's 6 million residents. Specifically, it recommended that (1) no new commercial facilities (e.g., a proposed shopping center) be permitted near the Pacific Coast Highway; (2) transportation needs be met through expanded bus service rather than widening of the highway; (3) private development on the beaches be prohibited; (4) physical access from the highway to the shore be assured every ½ mile, partially through the police power and partially through acquisitions; and, most important, (5) that a total of 420 acres of beaches and blufftops in eight locations be acquired to provide additional beaches, support facilities, and accessways. While quite expensive (estimated cost of $20 million), these acquisitions, unlike those at Sea Ranch, received the commissions' highest priority because they were designed to meet recreational needs near urban centers.[41]

In conclusion, the commissions appear to have done an admirable job of meeting the statutory requirement to provide "maximum visual and physical use and enjoyment of the coastal zone by the public," while at the same time balancing recreational uses with statutory injunctions to protect natural resources. The development of the plan did, however, involve some important compromises that are worth noting. First, the proposal made by the state staff in the intensity element to eventually establish (through acquisition, deed restrictions, and easements) a continuous band of public ownership along the entire 1100-mile coast, extending generally from the ocean to the nearest coastal road, was gradually whittled down to the much more modest proposal in the plan for a coastal trails system.[42] While the reasons for this substantial change—including enormous acquisition costs, conflicts with agricultural protection policies, and the vociferous objections of coastal property owners—were certainly understandable, the compromise was nevertheless considerable. Second, the heated opposition of coastal residents in many areas, e.g., Marin County and the northern Sonoma Coast, to the location of visitor facilities raised the possibility that extended stays in these areas would be infeasible for people unable or unwilling to stay in campgrounds.[43] Finally, as we saw in the case of Sea Ranch and Malibu, the plan relied heavily on acquisition to actually provide physical access to the coast. While this was understandable given the plan's status as a recommendation to the legislature, it merely postponed the question of what would be done if—as in the case of Sea Ranch—the legislature did not see fit to fund those proposed acquisitions.

B. The Protection of Wetlands

Although none of the Coastal Initiative's planning provisions dealt explicitly with wetlands protection, there can be little question that many of the general injunctions concerning the "preservation and management of the scenic and other natural resources of the coastal zone" and "the continued existence of optimum populations of all species of living organisms" pertained particularly to coastal wetlands. For, as we saw in Chapter 2, one of the principal motivations behind the massive effort that the Sierra Club and other environmental organizations put into passage and implementation of Proposition 20 was a desire to arrest the destruction of the state's wetlands by the construction of ports and marinas, dredging and filling for a variety of purposes, and sedimentation from logging, agricultural, and construction activities in upsteam watersheds.[44]

The coastal plan contained a number of policies dealing directly with the protection of wetlands and estuaries. In general, they required that (a) new development, including diking, filling, and dredging, in existing or restorable wetlands shall only be permitted when there is no net reduction in the functional capacity of the wetland, replacement areas are provided, there is no less environmentally damaging alternative, and the development conforms with an adopted comprehensive estuarine management plan; and (b) degraded wetlands capable of restoration (marshes or diked but unfilled former wetlands) shall be restored.[45] In addition, the plan (1) expressly forbid the dredging or filling of wetlands to accommodate new or expanded recreational boating facilities; (2) prohibited the creation of new ports, except for specialized facilities (e.g., LNG terminals); (3) required that the expansion of existing ports be based upon a comprehensive management plan approved by the successor coastal agency that would minimize adverse environmental impacts; (4) requied buffer zones and other controls on developments adjacent to wetlands and other fragile habitat areas; (5) required that mining of sand, gravel, and other minerals be strictly regulated; (6) proposed that environmentally significant areas, including wetlands, be given first priority (along with urban recreational needs) in any acquisition program; and (7) requested the legislature to establish an oil spill liability fund based on fees from petroleum production and imports.[46]

To control more indirect effects on wetlands (and other wildlife habitat), the plan proposed that comprehensive watershed management plans be prepared by a state agency, with the successor coastal agency having veto authority within the coastal resource management area (generally 5–10 miles upstream). These management plans would relate agricultural, logging, mining, pollutant discharge, dam construction, stream channelization, and other practices to the protection of water quality and would be required to contain, among other things, an erosion and silt-control ordinance. Pending the approval of these management plans and/or local coastal programs, the successor coastal agency would have review authority over all projects within 100 feet of coastal streams and all public works projects in or near such streams within the coastal resource management area; after certification, the coastal agency would be limited to requiring mitigation measures.[47]

In provisionally applying these policies to the area in and around Humboldt Bay, for example, the coastal commissions recommended that (1) development be prohibited on all marshlands and in the lands between Eureka and Arcata (except in a few already developed areas) in order to protect wildlife habitat and agricultural land; (2) a 450-acre open space easement be acquired in the Mad River Slough at the northwest corner of Arcata Bay in order to protect the habitat area; (3) maintenance dredging be permitted to ensure the economic viability of the harbor area, but that dredge spoil be disposed offshore rather than on lands adjoining the bay; and (4) existing off-road vehicle use on the spit south of the bay mouth be strictly regulated in order to protect the dunes (see Figure 5-1 in Chapter 5 for a map of the area).[48] But, perhaps because of the hostility of the

North Coast Regional Commission to watershed management schemes, the plan said virtually nothing about controls on the five major waterways leading into Humboldt Bay.

On the whole, however, the coastal plan appears to have done an excellent job of addressing the major factors adversely affecting coastal wetlands. Particularly noteworthy were the outright prohibition on expansion of marinas, the controls on commercial and industrial ports, the very strict regulation of dredging and filling, and the significant expansion in authority over development in coastal watersheds. While the state staff's original proposal for review authority over dams and other public works projects on any watercourse running to the coast was reduced from an apparently unlimited geographic scope to the 5–10 miles eventually included in the coastal resource management zone, this was nevertheless a considerable expansion of authority and would probably have been sufficient to control almost all adverse impacts on wetlands.

C. Reconciling Development with Environmental Protection

As has been noted several times, the 1972 California Coastal Zone Conservation Act had a strong "tilt" in favor of environmental protection and public access to the coast. About the closest the act came to requiring that these objectives be balanced against those of housing construction, energy production, and other economic development values was when it stipulated that the plan provide for "the orderly, balanced utilization and preservation, consistent with sound conservation principles, of all living and nonliving coastal zone resources" and when it required the plan to contain "a public services and facilities element for the general location, scale, and provision in the least environmentally destructive manner of public services and facilities including the possible siting of power plants in the coastal zone."[49] All but the most ardent environmentalists among commissioners and staff felt that the plan would have to accommodate orderly development along the coast while still providing access and mitigating adverse environmental impacts. Such balancing of often-conflicting values clearly was necessary if the commissions' proposals were to be perceived as "responsible"—and thus seriously considered—by the 1976 legislature. Moreover, if the commissions were to continue to receive funding under the 1972 *Federal* Coastal Zone Management Act, the plan would have to adequately consider the national interest in the siting of energy and other facilities.[50]

As we have already seen, the coastal plan incorporated numerous measures requiring new developments to provide physical access to the coast, not to impede scenic views, and to protect wetlands. In addition, the commissions proposed a number of policies designed to (1) concentrate development in existing developed areas, in part through restricting the conversion of agricultural land and the division of land in rural areas; (2) give priority to coastal-dependent development (e.g., ports) over any other developments, such as housing and shopping centers, with no particular need to be on the coast; (3) severely restrict new development in floodplains and in geologic hazard zones; and (4) protect the special character of communities like Venice and Mendocino.[51]

In addition to these sections applicable to all proposed developments, the coastal plan contained 32 policies specifically related to energy. Remember that the plan was developed in the context of the "energy crisis" following the 1973–1974 Arab oil embargo and the Nixon/Ford emphasis on reducing American dependence on foreign petroleum through, among other things, substantially increasing petroleum production on the continental shelf off California, significantly increasing the number of nuclear power plants, converting gas-fired power plants (such as those in Los Angeles) to coal, and transporting Alaskan oil to the Midwest, much of it through Long Beach and other California ports.[52] Because these proposals would have had a substantial impact on Southern California and, in the extreme, would have transformed many areas of California's coastline into large energy parks, the commissions decided to examine a much broader portion of the energy situation than actually required by the 1972 Coastal Act.[53]

On the one hand, the coastal plan recommended a detailed program of conservation—including restructuring of utility rates, efficiency standards for appliances, and encouragement of solar heating and cooling systems—to be applied within the coastal zone in the event that the California Energy Commission failed to meet its own July 1977 legislative deadline for implementing a statewide energy conservation program.[54] The plan also contained a series of detailed policies relating to offshore and onshore petroleum drilling and to the construction of energy facilities such as electrical generating plants, tanker terminals, and refineries and other petroleum facilities. In general, the latter recommended that (1) development be permitted only if part of a comprehensive, balanced national program of energy conservation and development, including a thorough analysis of alternative measures; (2) stringent safeguards be required to protect wildlife habitat and public safety, to minimize visual impacts, and, where possible, to provide some public access to the coast; and (3) the construction of all energy facilities within the coastal zone be subject to veto by the successor coastal agency.[55] These proposals raised howls of protest from other state and federal agencies with siting authority over power plants, as well as a (partially successful) suit by the petroleum industry seeking to block future grants under the federal coastal zone management program pending modification of state coastal policies.[56]

In general, the commissions went about as far as could be expected in restricting development in rural areas and, at least with respect to energy facilities, in meeting the statutory mandate to provide for such facilities "in the least environmentally destructive manner." In fact, as we shall see in the next chapter, the coastal plan went considerably further than the 1976 legislature was willing to countenance. Finally, the commissions went well beyond their statutory mandate in seeking to promote energy conservation as a means of environmental protection.

D. Proposed Institutions to Implement the Plan

If one returns to the conceptual framework of implementation outline in Chapter 1, the 162 substantive policies that constituted the bulk of the plan dealt with only two of the seven statutory variables affecting implementation: (1) the clarity and consistency of policy objectives and, to a lesser extent, (2) the adequacy of the underlying causal theory. The remaining five variables (and much of the causal theory component) were addressed almost solely in the section specifically devoted to plan implementation. Given the abundant evidence that even rather clear policies must allow some room for administrative discretion and thus the possibility of substantial noncompliance if assigned to unsympathetic implementing officials, probably the most crucial decisions in the entire plan involved the distribution of authority for plan implementation among the successor coastal agency, local governments, and other state agencies.[57]

Given the importance of this topic and the admirable manner in which the framers of the Coastal Initiative had structured implementation during the 1973–1976 interim period, it is somewhat surprising that they provided virtually no guidance about post-1976 institutional arrangements. Instead, the 1972 act simply noted that the plan should contain recommendations concerning "the governmental policies and powers" required for Plan implementation.[58]

As we have already briefly discussed the development and major features of the government powers and funding element, perhaps the best way to analyze this critical section of the plan is to examine it in terms of the statutory variables in our framework: (1) the clarity and consistency of policy directives, (2) incorporation of a sound causal theory identifying the major factors affecting those directives and specifying necessary points of intervention, (3) support for policy objectives among implementing officials, (4) the extent of hierarchical integration within and among implementing agencies, (5) supportive decision rules, (6) opportunities for intervention by supportive interest groups and sovereigns, and (7) adequacy of financial resources provided to implementing agencies. Given that the plan incorporated quite detailed policies to protect natural resources, provide public access, concentrate development in already developed areas, and give priority to coastal-dependent development,[59] the question now becomes the skill with which it structured the subsequent implementation process to achieve these objectives.

As indicated in Chapter 1, a valid causal theory is one in which the important causal factors affecting achievement of statutory goals are understood and subject to the control of (sympathetic) implementing agencies. Let us assume for the moment that the principal factors affecting wetlands and other habitats, access, and development location were understood. Did the plan give the successor coastal agency jurisdiction over those factors? On the one hand, the 1975 plan continued the Proposition 20 tradition of assuming that coastal resources could be protected via acquisitions and regulatory controls on the type and location of development—while doing very little to affect the demand for those resources, particularly for housing and general commercial activity. For example, the plan made very little effort to reduce the pressures created by the property tax system for more intensive use of coastal land.[60] On the other hand, the plan explicitly included within the successor coastal agency's geographical jurisdiction (i.e., within the "coastal resource management area") not only the beaches, dunes, coastal wetlands, and other nearshore areas subject to permit review under Proposition 20 but also agricultural lands influenced by coastal climate, areas where development could indirectly affect coastal recreation areas, and portions of the upstream watersheds of coastal wetlands. The result of this conscious effort to relate geographical jurisdiction to statutory values was an enormous expansion of the commissions' permit review authority from the 1000-yard boundary under Proposition 20 to almost the entire coastal zone (often extending 5–10 miles inland) in the four northern regions, with a highly irregular line in the South Coast and San Diego regions generally extending 1–2 miles inland, but including 5 miles inland in the Malibu area.[61]

This expansion in the geographical scope of the commissions' jurisdiction was accompanied by limitations on their ability to exercise absolute veto power within their domain. The commissions would now have to share permit review authority with local governments and other state agencies. In brief, local governments within the coastal resource management area would be required to bring their general plans and implementing ordinances into conformity with coastal plan policies, subject to certification by the successor coastal agency, within 4 years of the passage of the 1976 Coastal Act. After approval of the local implementation programs, the regional commissions would expire and the successor coastal agency's jurisdiction would be limited to hearing appeals from local decisions and reviewing proposed changes in the local programs.[62] The coastal agency would, however, continue to exercise exclusive permit jurisdiction over areas seaward of the mean high tide line, over energy projects, and, to a limited extent, over major public works and watershed projects; but the basic objective was to force special districts (such as port authorities) and public works agencies to have their projects go through the local certification process.[63]

On the basis of past experience, local and state agencies would show wide variation in their willingness to faithfully implement plan policies. Given that the basic policy orientation of the successor coastal agency would probably be very similar to that of the state commission under Proposition 20,[64] the critical issue was the review authority of the coastal agency over local (and other state) agencies—i.e., the degree of hierarchical integration among implementing institutions. Prior to local program certification, this would not be a problem since the coastal agency would have original jurisdiction over all development within 1000 yards of the coast and over major energy and public works projects within the resource management area.[65] After certification, within the resource management area the coastal agency had authority to revoke certification in the case of local violations, to amend coastal plan policies, to review proposed changes in local programs, and, most important, to hear appeals from local decisions involving (1) variances, conditional use permits, and other exceptions to local ordinances; (2) residential structures of more than four units or other developments of more than 10,000 square feet of floor space; (3) development on a beach, sand dune, floodway, or prime agricultural land, or within 100 feet of a wetland or coastal stream or 100 yards of a coastal bluff; (4) energy or public service facilities serving more than the local jurisdiction.[66] Given the apparently loose grounds for revocation and the liberal scope of appeals, the development review system would thus remain highly integrated (centralized) even after certification of local programs.

In addition, a new agency—the Coastal Conservation Trust—would be created to provide many of the critical management and acquisition capabilities concerning accessways, land banking, and restoration that the coastal commissions had so critically lacked under Proposition 20. Since the trust would likely be more sympathetic to plan objectives than local or state agencies authorized to perform the same functions and would involve fewer veto points, its creation was also conducive to a more integrated implementation system.[67]

Turning briefly to the other aspects of the framework, most of the supportive decision rules—particularly the two-thirds vote requirement on important permit cases—would be retained. In addition, the probability that potentially nonconforming local decisions after certification would be appealed was improved by a new provision allowing appeals to be brought not only by "aggrieved parties"—the liberal rule of standing in Proposition 20—but also by any two members of the state coastal agency.[68] Similarly, participation by environmental groups and other supporters was likely to be enhanced by the new provision permitting recovery of attorney's fees in litigation to prevent or halt violations of the coastal act.[69] On the other hand, the virtual immunity from legislative oversight that the Initiative statute had provided the commissions during the 1973–1976 period would no longer be possible, and the plan made no attempt to centralize oversight in favorable hands. Finally, the plan recommended supposedly adequate funding to allow the coastal agency to complete the expanded task of reviewing detailed local implementation programs over a substantially greater geographical area than had concerned the Proposition 20 commissions.[70]

In short, the coastal plan (if enacted by the 1976 legislature) would have structured the implementation process at least as effectively as had the 1972 Coastal Initiative. The much clearer and more detailed policies incorporated in the plan represented a major improvement over the "tilt" of Proposition 20. The larger geographic jurisdiction linked explicitly to statutory objectives was likewise an important improvement. On the other hand, the implementation system would be less hierarchically integrated than under the Initiative, as the coastal agency's primary jurisdiction in development review would be limited to areas seaward of the mean high tide line. Although the successor agency would have adequate legal authority to veto and/or amend local implementation programs not conforming to plan policies, to hear appeals on critical matters after certification, and even to revoke certification, there was still some question as to whether the state and regional commissions would have sufficient time and resources actually to monitor the development and implementation of detailed local programs for the 15 counties and 45 incorporated cities along the coast; this would be particularly problemmatic after the regional commissions expired, leaving the state coastal agency (and its supporters in environmental groups) with the enormous task of monitoring local compliance throughout an approximately 5-mile-wide stretch along the 1100-mile-long California coast. Despite these potential difficulties, the coastal plan still envisioned a larger state role in protecting natural resources and public access along the coast than that in any of the other coastal states.[71]

IV. Summary

The evidence presented suggests that the 1975 coastal plan substantially conformed to the Coastal Initiative's general objectives of maximizing the public's physical and scenic access to the coast (consistent with resource protection), protecting wetlands and other important wildlife habitat, and impeding the conversion of agricultural land to urban uses. This conclusion flows not only from our review of approximately 75% of the plan's 206 policies but is also suggested by the survey data presented in Table 7-1 indicating that about 79% of the informed activists who had voted for Proposition 20 supported the plan, while 81% of those who had opposed the Initiative likewise opposed the plan.

These survey data also suggest, however, that the commissions did not have much success in reducing opposition from the development community, many state agencies, and people concerned with the protection of private property rights. While changes in the political complexion of local governments—arising partially from the political organizations spawned during the Proposition 20 campaign—apparently made some local officials more receptive than in 1972 to a state role in coastal land use decisions, the County Supervisors' Association and the League of Cities felt that the commissions' proposals for carrying out the plan did not delegate sufficient authority to local governments.

In short, by faithfully implementing the substantial changes in coastal re-source utilization mandated by Proposition 20, the coastal commissions essen-tially assured that the coastal plan would arouse at least as much controversy in the 1976 legislature as had the coastal bills 4 years previous.

NOTES

1. For a discussion of this topic, see Robert Healy et al., Protecting the Golden Shore: Lessons from the California Coastal Commissions, (Washington, D.C.: Conservation Foundation), chap. 2. As we shall see later in this chapter, the permit review process had a particularly strong influence on at least one of the nine elements that went into the coastal plan, that dealing with intensity of development.
2. The coastal plan was formally adopted by the state commission in September 1975. Although the Outline for Coastal Zone Planning approved by the state commission in June 1973 indicated (p. 10) that "as elements of the Coastal Conservation Plan are adopted—even tentatively—they will become the basis for subsequent permit decisions," there is not a great deal of documented evidence that this actually occurred. In our analysis of (a random sample of) permit appeals acted on by the state commission—which included a content analysis of the entire record, including the staff recommendation and the minutes of the state commission's deliberations—on only 28% of the 46 appeals voted on from January through June 1975 was explicit reference made to plan policies, even though seven of the nine elements had been previously approved by the commis-sion. This finding should, however, be treated with some caution because of the small number of cases in the sample and the fact that this was not a major concern in the coding.
3. For the statutory provisions relating to plan preparation, see California Public Resources Code, Section 27300–27320 (1974).
4. The "coastal zone"—the planning jurisdiction of the commissions—was defined by Section 27100 as the area from the seaward boundary of the state's jurisdiction to "the highest elevation of the nearest coastal mountain range . . . or five miles inland from the mean high tide line, whichever is the shorter distance."
5. This discussion is based largely upon interviews with Jack Schoop (chief planner), 21 May 1975, and with Bill Travis (assistant chief planner), 4 November 1974, as well as Joe Bodovitz, "The Coastal Zone: Problems, Priorities, and People," Address to the Conference on Organizing and Managing the Coastal Zone, Annapolis, Maryland, June 13–14, 1973.
6. Recall that the chairman, executive director, and chief planner of the state commission had all previously held similar posts with BCDC. For discussions of the BCDC planning effort, see Rice Odell, The Saving of San Francisco Bay (Washington, D.C.: Conservation Foundation, 1972), and E. Jack Schoop and John Hirton, "The San Francisco Bay Plan," Journal of the American Institute of Planners (January 1971), pp. 2–10.
7. This discussion is based on (1) interviews with Schoop and Travis, (2) the Bodovitz paper at the Annapolis Conference, and (3) the memorandum from Bodovitz to regional commissioners and staff, "Outline for Coastal Zone Planning," June 14, 1973. For an analysis of the apparently minimal use of scientific information by the commissions in the planning effort, see Healy et al., Protecting the Golden Shore, chap. 5.
8. Bodovitz's June 1973 memo on the planning process, as well as interviews with Schoop and Travis.
9. As might be expected, suspicion of the state staff was most vocal in the North Coast region (interviews with Jack Lahr, executive director, 24 September 1974, and John Mayfield, chair-man, 27 September 1974), where the regional staff prepared its own background technical document for each element. But similar charges were also voiced by North Central staff (inter-

views with Dave Dubbink, chief planner, 11 November 1974, and Michael Fischer, executive director, 26 August 1975), as well as by CCEEB officials (interview with Mike Peevey, 31 October 1975). On the other hand, the whole issue of staff domination was vigorously disputed by several state staff and commissioners, who pointed to the extensive formal contacts and the similarity of views of the two groups (interviews with Travis and Bob Mendelsohn).

10. For a more detailed discussion, see Paul Sabatier, ed., *The Development of the California Coastal Plan* (Unpublished manuscript, University of California, Davis, 225 pp). Changes in each of the policies in the three elements were coded at different stages in the process, i.e., the original state staff recommendations, the hearing drafts in each region, the policies adopted by the two regional commissions, the synthesis of the regional recommendations prepared by the state staff, and the policies ultimately adopted by the state commission. Each policy change was then coded as to its *magnitude* (e.g., no change, minor editorial change, substantive change, or major innovation) and its *direction* on a number of different accounts: (1) environmental protection, (2) state versus local control, (3) effect on community economic welfare, (4) public versus private control of resources, and (5) coastal commission versus other state agencies. Finally, on the basis of the documentary record and interviews with key staff, commissioners, and interest group leaders, an attempt was made to identify the people most responsible for the changes, as well as the underlying dynamics of the planning process.

11. Daniel Ray, "The Coastal Land Element," in *The Development of the California Coastal Plan*, ed. Paul Sabatier, chap. 2.

12. Even the two cases in which the regional recommendations were coded as providing less environmental protection were somewhat marginal, as they involved (1) the deletion of the state staff's proposed policy protecting rare and endangered species as a "motherhood issue," and (2) the deletion of several policies relating to air pollution as essentially irrelevant to the windswept coastal portions of this region. Ibid.

13. David Lambert, "The Intensity of Development Element," in *The Development of the California Coastal Plan*, ed. Paul Sabatier, chap. 3.

14. This contrasts with the conclusions reached by Healy (*Protecting the Golden Shore*, chap. 2) concerning the rather minimal effect of permit review on the formulation of plan policies. That may have been true for most elements, but certainly not for intensity.

15. In part because of the minor role played by commissioners in the development of the first three elements, the North Central Commission began assigning two or three commissioners to work closely with staff from the very beginning of the planning process. On some elements, such as this one, these "commissioner-consultants" played a major role.

16. This discussion is based upon Paul Sabatier, "The Government Powers and Funding Element," *The Development of the California Coastal Plan*, ed. Paul Sabatier, chap. 4.

17. It is important to realize that much of the anger expressed was probably a result of more general frustrations with (1) the high unemployment in the timber industry because of the national recession in the housing industry, (2) recent decisions by state courts and agencies requiring environmental impact reports on timber harvest plans, and (3) recent regional water quality control board decisions involving more stringent septic tank regulations. In short, the commission served as a focal point for local residents' frustrations with enviornmental regulation in general.

18. Under existing California law, state agencies were not required to conform to local general plans—a loophole that often seriously disrupted local planning efforts.

19. After certification of local programs, appeals were limited to (1) variances or conditional use permits, (2) large developments, (3) any development within 100 yards of a coastal bluff or 100 feet of a wetland or coastal stream, or (4) energy or public works projects serving an area beyond that of the local government involved. The entire certification and appeals process was limited to a Coastal Resource Management Area to be mapped by the regional commissions. Designed to

include significant recreational and natural resources, as well as coastal-dependent agriculture, it wound up including virtually the entire coastal zone in the four northern regions and Malibu, and about 1–2 miles inland in the remainder of the South Coast and San Diego regions. Pending certification of the local plans, however, the commissions' permit jurisdiction was limited to the 1000-yard area as well as the conversion of agricultural land and major public works projects throughout the Coastal Resource Management Area. In general, see California Coastal Commissions, *California Coastal Plan*, December 1975, pp. 179–190, 278–285, for delineation of Coastal Resource Management Area.

20. For discussions of the major activities in each region, see Healy *et al.*, *Protecting the Golden Shore*, pp. 35–37. Interviews with state staff indicated they felt that the North Central region had done the best job of providing a wide variety of forums for participation, while the South Coast and South Central Commissions had been the least effective. This was unfortunate given the large number of legislators from the latter regions.

21. Rosenbaum's study involved an August 1976 survey of 121 randomly selected participants (84% response rate) in the planning process drawn from the state commission's mailing list. There were similar studies of Hawaii's 5-year boundary review (1974), Oregon's statewide planning goals program (1974), and North Carolina's coastal management program (1974–1976). Following are the relevant responses:

	California	Oregon	Hawaii	North Carolina
Was information made easily available to all affected citizens?				
Yes	53%	58%	32%	30%
No	29%	27%	48%	60%
Don't know	18%	15%	20%	9%
	100%	100%	100%	99%
Were affected citizens provided with enough opportunities to express their preferences to decision-makers?				
Yes	59%	61%	56%	44%
No	23%	27%	36%	41%
Don't know	18%	12%	8%	15%
	100%	100%	100%	100%

See Nelson Rosenbaum, *Report on the Citizen Participant Survey* (Washington, D.C.: Urban Institute, 1977).

22. The survey was administered in the fall of 1977 and involved random samples of respondents from four populations: (a) the master mailing list of the commissions, (b) the list of "very concerned" participants in the development of one or more elements of the plan, (c) permit applicants whose cases were appealed to the state commission, and (d) commissioners and staff from all seven commissions, as well as legislators from the two natural resources committees. The overall response rate was 53%. For a more complete discussion see Appendix A.

23. In fact, given the extensive opportunities for participation, it seems quite probable that the expressions of dissatisfaction from businessmen and property owners should be interpreted as a perception that their arguments were not being seriously *considered* by the commission.

24. The Sierra Club and other environmental groups testified on each of the elements at the regional and state levels in the study cited previously (Sabatier, ed., *The Development of the California Coastal Plan*, chaps. 2–4). Moreover, of the randomly selected activists in the planning process (exclusive of commission officials) who responded to our 1977 evaluation survey, 72% reported they had voted for Proposition 20, 20% listed themselves as representatives of enviornmental groups, and 60% supported the coastal plan. Similarly, about 70% of the California respondents in Rosenbaum's study favored efforts to expand the role of government in land use planning and to weigh environmental protection more heavily than economic development (Rosenbaum, *Report on the Citizen Participant Survey*, Table B-2). In short, about two-thirds of the active participants in the planning process could loosely be considered "environmentalists."

25. For example, at the statewide hearing on government powers and funding, organizations such as the Western Oil and Gas Association (WOGA), the California Chamber of Commerce, the California Coordinating Council (a coalition of property rights groups), realtors, and several segments of the construction industry opposed any successor coastal agency and urged that the plan merely be used as guidelines by existing state and local agencies. Their position was dramatically summarized by the WOGA spokesman: "How should this commission's plan be carried out? It should be carried out the door on a stretcher for drastic surgery!" See State Commission, *Minutes*, September 3, 1975, pp. 14–25; also CCEEB, *Summary Review and Analysis of Government Powers and Funding*, 1975, pp. 1–5.

26. In the spring of 1975, members of the Coastal Alliance representing various environmental groups began meeting privately with representatives from CCEEB, local government associations, and others in an effort to explore differences and try to reach agreements on some issues. Generally, somebody from the state commission or its staff would be present as an observer. Representatives from the Sierra Club were also invited but did not participate very actively. While these meetings didn't reach any momentous decisions, they did open a dialogue that would prove to be very useful during the 1976 legislative session. (Interviews with Janet Adams, executive director of the Coastal Alliance, 2 December 1975, and Tim Leslie, legislative advocate for CSAC, 14 October 1975).

27. It is assumed that attitudes toward Proposition 20 had both a direct effect on support for the 1976 plan and an indirect effect via evaluation of the opportunities provided for participation; in other words, those who had supported Proposition 20 were likely to view the commissions' participation efforts positively, thereby enhancing their evaluation of the 1976 plan. These direct and indirect effects can be separated through the use of causal modeling or path analysis [see Herbert Asher, *Causal Modeling*, Sage University Papers on Quantitative Applications in the Social Sciences (Beverly Hills: Sage, 1976]. Using our survey of 117 "very active" participants in the planning process (exclusive of commission officials) as a data base, we find the following path coefficients:

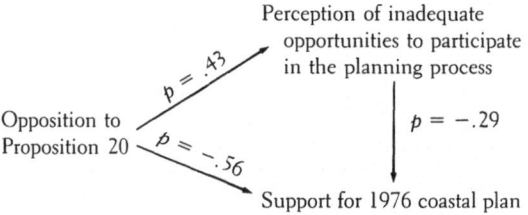

Given these coefficients, we can then estimate the direct and indirect effects of attitudes toward Proposition 20 and toward the commissions' participatory effects on respondents' attitudes toward the 1976 plan:

| | Effects on attitudes toward 1976 plan | | | | | Proportion of |
Predictor variable	Direct + effect	Indirect effect	=	Total effect (r)	R^2	explained variance
Prop. 20	$-.56$	$(.43)(-.29) = -.12$		$-.68$	.47	.87
Eval. of partic.	$-.29$	none		$-.29$	.07	.13
					.54	1.00

In other words, this model explains 54% of the variance (R^2) in respondents' support for the 1976 plan. Of this explained variance, 87% was accounted for by the direct and indirect effects of attitudes toward Proposition 20, while only 13% could be attributed to the effect of respondents' evaluation of the participatory efforts (independent of respondents' attitudes toward Proposition 20).

28. The planning provisions are contained in California Resources Code, Section 27300-27320 (1974).

29. Because different species are often engaged in a zero-sum competition for scarce resources, it is simply impossible to provide for "the continued existence of optimum populations of *all* species of living organisms" (emphasis added). While it is possible to optimize population levels of different species within a given area if one carefully specifies the constraints under which the optimization is to occur, this requires detailed data far beyond the capabilities of the commissions during the 1973–1976 period and assumes a much more developed body of ecological theory than actually exists [see W. B. Drury and I. C. T. Nisbet, "Succession," *Journal of Arnold Arboretum* 54 (1973), pp. 331–368].

30. Because of the tremendous length of the plan itself—with 205 policies covering 168 single-spaced, double-columned pages—our discussion of the policies covering access, wetlands, development, and future implementing institutions will be based primarily upon the 20-page summary developed by CCEEB and by the staff of the *State Coastal Report* (CCEEB, *Summary of the California Coastal Plan*, San Francisco, Fall 1975). In a 23 December 1975 letter to Michael Peevey, CCEEB's executive director, Joe Bodovitz indicated that the state commission staff had found only five minor errors in the summary of the 205 policies and, in general, found it to be "remarkably accurate." We have also relied upon summaries of the plan prepared by the commission itself (California Coastal Commission, *Coastal Plan*, p. 1–22) and by Peter Douglas and Joseph Petrillo in their excellent article, "California Coast: The Struggle Today—A Plan for Tomorrow," *Florida State University Law Review* 4 (October 1976), pp. 315–349, especially pp. 319–328. On certain points, however, particularly in the discussions of Sea Ranch and Malibu, we have relied on the text of the plan itself.

31. California Coastal Commissions, California Coastal Plan, Policies 44–56, 152.

32. Ibid., Policy 104.

33. Ibid., Policies 121–130, 158.

34. Ibid., Policies 155, 131–148.

35. Ibid., Policies 99–111.

36. Ibid., Policies 101, 102, 104.

37. Ibid., Policies 125–126, 58. The state commission first decided to force large developments displacing elderly persons to provide moderately priced units for the elderly in July and November 1973; see Douglas and Petrillo, "California's Coast," pp. 230–231. Other major decisions involving the protection of low- and moderate-income neighborhoods involved a San Diego barrio and the Venice area of Los Angeles.

38. Rising tax assessments were particularly a problem in Venice, where the Los Angeles City Council had encouraged (e.g., through zoning ordinances) the conversion of the area from single-family residences to higher-density, more expensive residential and commercial development, which would, of course, generate substantially greater revenues from property and sales taxes. The problem was that the process of conversion, once begun, was very difficult to arrest because the tax assessor was basing his assessment on what the property would bring on the open market rather than on its present use (see State Commission, *Minutes:* June 3, 1975, pp. 9–11). Moreover, there is little question that the restrictive policies pursued by the commissions (1) raised property values (and thus tax assessments) of developed residential parcels within the permit zone and (2) by limiting the size of visitor-serving facilities, increased the prices paid by consumers [see the survey of county assessors reported in "Impact of Proposition 20 on Coastal Property Mixed," *State Coastal Report* 3 (December 31, 1975), pp. 13–14]. For criticisms of the distributive consequences of the Coastal Plan, see Eugene Bardach *et al.*, *The California Coastal Plan: A Critique* (San Francisco: Institute for Contemporary Studies, 1976), pp. 39–50, 69–92. For more balanced discussions, see Dean Misczyncki, "The Awkward Economics of Coastal Planning," *Southern California Law Review* 49 (May 1976), pp. 737–748; Healy *et al.*, *Protecting The Golden Shore*, chap. 4; and Kneisel, *Economic Impacts of Land Use Control with Special Reference to the California Coastal Zone Commissions* (Davis, California: Institute of Governmental Affairs, 1979).

39. Definitive decisions concerning the application of plan policies would, of course, have to await the 1976 legislature's reformulation of those policies and the subsequent development of local coastal programs under the 1976 act.

40. For the provisions relating to Sea Ranch, see California Coastal Commissions, *California Coastal Plan*, pp. 210, 316–317. Under Policy 155 of the plan, however, these proposed acquisitions would be accorded second priority by the commissions.

41. The 420-acre figure comes from the final acquisition list of March 1976: California Coastal Zone Conservation Commissions, "Recommended Coastal Properties for Public Acquisitions," March 1976, pp. 35–36.

42. Policy 11 of the state staff's original (September 10, 1974) proposals on intensity of development indicated that the long-term goal should be public ownership and use of a band of land paralleling the coast, the width of which should vary according to local conditions but should be large enough to permit significant opportunities for public use and enjoyment (e.g., from several hundred feet to several hundred yards wide in major urban areas, and, to the extent possible, including all the area between the first public road and the mean high tide line. In rural areas coastal meadows could be leased out for view-protecting agricultural use). At the recommendation of several regions, this proposal was trimmed back considerably by the state commission in its meeting of April 8–9, 1975; see Joan Sweeney, "Plan for Public Land on Coast Curtailed," *Los Angeles Times*, 10 April 1975.

43. While there is little doubt that local residents vehemently opposed visitor-serving facilities in their communities, there is some evidence that a previous survey of coastal users (the Tomales Bay Study) revealed that users also preferred to leave the area in as natural a state as possible; at any rate, the North Central Commission essentially acquiesced to the wishes of local residents, both on the intensity element and in denying a very well-designed small inn along the Marin coast (see North Central Regional Commission, *Minutes*, October 24, 1974, pp. 4–8, and November 14, 1974, pp. 12–21).

44. As noted in chapter 2, California Fish and Game officials estimated a 67% reduction in coastal wetland habitat over the past hundred years because of the construction of boating facilities, filling, and sedimentation. With respect to the latter, in a Eureka restaurant there is a marvelous photograph of Humboldt Bay taken about 1910 showing the surrounding hills considerably higher, and the bay considerably larger, than at present. For a brief review of the literature on

the value of wetlands, see William Odum and Stephen Skjei, "The Issue of Wetlands Preservation and Management: A Second View," *Coastal Zone Management Journal* 1 (Winter 1974), pp. 151–163, 227–253.

45. This is a verbatim account of most of Plan Policy 15 taken from CCEEB, *Summary of the California Coastal Plan*, pp. 2–3. This policy is expanded in Plan Policies 17–18.

46. California Coastal Commissions, *California Coastal Plan*, Policies 146–147, 116–118, 28, 155, 12–13, and 41.

47. Ibid., Policies 14, 21–24, and Powers and Funding Policy 24.

48. Ibid., pp. 204, 296–297.

49. California Public Resources Code (1974), Sections 27302(c) and 27304(e)(6). On the other hand, the act also required that the plan contain a component specifying "the ecological planning principles and assumptions to be used in determining the suitability and extent of allowable development" (Section 27304b). While such principles were eventually published in the *Coastal Plan* (p. 19), they were developed at a very late date (i.e., they did not appear in the preliminary plan) and were not, for example, included in the state staff's original policy recommendations on the intensity element, or in the recommendation of any of the regional commissions to the state commission.

50. The Coastal Zone Management Act of 1972 (P.L. 92-583) is found in U.S. Code, Title 16, Section 1451-1464 (1973). Section 306(c)(8) made administrative grants to state agencies contingent upon the finding that "the management program provides for adequate consideration of the national interest involved in the siting of facilities necessary to meet requirements which are other than local in nature;" this was amended in 1976 to specifically mention energy facilities, although the implication was clear from the beginning. Moreover, Section 307(b) stated that "the Secretary of Commerce shall not approve the management program submitted by a state pursuant to Section 306 unless the views of Federal agencies principally affected by such program have been adequately considered. In case of serious disagreement between any Federal agency and the state, the Secretary, in cooperation with the Executive Office of the President, shall seek to mediate the differences." Given the Nixon/Ford emphasis on increasing energy production as part of Project Independence, it was clear that the coastal plan would have to carefully consider the siting of energy facilities.

51. California Coastal Commissions, *California Coastal Plan*, Policies 57–70, 30–39.

52. U.S. Senate, Committee on Interior and Insular Affairs, *Hearing on Project Independence*, 93rd Cong., 2nd Session, November 21, 1974.

53. Douglas and Petrillo, "California's Coast," pp. 323–324; California Coastal Commissions, *California Coastal Plan*, pp. 91–138.

54. Ibid., Policies 71–75.

55. Ibid., Policies 74–98, 12–13, and Powers and Funding Policies 11 and 23. For example, Policy 83 recommended that offshore and onshore oil drilling require (1) the least environmentally hazardous and aesthetically disruptive well sites; (2) assurance of the geologic safety of all sites; (3) the consolidation of drilling, production, and processing sites; (4) the use of submerged production systems where feasible and environmentally safe; (5) a preference of platforms over islands, with their impacts minimized; (6) minimal impacts from onshore petroleum facilities; and (7) the prohibition of liquified and gas extraction that would cause subsidence. In addition, related policies recommend separate permit review for petroleum exploration and development, the submission of all exploratory and production data from surveys or the drilling of wells to the Division of Oil and Gas within 60 days, and nine requirements for the approval of federal OCS petroleum development. Parenthetically, the requirements for the siting of power plants (Policies 77–79) were at least as detailed and restrictive.

56. In response to a September 1977 suit brought by the Western Oil and Gas Association (WOGA) seeking to block federal approval of California's coastal zone management program and distribu-

tion of $2.8 million to the coastal agency and local governments, in January 1978 the parties agreed—pending resolution of the merits of the suit—to a release of the funds if the state waived its right under the Coastal Zone Management Act to request federal agencies to conform to state policies regarding oil production in the outer (i.e., federally owned) continental shelf (California Coastal Commission, *Coastal News*, January 1978, pp. 1–2). The court subsequently ruled in favor of the commerce department, but as of this writing the case is on appeal by WOGA.

57. See the literature on policy implementation reviewed in chapter 1, particularly the studies of substantial noncompliance with relatively clear policy directives forbidding the use of prayer in schools and the existence of racially segregated dual school systems.

58. California Public Resources Code (1974), Section 27304e.

59. California Coastal Commission, *California Coastal Plan*, Policy 1.

60. It did, however, recommend a substantial revision in the manner by which timber was taxed, as well as a variety of means (including tax assessments based on present use rather than on market value) to impede the conversion of agricultural lands; ibid., Policies 31 and 39. But these measures would not ameliorate the effects of the property tax in existing urban areas.

61. The "coastal resource management area" was to include significant coastal resources such as beaches, dunes, wetlands and their immediate drainage basins, agricultural lands influenced by coastal climate, acquisition areas, special coastal communities, and areas where development could directly or cumulatively affect coastal recreation area access. Ibid., pp. 180 (definition) and the plan maps on pp. 280–420.

62. Ibid., Powers and Funding Policies 4–21.

63. Ibid., Powers and Funding Policies 21–25.

64. The composition of the state and regional commissions would remain unchanged (in terms of state vs. local appointees) during the interim period pending certification of the local coastal programs. After the dissolution of the regional commissions, the state coastal agency would be composed entirely of appointees of the governor, Assembly speaker, and Senate Rules Committee (Ibid., Powers and Funding Policies 16–17).

65. Ibid., Powers and Funding Policy 19.

66. Ibid., Powers and Funding Policies 11–13, 20, 26, 29.

67. The Coastal Conservation Trust was to have a governing board composed of the chairman of the coastal agency, representatives from the Resources Agency and the Department of Finance, and two gubernatorial appointees. It was given a wide range of acquisition and management powers, with its acquisition proposals exempt from review by the Public Works Board. Ibid., Powers and Funding Policies, 40–43.

68. Ibid., Powers and Funding Policy 11.

69. Ibid., Powers and Funding Policy 35.

70. The plan (Powers and Funding Policy 44) recommended a $2–3 million annual appropriation from the legislature, as well as other funds to service a coastal acquisitions bond issue; in so doing, it suggested a wide variety of state and federal sources, including tideland oil revenues.

71. For discussion of the distribution of authority among local governments, the state coastal agency, and other state agencies in other coastal states, see Jens Sorensen, *State–Local Collaborative Planning: A Growing Trend in Coastal Zone Management*, Prepared for Office of Coastal Zone Management and Office of Sea Grant (Washington, D.C.: Government Printing Office, 1978); and Office of Coastal Zone Management, U.S. Department of Commerce, *State Coastal Zone Management Activities 1975–76* (Washington, D.C.: Government Printing Office, December 1976).

8

EVALUATION AND REFORMULATION OF THE COMMISSIONS' MANDATE
The 1976 Legislative Session

In the opening chapter we suggested that the implementation of a regulatory statute be viewed as a cyclical process involving a number of stages: (1) the policy outputs (decisions) of the implementing agencies, (2) target group compliance with those decisions, (3) the *actual* impacts (both intended and unintended) of those decision, (4) the *perceived* impacts of agency policies as seen by the general public and relevant political elites, and (5) the political system's evaluation of the implementation process through revisions (or attempted revisions) in the basic statute. If extensive revisions emerge from the legislative process, the cycle simply begins anew.[1]

The 1972 California Coastal Initiative created a system of new agencies and charged them with two tasks. The first was to regulate all development within the 1000-yard permit zone during the 1973–1976 period. Chapters 5 and 6 examined the nature of the coastal commissions' permit decision (including their conformity with statutory objectives), target group compliance with those decisions, and the actual impacts of permit review on beach access, wetlands protection, and housing costs. The second task—discussed in Chapter 7—was the preparation of a coastal plan containing the commissions' recommendations concerning the policies and institutions to govern the future utilization of coastal resources. In forcing the commissions explicitly to address the adequacy of their statutory mandate and in creating the commissions as temporary bodies whose mandate would expire at the end of 1976, the Coastal Initiative formalized the cyclical nature of the implementation process to a greater extent than is normally done.

This chapter examines the reformulation of the commissions' mandate by their legislative and executive sovereigns in 1976. While the deliberations of the legislature and Governor Brown were essentially organized around the coastal plan (as introduced in bill form), it was clear from the outset that their decisions would be based in part upon their overall evaluation of the commissions' performance during the 1973–1975 period. As Joe Bodovitz[2] observed: "In addition to the Coastal Plan per se, what was on trial in the 1976 Legislature was coastal zone management as carried out by the commissions under Prop. 20 over three years. This is borne out by the fact that the legislative hearings on the Plan were often dominated by the so-called horror stories where anybody who was ever denied a permit would get up and tell the committee what had happened to him." In effect this was probably one of the first examples of what can be expected when a controversial regulatory agency is confronted with a formal "sunset" provision.

The reformulation of the commissions' mandate could be expected to be a function of (1) legislators' perceptions of public priorities, (2) their perceptions of the commissions' impacts on important values during the 1973–1976 period, (3) the ability of competing interest groups to present their cases, and (4) the skill and commitment of legislative and executive proponents of different views. In terms of the set of conditions presented in the initial chapter, this would imply that new legislation consistent with the Proposition 20 mandate was likely to emerge to the extent that (1) changing socioeconomic conditions and other factors had not substantially altered public priorities since November 1972; (2) the commissions were perceived as having acted fairly and having protected access, scenic views, and wetlands without unreasonable dislocations of housing markets, energy production, and other competing values; (3) environmental groups and other supporters were able to convince legislators that they represented a legitimate public need and/or a credible political threat; and (4) one or more committed and skillful legislative (or executive) supporters of coastal legislation emerged. Of course, the legislative process on any specific bill is also dependent upon personal relationships and other idiosyncratic features.[3] Thus, these conditions are only suggestive of some of the more important factors, rather than being determinative in any strict sense.

In examining the response of the legislature to the coastal plan and to the commissions' overall performance, we shall first set the context of the 1976 legislative session in terms of public and elite opinion concerning the commissions, as well as the resources of supportive constituency groups and key legislators. Next, we shall briefly review the tortuous path of coastal legislation during the session and try to isolate some of the more important factors contributing to the eventual passage of the 1976 Coastal Act. Finally, we shall compare the provisions of the 1976 act with the 1972 initiative and the 1975 plan.

I. The Political Context of the 1976 Legislative Session: Public and Elite Opinion and Resources

In setting the stage for discussion of the 1976 legislature's consideration of coastal bills, we shall go from those general contextual factors likely to have the most indirect and diffuse effects on legislators' deliberations (i.e., changes in socioeconomic conditions and public opinion) through the attitudes and resources of the interest groups and agencies concerned with coastal policy and finally to attitudes and resources of important legislative and executive officials on the eve of the 1976 session.

A. Socioeconomic Conditions and Public Opinion

As will be recalled from Table 3-1 and Figure 3-1, Chapter 3, California's economy was hard-hit by the nationwide recession of 1974–1975, as unemployment statewide rose from 7.0% in 1973 to 9.9% in 1975. The construction industry was particularly affected, as employment in this sector actually *declined* by about 12% both statewide and in coastal counties during this period. Fortunately for the fate of coastal legislation, however, 1976 marked a reversal of this trend, as overall unemployment improved slightly to 9.2% and employment in the construction industry rose about 5% from its 1975 nadir. Nevertheless, the legislature could still be expected to be cautious about imposing regulatory controls with potentially adverse economic consequences.

We also saw in Chapter 3 that public support for environmental protection declined somewhat during these years, as people became concerned with the potentially adverse effects on jobs. In fact, a January 1975 poll sponsored by California Council for Environmental and Economic Balance (CCEEB) indicated that fully 64% of California adults assigned greater priority to jobs than to environmental protection. That same poll revealed, however, that Californians may have viewed the coast as a special place, as 37% supported a moratorium on coastal construction while 36% opposed it (with the remaining 25% undecided).

In November 1975 CCEEB sponsored another statewide opinion poll, this time directed specifically at coastal regulation.[4] Some of the more important results are shown in Tables 8-1 and 8-2. Given the rather dismal state of the state's economy at the time, the poll indicated a surprising degree of support for the regulation of coastal development and for the continuation of a coastal agency. For example, part A of Table 8-1 deals with responses concerning the proper distribution of authority for a wide range of regulatory activities among local governments, the state coastal commission, other state agencies, and no agency at all. The results reveal, first of all, a substantial consensus in favor of the regulation of a wide variety of activities along the coast, as the percentage of

TABLE 8-1

PUBLIC OPINION CONCERNING THE PROPER LOCUS AND SCOPE OF COASTAL REGULATION, NOVEMBER 1975
$(N = 763)$[a,b]

PART A: SUBSTANTIVE JURISDICTION

Subject	Preferred locus, if any, of regulation					
	Local governments	State coastal commission	Other state agencies	No agency	Other/don't care	Total
Coastal recreation and views						
Tidal area	24%	56%	7%	3%	10%	100%
Preserving scenic views	30%	53%	10%	2%	4%	99%
Coastal recreational facilities	30%	50%	13%	2%	6%	101%
Beaches	38%	49%	4%	4%	5%	100%
Access to beaches	46%	34%	7%	7%	5%	99%
Major energy and other developments[c]						
Offshore oil development	23%	46%	20%	3%	9%	101%
Energy facilities	26%	42%	25%	1%	5%	99%
Ports	29%	43%	20%	1%	7%	100%
Highway construction	22%	40%	32%	2%	5%	101%
Water and sewage facilities	53%	29%	14%	0%	5%	101%

	Anywhere between ocean and bluff or first road or row of houses (approx. 1000 feet)	Anywhere between ocean and start of uplands (approx. 1/2 mile)	Anywhere between ocean and coastal mountains (approx. 5 miles)	Other	Don't know	Total
General development[c]						
Construction of high-rise apartments	52%	31%	10%	3%	5%	101%
Limitations on housing and development	50%	31%	11%	2%	6%	100%
Design and architectural features	45%	25%	15%	8%	6%	101%
Rezoning agricultural land to other uses	39%	30%	18%	5%	9%	101%
Other[a]						
Air quality in coastal area	22%	38%	28%	2%	10%	100%

PART B: GEOGRAPHICAL JURISDICTION

	Anywhere between ocean and bluff or first road or row of houses (approx. 1000 feet)	Anywhere between ocean and start of uplands (approx. 1/2 mile)	Anywhere between ocean and coastal mountains (approx. 5 miles)	Other	Don't know	Total
Extent of coastal area that should be included in these regulations[d]	10%	20%	39%	26%	2%	97%

[a] Given a sample size of 763, the confidence intervals are ±3–4%.

[b] Source: Public Response Associates, A Survey of Attitudes Toward the California Coastal Area, A study conducted for the California Council for Environmental and Economic Balance, San Francisco, December 1975.

[c] These subject headings grouping individual topics were added by the authors and thus were not part of the questionnaire presented to respondents.

[d] It is unclear from the question precisely which agency or level of government should have jurisdiction within this area. However, the 96 respondents who did not think that the coastal commissions or other state agencies should have jurisdiction over any of the 15 substantive topics were not asked this question.

people favoring no regulation at all never exceeded 7% for any of the 15 substantive topics. Second, a substantial plurality of respondents felt that the state coastal commission—rather than local governments or other state agencies—should have jurisdiction over projects affecting coastal views and recreation and over major energy and other developments, while at the same time preferring that general jurisdiction over development remain in the hands of local governments.[5] Third, part B of Table 8-1 indicates that about 59% of respondents felt that the zone of coastal regulation should extend at least ½ mile inland (i.e., approximately the permit area under Proposition 20), with 39% apparently willing to retain the 5-mile coastal zone envisaged in the coastal plan. It is unclear from the general nature of the question, however, precisely which agency should have jurisdiction over which functions within this zone.[6]

Fourth, the data presented in Table 8-2 indicate that the public, by almost a 2 to 1 plurality, felt that the commissions had done a good job of balancing coastal protection and development during the 1973–1975 period. There were also slight pluralities in favor of a moratorium along the coast and a $250 million bond issue to acquire additional coastal beaches and parks. On the other hand, an overwhelming 67% of respondents felt that the state should purchase parcels on which state regulation precluded development. Finally, there was little variation in responses to most of the questions by people in different areas of the state and between those in coastal and inland counties.

Interpreting the importance of these survey results is complicated by two factors. First, most of the questions were fairly general and did not ask respondents to suggest precise delineations of substantive and geographical jurisdiction, or to determine the precise line between legitimate regulation and a "taking" of property. While the results indicated substantial satisfaction with the commissions' past performance, support for continued coastal commission jurisdiction in the nearshore area and over major energy projects, and, conversely, strong support for private property rights, they provided only general guidance to those responsible for the development and passage of coastal legislation. Second, this was a privately funded poll whose results were not publicly released during the 1976 legislative session. Although the results certainly influenced some of CCEEB's positions with respect to the feasibility of a coastal bond act and the geographic scope of the commissions, they were discussed in only a very general way with a few legislators and other actors, and thus their effect on the subsequent course of legislation is impossible to ascertain.[7]

Nevertheless, several conclusions do seem reasonable. The substantial support for the coastal commissions probably encouraged CCEEB to continue its pursuit of a reasonable compromise with environmental groups and may have indicated that a successful initiative based upon the general thrust of the plan was not out of the question were the legislature to become deadlocked and not renew the commissions' mandate. In addition, the general thrust of public opinion

TABLE 8-2
PUBLIC OPINION CONCERNING COASTAL REGULATION AND ACQUISTIONS, BY AREA AND PROXIMITY TO COAST, NOVEMBER 1975
(N = 763)[a,b]

| | Statewide total (N = 763) | By area | | | | | By proximity to coast | |
		SF Bay Area (N = 184)	LA/ Orange (N = 341)	Rest of So. Cal. (N = 126)	Central Valley (N = 112)		Coastal County (N = 522)	Inland County (N = 241)
1. Everything considered, I think the Coastal Commission has done a good job of balancing protection and development of the coastline.								
% Agree	42%	36%	44%	47%	45%		44%	37%
% Ambivalent	12%	11%	13%	11%	13%		13%	12%
% Disagree	23%	15%	29%	24%	17%		27%	14%
% No opinion	22%	38%	13%	18%	25%		15%	37%
	99%	100%	99%	100%	100%		99%	100%
2. I think the state and regional coastal commissions should stop entirely any type of construction along the California coastline.								
% Agree	44%	39%	46%	38%	49%		46%	40%
% Ambivalent	11%	14%	10%	13%	7%		10%	12%
% Disagree	37%	40%	37%	46%	28%		38%	37%
% No opinion	8%	7%	6%	4%	16%		6%	11%
	100%	100%	99%	101%	100%		100%	100%

[a]Given a sample size of 763, the confidence intervals are ±3–4%.
[b]Source: Public Response Associates, *A Survey of Attitudes Toward the California Coastal Area, A Study Conducted for the California Council for Environmental and Economic Balance,* San Francisco, December 1975.

(continued)

TABLE 8-2 (Continued)

| | Statewide total (N = 763) | By area | | | | | By proximity to coast | |
		SF Bay Area (N = 184)	LA/ Orange (N = 341)	Rest of So. Cal. (N = 126)	Central Valley (N = 112)	Coastal County (N = 522)	Inland County (N = 241)
3. I agree with private property owners that if the state won't let them develop a parcel, the state should buy it at its fair market price.							
% Agree	67%	70%	66%	62%	69%	66%	68%
% Ambivalent	10%	13%	11%	10%	5%	12%	7%
% Disagree	13%	11%	13%	18%	17%	13%	16%
% No opinion	9%	7%	10%	10%	9%	9%	9%
	99%	101%	100%	100%	100%	100%	100%
4. I think we already have sufficient recreational areas throughout the state and the government should develop them before buying any more land in the coastal area.							
% Agree	39%	29%	41%	42%	47%	39%	38%
% Ambivalent	8%	10%	6%	10%	7%	8%	9%
% Disagree	47%	55%	46%	44%	40%	47%	48%
% No opinion	6%	5%	7%	3%	7%	6%	6%
	100%	99%	100%	99%	101%	100%	101%
5. Would you support a $250 million bond issue to be used for preservation and enhancement of coastal resources and to acquire, develop, and maintain additional coastal beaches and parks?							
% Support	49%	52%	49%	48%	45%	51%	44%
% Oppose	34%	23%	35%	40%	41%	34%	34%
% Don't know	17%	25%	15%	13%	14%	14%	22%
	100%	100%	99%	101%	100%	99%	100%

probably became known to many legislators from at least coastal counties, as coastal legislation was certainly one of the most publicized issues of the 1976 session and probably of sufficient salience to people in coastal counties that legislators would make some effort to ascertain the general orientation of their constituents.[8]

B. The "Attentive Public": Agency and Interest Group Perceptions of the Commissions' Impacts and Evaluation of the Commissions' Performance

Whatever the effects of general public opinion on the legislature's deliberations, research has repeatedly demonstrated that "the attentive public"—i.e., agency officials, interest group leaders, journalists, and other citizens knowledgeable about the commissions' performance—were likely to play a much more important role in providing information and advice to legislators.[9]

The importance of these sources was heightened by the paucity of good information on the activities and impacts of the coastal commission during the 1973–1975 period. True, the coastal plan comprised a comprehensive set of recommendations to the legislature. But there was little systematic, disinterested information available on the permit activities of the commissions, other than the commissions' oft-trumpeted and somewhat misleading (as we saw in Chapter 5) claim that they had denied only 3% of permit applications during those 3 years.[10] As for the impacts of the commissions' activities, the 1975 legislature commissioned several studies that (1) documented the enormous difficulties of doing competent economic analyses of the probable effects of coastal plan policies, (2) made some tentative predictions concerning the plan's effects on commercial and residential development, and (3) estimated the administrative costs on state and local agencies of carrying out the proposed plan.[11] A previously published survey of county assessors along the coast cautiously concluded that the commissions' permit activities had increased the value of developed parcels and depressed the value of undeveloped parcels within the permit zone, with very uncertain net effects on local tax bases; these findings were generally confirmed by several other studies, although attempts to estimate the impacts of plan policies on employment proved considerably more problematic.[12] Unfortunately, the type of rigorous, systematic analysis on housing impacts over a significant portion of the coast (such as that presented in Chapter 6) only became available 2 years after the legislature's decision on the 1976 Coastal Act.

Because of this paucity of reliable information during the 1976 session on the actual impacts of the commissions' permit activities or proposed planning policies on housing construction, beach access, wetlands protection, or any other politically significant objectives, legislators had to rely for information primarily upon the *perceptions* of supposedly knowledgeable observers of the commissions. Tables 8-3 and 8-4 report some of the results of our survey of a random sample of 486 members of the "attentive public" concerning their perceptions of the commissions' impacts and their evaluations of the commissions' performance during the 1973–1975 period. The data are based upon a questionnaire mailed to samples of (1) commissioners and staff, (2) people on the general mailing list of the seven commissions, (3) outsiders who had actively participated in the formulation of at least one of the plan elements, and (4) permit applicants whose cases had been appealed to the state commission.[13]

Table 8-3 tabulates the perceptions of seven categories of respondents concerning, first, the extent of the commissions' effects, and, second, respondents' evaluations of those effects on eight types of impacts ranging from park acquisitions through housing construction to preserving visual access of the ocean from the coastal highway.[14] Turning first to perceptions of the *extent* of commission impacts (measured in terms of group means from 5 = very great effect to 1 = no effect at all), almost all of the groups felt that the commissions had a substantial effect in preventing reductions in wetlands and other wildlife habitat and in giving priority to environmental protection and beach access over other considerations in permit decisions. In addition, business representatives, coastal property owners, and other groups to a lesser extent perceived the commissions as having substantially slowed housing construction in the permit area and as having restricted new development to urbanized areas. The commissions were also viewed by most groups as having moderate-to-substantial effects on preserving visual and physical access to the ocean and in generating public awareness of the cumulative effects of development decisions, with considerable disagreement concerning the extent of their effects on accelerating park acquisitions.

From a political standpoint, however, what was crucial was different groups' *evaluations* of these impacts. These judgments were measured on a curvilinear scale from 1 = went too far, through 3 = about right, to 5 = didn't go far enough. In general, coastal commission officials and environmental group leaders felt that the commissions had done about right or hadn't gone far enough in promoting these eight objectives, while business leaders and coastal property owners felt that the commissions had gone too far, particularly in areas like providing physical access to the beach, restricting new development to existing urbanized areas, and slowing housing construction in the permit area. Respondents in the other three categories (civic/community groups, interested citizens, and state/local agency officials) tended to have widely varying opinions. The commissions did, however, receive relatively high ratings from most groups concerning their performance in preserving visual access, accelerating park acquisitions, protecting wetlands, and generating awareness of the cumulative impacts of development decisions.

Differences among groups emerged more clearly on a number of more general items relating to the commissions' performance in permit review, plan development, and general procedural fairness. As can be seen in Table 8-4,

Table 8-3
Different Groups' Perceptions of the Impacts of the Coastal Commissions

Potential impact of coastal commissions	Commissioner/staff (N = 84)	Environ. group (N = 31)	Civic/community group (N = 22)	Interested observers/citizens (N = 68)	Local/state agency (N = 64)	Business (N = 34)	Applicant/property owner (N = 166)
				Group affiliation of respondents[a]			
1. Prevented reduction of wetlands and other wildlife habitat							
Extent of impact[b]	4.1	3.6	3.8	3.7	3.9	3.8	3.8
Evaluation of impact							
Went too far 1	5%	0%	15%	12%	7%	26%	23%
2	10%	4%	5%	12%	8%	19%	15%
About right 3	26%	14%	25%	48%	50%	29%	32%
4	25%	46%	35%	15%	22%	13%	12%
Didn't go far enough 5	34%	36%	20%	13%	13%	13%	18%
2. Gave priority to environmental protect. and beach access							
Extent of impact[b]	3.6	3.3	3.5	3.6	3.6	4.0	3.8
Evaluation of impact							
Went too far 1	6%	4%	29%	25%	16%	39%	56%
2	11%	4%	5%	7%	21%	29%	17%
About right 3	35%	18%	24%	30%	31%	16%	13%
4	24%	44%	14%	20%	16%	6%	6%
Didn't go far enough 5	24%	30%	29%	18%	17%	10%	8%

[a]Based upon respondents' categorization of the primary role in which they followed the coastal commissions' activities during the 1973–1976 period.
[b]This is the group mean on a scale from 5 (very great effect) to 1 (no effect at all).

(continued)

TABLE 8-3 (Continued)

Potential impact of coastal commissions	Group affiliation of respondents[a]						
	Commissioner/staff (N = 84)	Environ. group (N = 31)	Civic/community group (N = 22)	Interested observers/citizens (N = 68)	Local/state agency (N = 64)	Business (N = 34)	Applicant/property owner (N = 166)
3. Restricted new development to urbanized areas							
Extent of impact[b]	3.2	3.3	3.1	3.2	3.3	3.9	3.7
Evaluation of impact							
Went too far 1	10%	3%	25%	27%	23%	70%	57%
2	7%	10%	10%	8%	22%	15%	12%
About right 3	39%	21%	15%	23%	32%	6%	12%
4	22%	31%	25%	18%	13%	3%	6%
Didn't go far enough 5	23%	34%	25%	23%	10%	6%	12%
4. Slowed housing construction in permit area							
Extent of impact[b]	3.3	3.3	3.2	3.5	3.5	4.3	4.0
Evaluation of impact							
Went too far 1	9%	4%	29%	27%	30%	61%	67%
2	22%	0%	10%	14%	18%	23%	11%
About right 3	38%	30%	19%	22%	38%	6%	14%
4	15%	41%	24%	15%	10%	3%	1%
Didn't go far enough 5	16%	26%	19%	22%	3%	6%	7%
5. Preserved visual access to ocean from highway							
Extent of impact[b]	3.7	3.5	3.2	3.4	3.5	3.7	3.5
Evaluation of impact							
Went too far 1	8%	4%	25%	15%	11%	30%	44%
2	10%	0%	10%	17%	9%	17%	13%

About right 3	38%	43%	10%	38%	52%	33%	25%
4	24%	13%	20%	21%	20%	13%	7%
Didn't go far enough 5	21%	39%	35%	9%	9%	7%	10%

6. **Generated awareness of the problems of cumulative impacts of local decisions**

Extent of impact[b]	3.6	3.8	3.8	3.3	3.4	3/5	3.1
Evaluation of impact							
Went too far 1	4%	0%	10%	8%	6%	40%	35%
2	7%	0%	0%	8%	11%	13%	15%
About right 3	34%	41%	45%	32%	45%	23%	20%
4	31%	38%	20%	21%	14%	20%	13%
Didn't go far enough 5	24%	21%	25%	31%	23%	3%	16%

7. **Insured greater public access to beach**

Extent of commissions impact[b]	3.4	3.4	3.6	3.1	3.3	3.1	2.9
Evaluation of impact							
Went too far 1	6%	7%	26%	16%	16%	58%	42%
2	7%	7%	16%	14%	16%	6%	16%
About right 3	30%	34%	21%	33%	34%	12%	19%
4	21%	28%	26%	16%	12%	15%	11%
Didn't go far enough 5	36%	24%	10%	22%	22%	9%	13%

8. **Accelerated coastal park acquisitions**

Extent of impact[b]	3.5	3.4	2.8	3.2	2.7	2.3	2.8
Evaluation of impact							
Went too far 1	6%	4%	21%	17%	5%	22%	32%
2	5%	7%	10%	7%	16%	16%	13%
About right 3	38%	32%	21%	40%	25%	25%	17%
4	20%	43%	21%	20%	28%	16%	14%
Didn't go far enough 5	32%	14%	26%	17%	26%	22%	23%

commission officials, environmental groups and civic/community groups generally approved of the commissions' permit decisions, while business representatives and coastal property owners just as strongly opposed them, with "interested
citizens" and local/state agency officials quite divided in their judgments. While
somewhat the same pattern of responses emerged on the politically important
issue of whether the commissions had been sufficiently sensitive to the hardships
their decisions imposed on permit applications, there was nevertheless a discernible undercurrent of doubt; for example, whereas 75% of coastal agency officials
generally approved of the agency's permit decisions, only 57% felt that they had
adequately considered hardships on permit applicants. A similar configuration of
attitudes emerged concerning the coastal plan—or, more precisely, respondents'
perceptions of the content of the plan. Commission officials, environmental and
civic group leaders, and even a substantial majority of "interested citizens"
supported the general thrust of the plan, while business leaders and coastal
property owners were generally opposed and state/local agency officials on the
commissions' mailing list were ambivalent. Nevertheless, even some of those
opposed or ambivalent to the plan admitted that it was generally consistent with
statutory objectives. Conversely, even some plan supporters had doubts whether
it reflected the views of a majority of Californians.

Some of these undercurrents of doubt about the procedural and substantive
fairness of the commissions are further explored in part C of Table 8-4. The data
reveal quite clearly that, with the exception of businessmen and, to a lesser
extent, permit applicants, most observers of the commissions felt that the coastal
agency had provided permit applicants adequate opportunities to make their
case. On the other hand, there were substantial doubts among all groups—with
the exception of environmentalists and, to a lesser extent, commission officials—
about whether a fair balance of interests was represented on the state commission
and thus, implicitly, whether the state commission had really listened to proponents of development. As we saw in chapter 4, these doubts were not without
foundation, as over 90% of state commissioners (and staff) had voted for Proposition 20. Nevertheless, defenders of the commissions could argue that the imple-
menters of a regulatory statute with a definite policy orientation should reflect
that bias rather than a wide spectrum of views.

On the whole, however, the lineup of supporters and opponents to the 1975
coastal plan (and attendant legislation) was very similar to that on Proposition 20
4 years previous. Environmental groups were strongly supportive, as were a
majority of civic and community organizations (e.g., the League of Women
Voters) and individual citizens concerned with coastal management. In contrast,
most corporations and coastal property owners on the commissions' mailing lists
were opposed. Local and state agency officials tended to reflect a wide variety of
viewpoints. Additional evidence of the strong links in attitudes toward Proposition 20 and the coastal plan is provided if one makes the reasonable assumption

that the individuals on the coastal commissions' mailing lists roughly approximated the "attentive public" (exclusive of commission officials) who were knowledgeable about commission activities and concerned with the utilization of California's coastal resources. Of the random sample of 306 survey respondents on
the mailing lists, 70% of those who had voted for Proposition 20 supported the
coastal plan (while 18% were neutral and 12% opposed the plan); analogously,
77% of those who had voted against Proposition 20 likewise opposed the coastal
plan (while 17% were neutral and 4% supported the plan). Finally, a more
detailed analysis of this data set published elsewhere indicates that the vast
majority of variance in support for the coastal plan could be attributed to basic
policy predispositions, with evaluations of the commissions' impacts playing a
comparatively minor role. [15]

Nevertheless, attitudes toward the coastal plan (or any other legislation)
matter little unless advocates have the desire and the resources to see that their
positions are seriously considered by legislators. This generally involves, first, an
ability to make reasoned arguments through personal lobbying and, second, the
capacity potentially to influence electoral campaigns through evidence that one's
views are amply represented in the legislator's district and/or through the provision of resources (e.g., money, volunteers) useful in any campaign. [16]

In this regard, there was every reason to believe on the eve of the 1976
legislative session that the opponents of the coastal plan would have little difficulty in presenting their case. The land developers, oil companies, utilities,
realtors, construction trades unions, and other members of the development
community all had professional lobbyists in Sacramento and a history of substantial campaign contributions. Moreover, the CCEEB poll certainly indicated
strong public support for the protection of property rights and for returning some
control over coastal development to local governments. Finally, most local governments and state agencies—whatever their position on coastal legislation—had
legislative advocates in Sacramento. [17]

Thus, the critical question concerned the ability of the coastal commissions, environmental groups, and other supporters of the coastal plan to effectively present their case. In this respect, the Sierra Club, Coastal Alliance,
Planning and Conservation League (PLC), and League of Women Voters,
among others, were viable organizations with active members in most legislative
districts and professional lobbyists in Sacramento. [18] Moreover, Mel Lane and
Joe Bodovitz had good connections in the Republican and Democratic parties,
respectively, plus the experience of having successfully pushed the BCDC plan
through the legislature in 1969. In addition, Bill Press—a former PCL lobbyist
and active member of the Coastal Alliance—was director of the Office of Planning and Research in the governor's office and thus could be expected to effectively present the commissions' case to Governor Brown. Finally, the proponents
of coastal legislation had one important resource they had not possessed in the

TABLE 8-4

GENERAL EVALUATION OF THE COMMISSIONS' PERFORMANCE BY DIFFERENT GROUPS

	Commissioner/staff (N = 84)	Environ. group (N = 31)	Civic/community group (N = 22)	Interested observers/citizens (N = 68)	Local/state agency (N = 64)	Business (N = 34)	Applicant/property owner (N = 166)
				Group affiliation of respondents			
A. Permit review							
1. Permit decisions generally good							
% Agree	75%	73%	55%	40%	31%	9%	15%
% Ambivalent	17%	10%	27%	26%	41%	38%	18%
% Disagree	8%	17%	18%	34%	28%	53%	67%
2. Given conservationist principles of 1972 Initiative, commissions sufficiently sensitive to hardships on permit applicants							
% Agree	57%	77%	41%	47%	24%	3%	12%
% Ambivalent	24%	10%	23%	15%	20%	16%	6%
% Disagree	19%	13%	36%	38%	56%	81%	82%
B. Coastal plan							
1. Support for 1975 plan							
% Support	84%	97%	64%	61%	33%	18%	20%
% Ambivalent	10%	3%	9%	10%	45%	21%	18%
% Oppose	6%	0%	27%	28%	22%	62%	62%
2. Plan generally consistent with objectives of Prop. 20							
% Agree	84%	81%	77%	59%	38%	24%	29%
% Ambivalent	11%	10%	0%	21%	28%	15%	30%

% Disagree	41%	61%	34%	20%	23%	10%	5%
3. Plan *failed* to reflect views of majority of Californians							
% Agree	61%	79%	52%	38%	27%	0%	16%
% Ambivalent	22%	9%	18%	12%	18%	16%	12%
% Disagree	17%	12%	31%	49%	55%	84%	72%
C. Procedural fairness							
1. Inadequate opportunities for affected citizens to participate in permit review [with similar figures for planning]							
% Agree	49%	59%	28%	28%	14%	3%	7%
% Ambivalent	16%	12%	27%	13%	9%	7%	5%
% Disagree	35%	28%	45%	58%	77%	90%	88%
2. Fair balance of interests represented on state commission							
% Agree	15%	3%	29%	30%	38%	71%	56%
% Ambivalent	16%	24%	36%	35%	19%	14%	18%
% Disagree	69%	73%	36%	35%	43%	14%	24%
D. The starting point							
1. Support Prop. 20							
% Support	40%	35%	73%	73%	73%	100%	82%
% Oppose	60%	65%	23%	27%	27%	0%	18%
E. The result							
1. Support for the 1976 Coastal Act							
% Support	16%	9%	37%	55%	62%	90%	82%
% Ambivalent	18%	27%	43%	23%	14%	10%	8%
% Oppose	66%	64%	20%	22%	24%	0%	10%

1971–1972 legislative sessions—namely, the knowledge on everyone's part that they had successfully sponsored the 1972 Coastal Initiative and might well go to the electorate again if rebuffed by the 1976 legislature.[19]

C. The Actual Decision-Makers: The 1976 Legislature and Governor Brown

We come finally to the views of the people who would actually decide the fate of coastal legislation, namely, the members of the 1976 California legislature and Governor Jerry Brown. While always important, the attitudes and skill of legislators of different persuasions were likely to be particularly critical when, as in this case, both the outside supporters and opponents of the coastal plan were more or less evenly matched.

At this point, the independence from legislative and executive scrutiny provided the commissions by the 1972 Coastal Initiative began to have mixed effects. While the secure source of funding and the impediments to weakening amendments had been very helpful in allowing the commissions to complete their tasks on time without the normal involvement of sovereigns in important permit cases (see Chapter 3), some disadvantages were also entailed during the 1976 session. First, the commissions' autonomy from normal oversight meant that legislators—even those on the natural resources committees in the Assembly and the Senate—had had little experience with coastal issues since 1972 and were generally quite unfamiliar with the commissions' record on permit review or the provisions of the coastal plan.[20] This meant that legislators would have to undergo a substantial educational process on very long and complex legislation during the 7 months of the 1976 session.[21] Second, the lack of any major confrontation with the legislature (or the governor) during the previous 3 years meant that the commissions had very little experience dealing with the legislature and few close relationships with legislative supporters.

As a partial result of this autonomy, the commissions had no knowledgeable and influential "fixer" comparable to Senator Muskie on federal pollution control legislation or Assemblyman Frank Lanterman on California mental health issues. While Assemblyman Alan Sieroty was a long-term advocate of coastal legislation, he lost his chairmanship of the coastal subcommittee in the Assembly Natural Resources Committee at the beginning of the 1976 session, and there were strong tactical reasons for having a senator carry the major coastal bill. Assembly Speaker Leo McCarthy was a supporter of coastal legislation and extremely powerful. But he had to be concerned with a wide variety of bills; moreover, it was widely recognized that the fate of coastal legislation would primarily be decided in the Senate, where his influence was limited. Likewise, Senate Pro-Tem James Mills was a long-time advocate of coastal protection and, indeed, had taken some important steps during the 1973–1975 period to change the composition of the Natural Resources Committee so that it would no longer be the graveyard of strong coastal legislation.[22] Nevertheless, the widely dispersed nature of power in the Senate—often characterized as 40 individual fiefdoms—limited his overall influence.[23]

In sum, on the eve of the 1976 session it was assumed that the Assembly would generally be supportive of the coastal plan, just as it had passed the forerunners of Proposition 20 in 1971 and 1972. But approval would certainly not be automatic, in part because of the very complexity of the legislation, in part because the new chairman of the Assembly Natural Resources, Energy, and Land Use Committee, Charles Warren, indicated very early that he would have to be convinced of the need for a successor coastal agency. In the Senate, on the other hand, proponents of the coastal plan could only count about 13 "sure" votes out of 40 senators, with the Natural Resources Committee rather evenly divided. Moreover, since the legislation would involve the future expenditure of funds, it would have to go through the normally conservative Finance Committee. In part because of the crucial role of that committee, proponents decided to ask the finance chairman, Senator Anthony Beilenson, to carry the major coastal bill. While Beilenson had no history of involvement with coastal legislation, he was widely respected within the upper house and was generally supportive of the coastal plan.[24]

As for Governor Jerry Brown, it was assumed that—unlike former Governor Reagan—he would eventually support coastal legislation of some sort. While he had generally been supportive of environmental legislation and had included an appropriation for a successor coastal agency in his FY 1977 budget proposal, he was also very skeptical of comprehensive planning. Moreover, his general strategy of letting the legislature take the lead on controversial topics was reinforced in this case by the objections of numerous agencies within his administration to various segments of the coastal plan. Thus, the general expectation was that the governor would probably support the commissions' legislation after it had been modified to meet some of the objections of state agencies and other groups.[25]

In conclusion, the fate of coastal legislation was very uncertain at the beginning of the 1976 session. California, like the rest of the nation, was emerging from a serious recession, which had hit the construction industry particularly hard. And, in commissioning a study of the economic consequences of the coastal plan, the legislature indicated that it would be very sensitive to such considerations. While public opinion was generally supportive of the commissions, the CCEEB poll also indicated a strong desire for returning some, and possibly even principal, authority for development review to local governments, as well as strong public support for the protection of property rights. Our survey of the "attentive public" indicated widely different evaluations of the commissions' performance, with environmental groups and many community organizations strongly supportive, business groups and coastal property owners just as strongly opposed, and state and local agencies more or less evenly split. And the

1976 legislature itself was split between the generally supportive Assembly and the very uncertain Senate. Nevertheless, there had been at least three developments since Senate rejection of coastal legislation in 1972 that promised a more even match this time: (1) the knowledge in everyone's mind that another initiative was a possibility, although explicitly advocated by no one; (2) the more even composition of the Senate Natural Resources Committee; and (3) the change from a hostile to a probably supportive person in the governor's office.

II. The 1976 Legislative Session

In early 1976 three major pieces of legislation incorporating the principal features of the coastal plan began their tortuous path through the legislature. They are briefly summarized in Table 8-5. The most important bill, carried by Senator Beilenson, sought to establish the basic policies to guide future utilization of coastal resources and to distribute authority for the implementation of those policies among the successor coastal agency, other state agencies, special districts (e.g., ports), and local governments. While it did not include all of the policies in the coastal plan, it contained substantial portions of most of them and, in many places, referred to specific plan policies. In addition, the bill required that "all public agencies consider and be guided by the Coastal Plan in carrying out their duties . . . affect[ing] coastal resources."[26] This fidelity to the coastal plan thus assured that the Beilenson bill would be a long (106 pages), complex, and extremely controversial piece of legislation, which would demand a considerable portion of its author's and the legislature's time and energy.

The other two bills were considerably simpler. One would establish a five-member Coastal Conservancy to acquire partial interest in agricultural lands (for which police power controls were inadequate to maintain in agricultural production) and to provide grants to state and local agencies for the acquisition of public accessways, the restoration of degraded areas, and the protection of wildlife habitat areas. In short, the conservancy was to help provide some of the management functions necessary to supplement the police powers incorporated in the Beilenson bill. The third piece of legislation was a $290 million bond issue to be placed on the November 1976 statewide ballot to acquire park and valuable habitat areas along the coast designated as priority acquisitions by the coastal commissions.[27]

In this section, we shall first review the very convoluted path of the Beilenson bill (and its successor) through the 1976 session and then briefly review the fate of the conservancy and coastal bonds bills. We shall then assess some of the major factors involved in the legislature's reformulation of the coastal plan.

Table 8-5

Major Legislation Sponsored by Coastal Commissions to Implement 1975 Coastal Plan

	SB 1579	AB 3544	AB 2948
Major bills and principal function	Incorporate most plan policies and divide authority between successor coastal agency and other state and local agencies	Create coastal conservancy	Authorize bonds for coastal acquisitions
Bill number Legislative sponsor	Senatory Anthony Beilenson (Dem-Los Angeles County) chairman, Senate Finance Committee	Assemblyman Michael Wornum (Dem-Marin County) chairman, Coastal subcommittee of Assembly Land Use and Natural Resources Committee Freshman Former chairman, North Central Regional Commission	Assemblyman Gary Hart (Dem-Santa Barbara County) Freshman Former member, South Central Regional Commission
Major provisions	General policies Protect wetlands, water quality, wildlife habitat, and agricultural land Protect viewsheds and scenic areas Generally restrict new development to existing urbanized areas and give priority to coastal-dependent uses Determine need for new transportation and energy facilities and minimize their adverse environmental effects Maximize public access and recreation consistent with resource protection	Establish Coastal Conservancy Members 1. Secretary Resources Agency 2. Director, Dept. of Finance 3. Chairman, Coastal Comm. 4–5. 2 appointees of governor Functions: Acquire interest in agricultural lands Provide grants to local agencies to acquire scattered lots and restore degraded areas Provide grants to other state depts. for acquisition of public accessways, re-	Place $290 million bond issue on Nov. 1976 ballot to purchase coastal lands designated by coastal commissions $150 million for state parks along coast $55 million for other acquisitions $50 million in grants to local govts. for coastal parks $10 million for purchase of wetlands $25 million for Coastal Conservancy

(continued)

TABLE 8-5 (Continued)

Major bills and principal function

Incorporate most plan policies and divide authority between successor coastal agency and other state and local agencies	Create coastal conservancy	Authorize bonds for coastal acquisitions
Maintain state and regional commissions until certification of local coastal programs; after certification, regional comms. dissolved and regional reps. replaced by state appointees on state commission	sources protection zones Fund state and local programs for habitat protection and restoration	
Distribution of authority for development review		
Prior to certification, coastal agency has permit authority within permit area and over public works and energy projects in larger area		
Local governments must revise plans and ordinances to conform to Coastal Act policies within coastal resource management area prior to 1981, after certification, appeals to coastal agency same as proposed in coastal plan		
Even after certification, coastal agency retains permit jurisdiction seaward of coast and apparently over public works and energy projects		
Provision for citizen suits and recovery of attorneys fees, while prohibiting bonds		

A. A *Cliff-Hanger*

The major groups of actors and their strategies are outlined in Table 8-6. The proponents of the Beilenson bill—the coastal commissions, environmental groups, and a few state agencies—sought to maintain the central thrust of the bill in terms of its emphasis on environmental protection, public access, and a strong coastal agency, while at the same time accepting the compromises needed to obtain passage. Their basic strategy was to try to get an acceptable bill through the Senate, strengthen it if necessary in the Assembly, and then resolve differences between the two versions in conference committee. If an acceptable bill did not emerge, they would try for a 1-year extension of the commissions' existing authority and perhaps eventually prepare another initiative. But the initiative was definitely seen as a measure of last resort because of the complexity of the issues involved, the perceived decline in public support for environmental protection, and the difficulty in avoiding at least a 1-year hiatus in the commissions' authority. [28]

Under this general rubric, the initial proponents of the Beilenson bill developed a negotiating strategy involving two elements. [29] First, as SB 1579 was in many respects a coastal commission bill, a few members of the state commission staff played a major role in almost all negotiations, assisted at times by lobbyists from the Sierra Club and the PCL, with the final decisions generally in the hands of Senator Beilenson (or, as it turned out, often his legislative assistant). Second, it was decided to negotiate separately with specific opponents more or less one at a time rather than try to anticipate the concerns of large numbers of groups very early in the process. Moreover, the first set of negotiations were focused primarily on state agencies, as satisfying their concerns would be critical to obtaining the governor's support and to avoiding the possibility of an administrative bill as an alternative to SB 1579. It was also acknowledged that the Beilenson bill would have to obtain the support of at least one of the local government associations (the League of California Cities or the County Supervisors Association of California) and some of the development groups originally opposed to SB 1570.

Table 8-6 also identifies three categories of actors with objections to the (original) Beilenson bill. On the one hand were many state agencies, which accepted the general outlines of the bill but sought rather limited and specific amendments, chiefly to reassert their functional authority within the coastal zone while at the same time acknowledging the need for some sort of concurrent jurisdiction by the successor coastal agency. [30] At the other extreme were the building trades unions, property rights organizations, some business groups, and a few cities (notably Long Beach and the Los Angeles City Council), which essentially rejected any real role for a successor coastal agency and wished to return to the days before Proposition 20. Their initial strategy was an initiative

TABLE 8-6

GENERAL POSITION AND STRATEGY OF MAJOR ACTORS WITH RESPECT TO THE BEILENSON BILL (SB 1579)

	Category 1	Category 2	Category 3	Category 4
Basic positions	Support Beilenson bill as introduced or with very minor amendments	Agree with general thrust of Beilenson bill, but with some limited and fairly specific objections	Accept concept of successor coastal agency, but major revisions needed to delegate more authority to local governments and other agencies and/or to provide more balance with economic development	Basically implacable opponents: State provide guidelines but actual authority should be in hands of local governments and existing agencies, assign more emphasis to economic development and compensation to property owners
Principal advocates	Coastal commissions Environmental groups Dept. of Fish and Game Federal Office of Coastal Zone Management	Dept. of Water Resources Water Resources Control Board Div. of Forestry Energy Commission	CCEEB League of Cities County Supervisors Assoc. Most labor unions Recreational boating groups	Chamber of Commerce Construction trades unions Realtors Property rights groups Ports

Air Resources Board State Dept. of Agriculture	State dept. of recreational boating (DNOD) Dept. of Parks and Recreation League of Women Voters State highway dept.	Farm Bureau	Association of General Contractors State Land Commission Oil Companies Utilities Los Angeles and Long Beach Sea Ranch Association
Basic strategy	Support Beilenson bill; accept compromises necessary to pass Senate and then try to strengthen it in Assembly	Negotiate specific compromise with Beilenson	Try to negotiate major compromises with Beilenson If unsuccessful, support Keene bill (AB 3875)	Support local control initiative Support Carpenter bill (SB 1919) When those failed, support Cullen-Ayala bill Late in session many supported Keene bill as only alternative to Smith bill

aSources: (1) Testimony presented during March–June 1976 at hearings of Assembly Subcommittee on Coastal Zone Planning, Senate Natural Resources Committee, and Senate Finance Committee; (2) articles in the *State Coastal Report*; and (3) interviews with Bob Testa (chief staff to Senate Natural Resources Committee), Jim Carroll and Bill Geyer (CCEEB), Mel Lane, Gail Osherenko (Beilenson's legislative assistant), Preble Stolz (governor's office), Peter Douglas (Assembly Land Use and Energy Committee staff), Joe Petrillo (former staff counsel to state commission and then legislative assistant to Senator Jerry Smith), and Larry Moss (lobbyist for Planning and Conservation League).

statute returning all regulatory authority over coastal land use to local governments. When it became obvious by late April that this initiative would not obtain
sufficient signatures to qualify for the November ballot and that some form of
coastal legislation was likely to emerge from the 1976 session, most of these
groups began supporting a bill by Assemblyman Mike Cullen and Senator
Ruben Ayala that would have required local governments to prepare coastal
programs within a 1000-yard coastal area to be reviewed by a Compliance Council composed primarily of representatives from existing state agencies; the
Cullen-Ayala bill would also have required that property owners be compensated
for loss and expense due to land use restrictions imposed by the council.[31]

In between the strong advocates of local control and the state agencies with
quite specific concerns were several critical groups—most notably CCEEB, most
labor unions, and local government associations—that accepted the principal of
a successor coastal agency but also sought major revisions in the Beilenson bill to
delegate more authority to local governments and/or to provide greater balance
between economic development and environmental protection. Although many
of these groups eventually had their concerns reflected in the Beilenson bill,
others supported a compromise measure drafted by Assemblyman Barry Keene
(Dem-Eureka) after the initial defeat of both the Beilenson and Cullen-Ayala
bills. It would have required local governments to revise their plans in a 1000-
yard coastal zone subject to the review of a restructured state coastal commission
under very tight time constraints. The Keene bill would also have (1) required
compensation be paid in case of down-zoning, (2) limited review of local permit
decisions (both before and after local certification) to appeals brought by the
commission's executive director or aggrieved parties, and (3) provided fairly
strong protection for 19 wetlands.[32]

Given the breadth of interests with objections to the Beilenson bill and the
numerous veto points within the legislature—each reflecting a slightly different
range of concerns—Senator Beilenson was faced with the formidable task of
obtaining the support (or at least neutralizing the opposition) of virtually all of
the actors in Category 2, most of those in Category 3, and a few of those in
Category 4 if he were to obtain a majority of votes in Senate Natural Resources (4
probable supporters, 1 swing vote, and 4 opponents), Senate Finance (5 supporters, 4 swing votes, and 4 opponents), and the Senate floor (only 13 sympathizers
out of 40 votes).[33] Unfortunately, this task was enormously complicated when,
10 days after introducing SB 1579, Beilenson decided to run for a recently
vacated congressional seat. This not only significantly limited the time he could
devote to mastering a complex piece of legislation and negotiating the necessary
compromises with various groups and other legislators, but also rendered him a
"lame duck," thereby probably diminishing his personal influence with his
peers.[34]

The long negotiating process began in the Senate Natural Resources Committee, where, over a 2-month period, the bill was revised to delete all references to the coastal plan, to satisfy the concerns of most state agencies, to provide for 50% representation for local authorities on the coastal agency after disbandment of the regional commissions, and to simplify the certification process for local coastal programs. These concessions were sufficient to win the grudging support of the committee's swing vote, Chairman John Nejedly, and SB 1579 was approved by a 5–3 vote on May 3. The same day the committee killed the Senate version of the Cullen-Ayala bill (SB 1920) by a 4–4 vote, with Nejedly voting to report out both bills.[35]

Because the bill had fiscal impacts (although it contained no appropriation), it then went to the Senate Finance Committee, where Beilenson was faced with the task of making sufficient concessions to win 2 of the 4 swing votes without so weakening the bill that he lost any of his anticipated supporters. He didn't succeed. In the process, however, he negotiated a number of compromises that ultimately were critical to the passage of coastal legislation. These included (1) a substantial reduction in the coastal resource management area from 5–15 miles in much of the northern two-thirds of the state to 2000 yards (approximately 1 mile) with "bulges" of up to 5 miles for wetlands, streams, and some agricultural lands; (2) further simplification of the local certification process and a reduction in the scope of postcertification appeals, as well as a provision for reimbursing local governments for the costs of preparing local coastal programs; (3) reduction of the scope of coastal agency review of energy and port facilities; and (4) revision of the policies dealing with the siting of energy and other industrial facilities to provide greater balance between economic development and environmental protection. These amendments were sufficient to win the support of the League of Cities (but not the County Supervisors Association), the ports and longshoremen, the Energy Commission, and, most important, 2 of the 4 swing votes (Grunsky and Holmdahl). On the other hand, the negotiations took so much of the committee's time that the bills of some of its members were lost in the shuffle, thereby probably losing the support of another swing vote (Alquist). In addition, the committee rejected the efforts of one of the bill's anticipated supporters, Senator David Roberti of an inner-city district of Los Angeles, to strengthen its provisions for public access and lower-income housing (by, for example, requiring access dedications on all new developments, including single-family residences, along the coast). Roberti was particularly upset about the commission's failure to obtain (what he considered to be) adequate access in wealthy residential areas like Malibu. Thus, despite the diligent efforts of the bill's supporters—who now included Governor Brown—SB 1579 was defeated in Senate Finance on June 11 when it could obtain only 6 votes on the 12-member committee.[36]

The history of the Beilenson bill is briefly summarized in Table 8-7A. But the bottom line was that, with only 2 months left in the 1976 session and the

Table 8-7A
History of Major Coastal Bill in 1976 Session: Amendments to, and Eventual Defeat of, SB 1579

Date	Locus/event	Principal effects on Beilenson bill (SB 1579)	Changes in organized support or opposition
February 20	Beilenson announces he is running for Congress	Reduces his ability to master and to manage negotiations on long and complex bill	
March 9–April 28	Senate Natural Resources Committee Hearings	1. Delete all references to coastal plan 2. Slight reduction in jurisdiction 3. Shorten policy section (from 42 to 19 pp.) and add some "balancing language" 4. After disbandment of regional commissions, half of state comm. will be local officials 5. Delete coastal agency jurisdiction over timber harvesting 6. Eliminate jurisdictional conflicts with State Water Resources Control Board 7. Reduce jurisdiction over energy facilities, especially power plants 8. Simplify local certification process 9. Delete ⅔ vote requirement for permits	Met concerns of most state agencies, with exception of Energy Commission and State Lands Commission
May 3	Senate Natural Resources Vote	Pass Beilenson bill by 5–3 vote after rejecting amendments to (1) delete ports from	

	coastal agency jurisdiction and (2) require posting of bonds in court suits Also rejected Ayala bill (SB 1920, as amended) on 4–4 vote		
May 26	Governor Brown endorses SB 1579		Brown administration
May 27–June 10	Senate Finance Committee Hearings	Amendments to bill 1. Reduce coast comm. inland (geographical) jurisdiction to 2000 yards plus "bulges" for wetlands, streams, and some agricultural lands 2. Streamlined local certification process and slightly reduced postcertification appeals 3. Significantly reduced jurisdiction over ports 4. Provide state funding for preparation of local coastal programs 5. Reduce authority over siting of energy facilities 6. Attorneys fees awarded to prevailing party in litigation	Gain support of 1. League of Cities 2. Ports 3. Energy Commission 4. Longshoremen's union
June 10–11	Senate Finance Committee Vote	After rejecting Senator Roberti's amendments to broaden access and low-income housing requirements, committee fails to pass SB 1579 thereby killing it (6–5 vote with 7 ayes needed)	

deadline passed for getting bills out of the house of origin, the coastal commissions' bill had been defeated. Moreover, on June 3 the Senate by a 19–17 vote deleted the appropriation for a successor coastal agency from the state budget (an action that was subsequently upheld in conference committee).[37] Thus, unless some way could be found to resurrect a coastal bill and provide state funding, the commissions would simply go out of existence on December 31, 1976.

After discussing a number of alternative strategies (including asking the governor to call a special session), the proponents of strong coastal legislation decided to amend the substance of the Beilenson bill into a bill (SB 1277) carried by Senator Jerry Smith that had previously passed the Senate and was pending in the Assembly Resources, Land Use, and Energy Committee. During the July recess, the bill was substantially rewritten by committee chairman Charles Warren in consultation with Senator Smith.[38] The revised bill incorporated some of the features of the Keene bill awaiting action in the Senate Natural Resources Committee; these included the addition of three state agency officials to the successor coastal agency and the limitation of "sensitive coastal areas" primarily to wetlands. Other important revisions involved (1) reduction of the commissions' inland jurisdiction from 2000 yards plus bulges to 1000 yards (less in urban areas), with bulges for wetlands and recreational resources but not streams or agricultural lands; (2) reorganization of the bill's policies and provisions dealing with state agencies; (3) provision under certain conditions for local governments to issue permits within the coastal zone (except in the nearshore area) pending certification of their coastal programs; (4) further streamlining of the local certification process, while also weakening the state's commitment to fund the preparation of local programs; and (5) deletion of the commissions' authority to issue cease-and-desist orders. The revised bill was generally considered to be better drafted than SB 1579 and now had the support of the Association of General Contractors—thereby somewhat undermining the contention of the building trades unions that the bill would wreak havoc on the construction industry.

On August 9, SB 1277 passed the environmentally sympathetic Assembly Resource Committee on a 10–3 vote. At virtually the same time, both the Keene bill (AB 3875) and a revived version of the Cullen-Ayala bill were being killed in the Senate Natural Resources Committee, in part because of a concerted effort by environmental groups, in part because Senator Nejedly was much more satisfied with SB 1277 than he had been with the Beilenson bill.[39] These defeats left the Smith bill as the only vehicle for major coastal legislation.

On August 11, the Smith bill was revised in the Assembly Ways and Means Committee to reincorporate some of the concessions to the League of Cities that had been deleted by Warren and to gain the support of the utilities, the Irvine Company (a moderate Orange County developer), and CCEEB. It then passed the committee (15–4) and, 2 days later, the Assembly floor (45–20) with considerable assistance from Speaker McCarthy.

The Smith bill then returned to the Senate for concurrence with the extensive amendments made in the Assembly. It was important to the bill's backers that it be approved by the Senate without amendment, as any changes would automatically send it to conference committee—with absolutely no guarantee that the committee would be dominated by supporters or would reach an agreement before the August 31 expiration of the session.[40] As a first step, on August 18, SB 1277 was approved 6–3 by the Senate Natural Resources Committee and sent directly to the floor (without referral to Finance).[41] There the struggle began to win the 21 votes necessary for passage. By this time, the Smith bill had the support of not only environmental groups but also the ports, the League of Cities, CCEEB, the longshoremen's union, all state agencies, the utilities, the general contractors association, and a few development firms. But it was still opposed by the construction trades and the AFL-CIO, property rights groups, the County Supervisors Association, the City Councils of Long Beach and Los Angeles, and some development firms. After 5 days of intense lobbying by both sides, supporters had 18 votes and a conference committee seemed imminent. But then Governor Brown intervened directly to negotiate a series of largely symbolic amendments that gained the support of the AFL-CIO, whereupon SB 1277 was approved 25–14.[42]

In order to avoid a conference committee, the amendments were incorporated into a trailer bill. Moreover, a special appropriation bill for the successor agency had to win a two-thirds vote in both houses. In a flurry of activity—and with enormous help from Speaker McCarthy—the two bills were finally approved on the last day of the 1976 session and (along with the Smith bill) were subsequently signed by Governor Brown.[43]

The history of SB 1277 is summarized in Table 8-7B. While it certainly was the major piece of coastal legislation during 1976, its regulatory focus was complemented by the management and acquisition functions provided by the conservancy and coastal bonds bills. It is to a brief history of these other two parts of the coastal plan package that we now turn.

B. The Coastal Conservancy and Coastal Bonds Bills

As indicated previously in Table 8-5, the conservancy bill (AB 3544) would have established a Coastal Conservancy composed of five appointed state officials with authority to acquire partial interest in agricultural lands to keep them in agricultural production, to provide grants to local agencies for land-banking and restoration, and to provide grants to state and local agencies for the acquisition and management of accessways, buffer zones around parks, and wildlife habitat.

TABLE 8-7B

HISTORY OF MAJOR COASTAL BILL IN 1976 SESSION: REVISION AND EVENTUAL PASSAGE OF SB 1277

Date	Locus/event	Principal effects on Smith bill (SB 1277)	Changes in organized support/opposition
June 18	Substance of SB 1579 amended into SB 1277 (Smith), which had previously passed Senate	Revived coastal legislation	
June 24–August 9	Assembly Resources, Land Use, and Energy Committee	Smith bill (SB 1277) completely redrafted to 1. Revise inland boundary of coastal zone to 1000 yards (somewhat less in urban area) plus "bulges" for habitat and recreational resources but not streams or agricultural land 2. Slightly weakened policy mandate for public access and protection of scenic views while somewhat strengthening those for low-income housing, recreational facilities, and natural habitat protection 3. Added 3 state agency officials to the state commission 4. Clarified relations with other state agencies 5. Allows local governments to exercise interim permit review except in nearshore area under certain conditions 6. Streamlined local certification process but weakened state funding of local programs	Generally considered to be better organized than SB 1579 Gained support of Association of General Contractors—thereby undermining opposition of construction unions But lost support of League of Cities because some of Beilenson's concessions withdrawn

Date	Body	Action	Support
		7. Provide for special treatment areas in timber harvesting	
		8. Delete commission's authority to issue cease-and-desist orders	
August 10	Assembly Resources Committee	Approve SB 1277, 10–3	
August 2–10	Senate Natural Resources Committee	Keene bill (AB 3875) killed 2–6 and (revived) Cullen-Ayala bill died on Aug. 10, leaving Smith bill as only coastal legislation	
August 11	Assembly Ways and Means Committee	SB 1277 amended to 1. Limit postcertification appeals and simplify certification process 2. Reestablish state funding for local program preparation 3. Limit authority over power plants (against Smith's objection) Approved 15–4, after rejecting amendment to require compensation for downzoning	Gained support of: 1. League of Cities 2. Utilities 3. Irvine Company (development firm)
August 13	Assembly floor	Approved SB 1277 45–29 without amendment	CCEEB State Lands Commission
August 18	Senate Natural Resources Committee	Concurred in Assembly amendments to SB 1277 by 6–3 vote	
August 23	Senate floor	On first vote, SB 1277 fell 3 votes short of passage Governor Brown then negotiated amendments to win labor support, whereupon SB 1277 passed 25–14	AFL-CIO

(continued)

TABLE 8-7B (Continued)

Date	Locus/event	Principal effects on Smith bill (SB 1277)	Changes in organized support/opposition
		Brown's amendments 1. Clarify mandatory urban exclusion under certain conditions 2. State commission has option of not activating regional commissions 3. Require legislative certification of "sensitive coastal areas"	
August 25–31	Both houses	Trailer legislation incorporating Brown's amendments (AB 2448) and providing funding for successor agency and $23 million for some coastal acquisitions; pass committees and floor of both houses	
September 29	Governor	Governor Brown signs coastal bills, while reducing appropriation for coastal acquisition from $23 million to $12 million	

The bill was introduced by Michael Wornum, a freshman Democrat from Marin County and former chairman of the North Central Regional Commission. Unfortunately, Wornum was not a very skillful legislator, and thus the bill was almost killed a couple times despite minimal organized opposition. In particular, Assemblyman Sieroty and Speaker McCarthy rescued it from preliminary defeat in the Assembly Ways and Means Committee, and Bill Geyer and Bob Meyer of CCEEB apparently saved it in Senate Finance by convincing several conservative Republicans that the bill was essentially designed to provide partial compensation to people—such as the owners of marginal agricultural land or of lots in "premature" subdivisions—for whom the restrictive policies in the Beilenson-Smith bill promised real hardship.[44] At any rate, the bill was eventually approved in late August with only minor amendments.

The coastal bond bill was a more interesting case. In fact, as indicated in Table 8-8, there were two competing bond bills: One, sponsored by Assemblyman Gary Hart, earmarked $290 million solely for park and wetland acquisitions along the coast that had been recommended by the coastal commission. The other, by Senator Nedjedly, allocated only $150 million of its $360 million for coastal acquisitions (not necessarily those recommended by the commission), with the remaining $210 million going for park, recreational, and wildlife habitat acquisitions and development *throughout the state*. Nejedly's bill, which was supported by CCEEB, was based on the premise that a statewide program had a much better chance of being approved by the voters in November.[45] Both bills were approved by the relevant committees and the floor of each house with only minor amendments (including a reduction of $40 million in Hart's proposal and the allocation of $15 million in Nejedly's bill to the conservancy), whereupon they went to conference committee for final resolution. As can be seen in the third column of Table 8-8, the conference committee generally accepted the allocation of funds in the Nejedly bill while also reducing the total amount from $360 million to $280 million at the insistence of the Brown administration.[46] The conference report was then approved by both houses just before the June 26 deadline for placement on the November ballot.

As indicated in chapter 6, the Nejedly-Hart State Urban and Coastal Park Bond Act of 1976 was approved by 51.5% of the electorate in November after a very low-key campaign in which CCEEB played a major role. The margin of victory, which came primarily for the Bay Area, may not seem very impressive until one realizes that both other bond issues on the ballot went down to defeat despite being strongly supported by Governor Brown.[47]

On the whole, however, of the $255 million recommended by the coastal commissions in the Hart bill for the acquisition and development of coastal parks, only $110 million found its way into the Nejedly-Hart Bond Act. In an effort to recoup a portion of this shortfall, AB 400 (McCarthy) allocated $23.5

TABLE 8-8
COMPARISON OF COASTAL-RELATED ACQUISITIONS LEGISLATION, 1976 SESSION[a]

Allocation of funds	Coastal bond bills			Additional purchases from general fund in AB 400 (McCarthy) as signed by Governor Brown
	As presented in Assembly Resources Committee Feb. 1976		Nejedly-Hart Bond Act as approved by Conference Committee and signed by Governor Brown	
	AB 2948 (Hart)	SB 1321 (Nejedly)		
In coastal zone				
State park system acquisition and development	$150 million	$150 million	$110 million	$12 million
Grants to local governments for park acquisitions	50 million	—	—	—
Acquisition or development of property for recreation or historical preservation	55 million	—	—	—
Fish and Game for acquisition and management of wildlife habitat	10 million	—	10 million	—
Coastal Conservancy funding	25 million	—	10 million	—
Construction of nearcoast hostel facilities	—	—	—	2 million
Statewide				
State park system for acquisition and development	—	70 million	34 million	—
Grants to local governments for acquisition and development of recreational facilities	—	100 million	85 million	—
Fish and Game for acquisition of wildlife habitat	—	10 million	5 million	—
State Water Project recreational development	—	30 million	26 million	—
Totals				
Coastal subtotal	$290 million	$150 million	$130 million	$14 million
Statewide subtotal	0	$210 million	$150 million	0
Total	$290 million	$360 million	$280 million	$14 million

[a]Sources: *State Coastal Report*, January 7, 1976, pp. 7–9; February 17, 1976, pp. 3–6; June 1976, p. 7–8; October 1976, pp. 1–2; November 1976, p. 1.

million from the state's general fund for coastal park acquisitions and an additional $6.3 million for the construction of nearcoast hostel facilities. While these sums were approved by the legislature, they were substantially cut by Governor Brown to $12 million and $2 million, respectively, before he signed the bill into law.[48] In short, only about half of the funds requested by the commissions for coastal parks were actually provided by the legislature and the governor in 1976.

C. Important Factors in the Passage of Coastal Legislation

As this review has indicated, the legislature made a number of changes in the legislation embodying the coastal plan before enacting it into law. While this topic will be addressed in greater detail later in this chapter, the historical summary provided thus far indicates that the 1976 Coastal Act embodied more sharing of authority with local governments and other state agencies than originally proposed by the commissions. In addition, the legislature provided only about half of the requested funds for coastal acquisitions. On the other hand, the 1976 Coastal Act retained much of the plan's emphasis on environmental protection, public access, and concentrating development in existing urbanized areas; it retained the state and (probably) the regional commissions; and it retained the basic process of handing over primary jurisdiction for development review to local governments once their plans and implementing ordinances had been revised to conform to statewide coastal policies, with the coastal commissions retaining some appellate review as well as principal jurisdiction in the interim period and over offshore development. In short, while the legislature certainly did not enact the coastal plan, it did retain the basic policy orientation and implementing institutions recommended by the plan. This view is substantiated by the respondents to our 1977 survey of people on the commissions' mailing lists: Of those who supported the coastal plan, 81% also supported the 1976 Coastal Act; conversely, 85% of those who opposed the plan also opposed the act.[49]

Given the formidable opposition (exacerbated by adverse economic conditions) to the Beilenson bill and the numerous veto opportunities provided by the legislative process, why did the eventual outcome bear a reasonably close resemblance to the commissions' original proposals? We would suggest a number of reasons.

First was the legacy of public support for coastal protection represented by the passage of Proposition 20 in 1972 and confirmed by the November 1975 CCEEB poll, as well as the general satisfaction with the commissions' performance by those who had supported Proposition 20. Many legislators—most notably Senator John Nejedly—felt that the legislature's capacity to perform its representative function was on the line. As he observed in December 1975,[50] "The Legislature failed to do anything sensible in this area in the past. So finally the people expressed a decision, a concern. It seems to me we should learn something from Proposition 20. We ought to accept the responsibility for our actions. If you turn your back on the people, some of us should be turned out at the next election." Moreover, the possibility, however uncertain, of another initiative apparently had a moderating effect on some past opponents of coastal legislation—most notably, the Association of General Contractors—and CCEEB's own poll helped convince it of the desirability of a successor coastal agency and of a coastal bond issue.[51]

Second, the Beilenson-Smith bills and other legislation incorporating coastal plan recommendations had the strong support of the Democratic majority leadership in both houses and, eventually, of Governor Brown. This created a general expectation that some form of coastal legislation involving a successor agency was very likely to be enacted. It also assured the commissions' bills of excellent strategic advice throughout the session and of the intervention of the more powerful people in state government at critical moments. Most notable were Speaker McCarthy's intervention in the Assembly Ways and Means Committee on behalf of both the Smith bill and the conservancy bill, the major role he played on the trailer legislation; and, of course, Governor Brown's dramatic intervention to assure passage of SB 1277 on the Senate floor.

Third, the major coastal bill benefited enormously from the drafting and negotiating skill of a number of legislators. While Senator Beilenson could probably be faulted for taking so long to redraft his original legislation (which so closely reflected the unwieldy coastal plan), he did arrange the critical compromises with the ports and the League of Cities. Senator Nejedly, as chairman of the original policy committee, forced the first major revisions to drastically compress the policy section and to meet the concerns of most state agencies. He also developed a more salable bond act than that originally proposed by the commissions. Assemblyman Charles Warren presided over the major redrafting of the Beilenson-Smith bill during the July recess, turning it into a much more coherent piece of legislation. And Senator Jerry Smith apparently did a superb job of mastering the bill, arranging the final compromises, and then "working" the bill with his colleagues.[52]

Fourth, coastal legislation benefited from the willingness to compromise and subsequent support of some groups that the commission (and particularly environmental groups) tended to regard as opponents. Almost everyone we interviewed, for example, mentioned the superb job that David Beatty of the League of California Cities did in arranging reasonable compromises and in educating his constituents (with the exceptions of Long Beach and Los Angeles). And CCEEB served as an important bridge between the development community and the coastal commissions, as well as playing critical roles in the eventual passage of the conservancy bill and the Nejedly-Hart Bond Act.

Fifth, the coastal commissions (particularly the state commission staff) helped develop a coherent strategy—emphasizing sequential negotiations with the wide variety of agencies and interest groups concerned with the major coastal bill—and managed to play a central role in most of the negotiations. While the strategy delayed some of the necessary compromises and perhaps contributed to the defeat of the Beilenson bill in Senate Finance, it also produced a process of orderly compromise and helped maintain the general coherence and stringency of the legislation.

Finally, environmental and community groups supportive of the coastal plan played an important role in legislative deliberations and particularly in generating constituency support for the Beilenson-Smith bill. Most people we talked to indicated that the professional lobbyists for the Sierra Club, the Planning and Conservation League, and, to a lesser extent, the Coastal Alliance and PACE did a good job of assisting the bill's authors in negotiating and drafting amendments and in orchestrating constituent support. There was also general agreement that the defeat of SB 1579 in Senate Finance served to galvanize supporters, many of whom had become complacent or had been working on other legislation (most notably, the Nuclear Safety Initiative in the June 1976 primary). When asked for specific examples of the effective (although not always successful) marshaling of constituency support, legislative staff and lobbyists mentioned senators Grunsky, Holmdahl, Alquist, and Roberti on the Finance Committee, Senator Nejedly, Assemblyman Robert Beverly (the Republican floor leader whose support kept SB 1277 from becoming a partisan issue), and numerous senators during the hectic days of August, of whom the most surprising was Jack Schrade, a very conservative San Diego Republican who had been instrumental in killing coastal bills in 1970 but who supported SB 1277 on the floor (partially because he thought it would help in a close reelection campaign).[53]

A more systematic indication of the efforts of environmental groups and other actors comes from our survey of the "attentive public" concerned with coastal regulation. As indicated in Table 8-9, about 60% of the members of the environmental and community groups contacted legislators, over a third made public presentations concerning coastal legislation, and over 60% helped organize grass roots efforts. In fact, the data suggest quite strongly that—in contrast to the conventional wisdom that the beneficiaries of regulation are less active in legislative reformulation than regulated groups—the supporters of the coastal plan were about as involved in a wide variety of activities as the plan's opponents.[54] In sum, SB 1277 would never have become the 1976 Coastal Act without the ability of environmental groups and other supporters to mobilize support from the districts, thereby confirming our emphasis throughout this study on the importance of organized constituency groups.

TABLE 8-9
ACTIVITIES ON 1976 COASTAL LEGISLATION BY GROUP AFFILIATION AND ATTITUDE TOWARD COASTAL PLAN

| | By group affiliation | | | | | | | By attitude toward coastal plan[a] | | |
| | Generally supportive of coastal plan | | | | Generally opposed to coastal plan | | Ambivalent about coastal plan | | | |
Activity	Commissioner/ staff (N = 84)	Environ. group (N = 31)	Civic/ community group (N = 22)	Interested observers/ citizens (N = 68)	Business (N = 34)	Applicants/ coastal property owners (N = 72)[a]	Local/ state agency officials (N = 64)	Supported coastal plan (N = 210)	Opposed coastal plan (N = 110)	Ambivalent toward coastal plan (N = 62)
Direct contact with legislators										
Personally write or contact legislators, their staff, or governor's office	42%	71%	59%	64%	44%	64%	25%	53%	64%	24%
Attend or testify before legislative hearings	33%	10%	23%	16%	24%	15%	16%	23%	21%	14%

Suggest amendments to bills									
Suggest amendments to coastal legislation	38%	23%	32%	20%	62%	39%	47%	31%	53%	29%
Participate in negotiations over sections of bills	12%	0%	18%	3%	9%	4%	14%	9%	10%	10%
Stimulate constituent activity									
Speak at public forums or write articles concerning coastal legislation	49%	42%	36%	26%	26%	25%	12%	36%	30%	19%
Help in grass roots efforts (letter writing, phone campaign, etc.) in support of your position	21%	61%	68%	33%	21%	53%	6%	36%	44%	14%
Aggregate activity									
Participate in two or more activities	51%	61%	59%	46%	65%	58%	30%	52%	62%	33%

aIncludes only property owners and applicants on the commissioners' two mailing lists (i.e., not those from our separate sample of permit applicants). Thus, their selection is by the same criterion as all other groups, with the exception of commissioners/staff.

III. The Reformulation of the Commissions' Mandate: A Comparison of the 1976 Coastal Act with the 1972 Initiative and the 1975 Plan

It is now time to examine in greater detail the manner in which the legislature and the governor reformulated the commissions' mandate. Table 8-10 compares the 1976 Coastal Act (as supplemented by the conservancy and the bond act) with the 1972 initiative and the 1975 plan in terms of the statutory variables used throughout the study, namely, (1) the clarity and consistency of objectives, (2) the adequacy of the overall causal theory implicit in the legislation, and (3) the set of variables affecting the statute's ability to effectively structure the implementation process.

As has been emphasized throughout this report, the 1972 Coastal Initiative had a strong general "tilt" in favor of environmental protection rather than economic development, but more specific objectives had to be inferred from those instances—involving proposed developments affecting wetlands, public access, viewsheds, agricultural land, and geologic hazards—in which a two-thirds vote was required and/or conditions (or denials) were specifically mandated. The 1975 coastal plan continued the strong environmental orientation of the Initiative but "fleshed out" its implications by adding numerous policies designed to protect wetlands, provide physical and scenic access to the coast, concentrate new developments in existing urbanized areas, and give priority to coastal dependent development such as ports and water-oriented recreation. In addition, it added an entirely new set of policies relating to the protection of special coastal communities and moderate income housing. The legislature's reaction to the plan was to retain the emphasis on wetlands protection, public access, priority for coastal-dependent development, and, to a lesser extent, concentrating development in urbanized areas (including strong protection for prime agricultural land). But it also placed greater emphasis than had the plan on meeting the housing and energy needs of the state, protecting the property rights of coastal residents, and involving local governments in the implementation process—as well as indicating in greater detail than did the plan the manner in which these values should be balanced against the statute's general goals of access and environmental protection. [55]

The comparison of objectives among the three pieces of legislation can perhaps best be seen by examining specific policies—for example, the requirements for public access in new developments. Proposition 20 simply required that "access to publicly owned or used beaches, recreation areas, and natural reserves be increased to the maximum extent possible by appropriate dedication" (Section 27403a). Following are the relevant policies from the plan and the 1976 Coastal Act.

1975 Coastal Plan (Policy 123)	1976 Coastal Act (Section 30212)
New developments shall provide access to the shoreline except when adequate access exists nearby, the development is too small to include an accessway, or access would endanger public military safety, fragile natural resources, or agricultural land. When access dedications are infeasible, in lieu fees shall be required to help provide access elsewhere.	Public access from the nearest public roadway to the shoreline shall be provided except when adequate access exists nearby or access would adversely affect public or military safety, fragile resources, or agricultural land. A dedicated accessway shall not be opened to public use until a public agency or private association agrees to accept responsibility for maintenance and liability.

As can be seen, the principal changes made by the legislature were the deletion of the requirement for in-lieu fees and the addition of a requirement (implicit in Policy 124 of the plan and explicit in Section 302746 of the original Beilenson bill) that accessways not be opened until maintenance and security could be provided. In effect, the 1976 Coastal Act retained the essential thrust of the plan policy, while also meeting some of the concerns of property owners that they not be responsible for providing maintenance and security and not be required to assist in providing access beyond the boundary of their property.

Given these objectives, the 1975 plan and the 1976 Coastal Act incorporated more sophisticated theories than the 1972 Initiative about how the goals could be achieved. As indicated in the previous chapter, Proposition 20 implicitly assumed that its objectives of protecting habitat and access could be achieved through the commissions' total control over development for 3 miles seaward and 1000 yards landward of the mean high tide line. While adequate to protect viewsheds from the coastal highway and to protect most wetlands, the absence of authority to purchase and to manage accessways and beach-front lots proved a serious impediment to actually *providing* access. In addition, the Initiative provided the commissions no authority to affect the demand for coastal resources brought about, for example, by the inducements of property and estate taxes for more intensive uses of land.[56]

The coastal plan addressed most of these deficiencies, and many of its innovations were incorporated into the 1976 Coastal Act. The plan's recommendation for broadening the commissions' geographical jurisdiction (to 5 miles inland in most places) provided better authority to protect wetlands (by including upstream areas), recreational areas (such as the Santa Monica mountains) near the coast, and agriculture dependent upon a coastal climate (e.g., artichokes). In

TABLE 8-10A

COMPARISON OF THE 1976 COASTAL ACT, THE 1972 COASTAL INITIATIVE, AND THE 1975 COASTAL PLAN: POLICIES AND OVERALL THEORY

	1972 Coastal Initiative[a]	1975 Coastal plan/original SB1579	1976 Coastal Act[b]
I. Clarity and consistency (or explicit ranking) of policy objectives			
A. General objectives	Avoid substantial adverse environmental effects; preserve, protect, and restore resources of coastal zone, no explicit balancing with economic objectives	Basic goals of (1) protecting natural resources, (2) giving priority to coastal-dependent development, (3) maximizing access to the coast consistent with resource protection, (4) concentrating development in existing urbanized areas, (5) protecting special coastal communities (Pol. 1)	Basic goals of (1) protecting natural and man-made resources, (2) assuring orderly, balanced utilization with regard to social and economic needs, (3) maximizing access and recreational opportunities consistent with resource protection and property rights, (4) giving priority to coastal-dependent development, and (5) encouraging state–local collaborative planning, conflicts to be resolved to protect significant coastal resources
B. Specific objectives/topics			
1. Wetlands and wildlife habitat	⅔ vote and conditions to protect wetlands, wildlife habitat, and water quality	Essentially prohibited development resulting in a net reduction in functional capacity of wetlands; clear preference for commercial over recreational boating	Allow certain types of development in wetlands (including expanded recreational boating) subject to environmental safeguards

2. Public access to coast and recreation areas	⅔ vote and conditions to provide maximum access to beach and other recreation areas	Require access dedications in new developments on the coast; give priority to commercial recreation near coast; and provide transportation to the coast—all consistent with resource protection	Maximize public access and recreation consistent with protection of property rights and natural resources from overuse by requiring access in most new developments and by giving priority to coastal-dependent recreation
3. Scenic resources and views	⅔ vote and conditions to protect viewsheds and scenic resources	Require new developments to be unobtrusive and not interfere with viewshed and highway	New development to protect scenic view and to be unobtrusive
4. Protect agricultural land; reduce sprawl	⅔ vote on conversion of agricultural land	Explicitly seek to protect agricultural land, concentrate development in existing areas, and restrict lot splits	Strong policies to protect prime agricultural land, concentrate development in existing areas, and restrict land divisions
5. Permissible development	Essentially silent except to require conditions on developments in floodplains and geologic hazard zones	Give priority to coastal-dependent developments, e.g., ports and water-related recreation; restrict new development in floodplains and geologic hazard zones; allow energy facilities under stringent environmental safeguards	Give priority to coastal-dependent industrial development and, secondarily, to water-related recreation; restrict new development in geologic hazard zones; allow energy (and other industrial) facilities under environmental safeguards
6. Protect special coastal communities and moderate-income housing	—	Protect communities of unusual aesthetic character; discourage reduction in moderate-income housing	Little mention of special coastal communities; protect and provide lower-cost visitor facilities and housing where feasible

(continued)

aFocus is on permit review, as most analogous to development review under coastal plan and 1976 act.
bIncludes provisions of SB 1277 and relevant portions of AB 2948 and AB 400.

TABLE 8-10A (Continued)

	1972 Coastal Initiative[a]	1975 Coastal plan/original SB1579	1976 Coastal Act[b]
II. Incorporation of a sound overall theory for achieving statutory objectives			
A. Knowledge about causal factors affecting objectives	Low/moderate	Moderate/high; varied with objectives	Moderate/high; varies with objectives
B. Jurisdiction over principal causal factors	Control over all development seaward and landward within 1000 yards of coast,[c] sufficient to protect scenic views and probably most wetlands, but lacked management and acquisition authority necessary to actually provide public access or ability to provide transportation to coast, moreover, practically no control over demand for land or mineral resources or pressures from property and estate taxes for more intensive land uses	Broadened geographical jurisdiction to include most land within 5 miles of coast (2 miles in S. Calif.),[c] sufficient to protect views, wetlands, potential recreational areas, and coastal-related agriculture, ability to actually provide access aided substantially by conservancy and acquisitions, but still minimal control over demands for land and mineral resources, although did recommend changes in tax assessments on agricultural land and timber and did recommend energy conservation measures to reduce need for oil drilling and energy facilities	Geographical jurisdiction now 3 miles seaward and 1000 yards landward, with bulges up to 5 miles inland on about 40% of the coast,[c] sufficient to protect views, habitat, and recreational resources from development, ability to provide access and habitat aided by conservancy and acquisitions, tho less than in plan, still minimal control over demand for land and mineral resources, although some changes in property tax assessments and major change in timber taxation

[c]Excludes San Francisco Bay, which remains under the jurisdiction of BCDC.

addition, the recommendations for extensive acquisitions, a Coastal Conservancy, and the provision of public transportation and parking areas near the coast demonstrated a much more sophisticated understanding of what was required actually to provide beach access. Finally, the plan made some effort to address the demand for coastal resources through its suggestions concerning revisions in the property tax system as applied to agricultural land and timber, as well as its innovative recommendations for energy conservation (thereby lessening the need for petroleum drilling and energy facilities along the coast). The 1976 legislature reduced the proposed geographical jurisdiction to approximately 1000 yards landward plus extensive "bulges," in the process deleting protection of coastal-dependent agriculture (but not prime agricultural land that happened to be near the coast) as a statutory objective. It retained the conservancy, while reducing the proposed recreational and wetland acquisitions by about half. It required tax assessors to take into account development restrictions imposed by the commissions in determining the taxable value of property; in addition, separate legislation completely revised the method of taxing timber.[57] Finally, the legislature deleted the energy conservation recommendations as duplicative of authority previously assigned to the Energy Commission. In sum, the coastal plan and, to a slightly lesser extent, the 1976 Coastal Act addressed far more of the important causal factors affecting statutory objectives than had the 1972 Coastal Initiative.

Assigning authority to implementing institutions to address these factors is not, however, sufficient to attain statutory objectives. The statute must also establish an implementation structure that maximizes the probability that implementing officials and the ultimate target groups will act consistent with those objectives. This involves a number of different variables. We have already indicated that both the plan and, to a lesser extent, the 1976 Coastal Act provided much clearer policy guidance to implementing officials than did the Coastal Initiative. On the other hand, Proposition 20 assigned implementation to more supportive officials, i.e., the coastal commissioners, whereas both the plan and particularly the 1976 act assigned principal responsibility to local governments after the certification of local coastal programs. Proposition 20 also provided a much more hierarchically integrated system, with the state commission responsible for almost all important decisions within the 1000-yard permit zone. This was not greatly changed in the plan, as the coastal commissions would retain primary jurisdiction over major public works and energy projects and would have extensive appellate review of local decisions, even after certification of local coastal programs (LCPs). The 1976 Coastal Act, however, provided less appellate review after LCP certification and more liberal criteria for the exclusion of urban areas from any review by the commissions; as a result, in urban areas the commissions' review authority after certification would essentially be limited to major energy and public works projects and to developments within 300 feet of a beach or 100 feet of a wetland.[58] In addition, the 1976 act went into consider-

TABLE 8-10B
COMPARISON OF THE 1976 COASTAL ACT, THE 1972 COASTAL INITIATIVE, AND THE 1975 COASTAL PLAN: STRUCTURING THE IMPLEMENTATION PROCESS

	1972 Coastal Initiative: 1972–1976	1975 coastal plan/original SB1579	1976 Coastal Act[a]
III. Support for policy objectives among implementing officials	By statute, regional commissions split between state appointees and local elected officials, with state commission more favorable; staff could be expected to be favorable, see Chap. 4 for actual attitudes	Pending certification of local coastal programs (LCPs) would remain unchanged from Prop. 20; after certification, principal authority for implementation assigned to local and state officials with widely varying attitudes; state coastal commission would be composed entirely of state appointees and thus probably favorable to statutory objectives	Same as in coastal plan, except state commission after LCP certification would be half state appointees, half local officials
IV. Extent of hierarchical integration within and among implementing agencies	Very high within permit zone, as commissions had absolute development review authority and state commission heard most important cases; rather low, however, when required cooperation of other agencies, e.g., on management of accessways, civil penalties generally sufficient to bring about target group compliance if commissions and attorney general willing to provide resources for prosecution	Pending certification of LCPs, even more integrated than under Prop. 20 because of commissions' wider geographic jurisdiction over energy and public works projects and the presence of sympathetic conservancy for some management functions; after certification, primary authority to local governments, but almost all significant projects could be appealed	Pending certification of LCPs, rather similar to Prop. 20 situation except that coastal commissions would have much less jurisdiction over power plants and timber harvesting; also precluded from setting more stringent standards of review than state agencies in their area of expertise; after certification of LCPs by commissions, appeals limited to (1) developments on sea or within 300 ft of beach or

			100 ft of any estuary, (2) non-principal permitted uses in counties, (3) public works and energy projects, and (4) developments in "sensitive coastal areas" designated by the legislature
V. Supportive decision rules	Burden of proof on permit applicants; require ⅔ affirmative vote on most important permit cases	Retain burden of proof and ⅔ vote provisions	*Delete ⅔ vote provision*
VI. Opportunities for intervention by supportive interest groups and sovereigns	Require public hearings on permit and planning decisions; liberal rules of standing for administrative and judicial appeals; Initiative precluded weakening amendments by legislature prior to 1976, when sunset clause went into effect	Similar requirements for public hearings and liberal rules of standing, with any two commissioners now also able to bring administrative appeals; encourage litigation by supporters through recovery of attorneys fees and portion of civil penalties, as well as prohibiting requirements to post bonds	Retain plan's provisions for public hearings, liberal rules of standing, authority for any two commissioners to bring appeals, and prohibition against bond requirements; delete provision for attorneys fees and civil penalties; silent about legislative review
VII. Adequacy of financial resources to implementing agencies	Guaranteed appropriation of $5 million for 4 years; also federal grants	Suggested variety of funding sources	Appropriated $1.47 million for first year and obligated state to offset local governments' costs of preparing LCPs, but all appropriations on annual basis; also expect federal grants, with 50% allocated to local governments

ᵃIncludes provisions of SB 1277 and relevant portions of AB 2948 and AB 400.

able detail to clarify the substantive scope of the commissions' jurisdiction and to make sure the commissions deferred to other state agencies in the latter's area of expertise. With respect to sewage treatment plants, for example, the coastal commissions would have jurisdiction over location, visual appearance, and scope of service area served by the facility, while decisions concerning the extent of treatment were expressly reserved to state water pollution control authorities. [59]

Turning to other aspects of statutory structuring, the legislature decided not to retain the requirement in Proposition 20 and the plan for a two-thirds vote on most major permits. But the 1976 act retained Proposition 20's provision to encourage citizen participation in agency proceedings and in juridical review, while rejecting the plan's suggestions for recovery of attorney's fees and civil penalties to litigants seeking to prevent violations of the act. Finally, the 1976 act did not renew the Initiative's provisions insulating the commissions from annual legal and financial scrutiny by the legislature. At the same time, however, it established the commissions as permanent agencies without any "sunset" clause.

In conclusion, if one assumes that Proposition 20 was instituted primarily to protect wetlands and other wildlife habitat and to provide physical and scenic access to the coast, it is clear that the coastal plan would have attained those objectives better than would the Coastal Initiative because of its clearer policy directives, more extensive geographical jurisdiction, and provisions for the ac-quisition and management of accessways and threatened wetlands. Just as clear-ly, the 1976 legislature rejected some of the more grandiose provisions of the 1975 plan with respect to the coastal commissions' authority over state and local agencies throughout the 5-mile-wide coastal zone. In comparison to the plan, the 1976 Coastal Act restricted the commissions' authority to the protection of specific coastal resources (especially wetlands and access) and incorporated pro-cedures designed to reduce the paper work burden on permit applicants, increase the role of (local) elected officials in the implementation process, and balance access and environmental protection against the housing and energy needs of the state (particularly in urban areas).

It is not at all clear, however, whether the 1972 Initiative or the 1976 Coastal Act did a better job of providing access and protecting wetlands. On the one hand, Proposition 20 did not have any balancing language; it provided a much more hierarchiacally integrated structure, with the state commission clear-ly able to override the decisions of other state and local agencies on development applications; and it insulated the commissions from annual legislative review and the attendant leverage of legislators on important permit cases. On the other hand, the 1976 Coastal Act provided clearer policy directives, with the balancing provisions probably having little effect on access and wetlands; the larger geo-graphical jurisdiction and the provisions for acquisitions and the conservancy have increased the ability to actually provide access and to manage wetlands; the revisions in tax assessment practices have probably had some dampening effect on pressures for more intensive utilization of coastal land; and, most subtly but perhaps most importantly, the creation of a "permanent" coastal agency and the (supposed) integration of coastal concerns into the decision structure of state and local agencies have perhaps altered everyone's expectations of what constitutes acceptable development along the coast, thereby reducing the enormous opposition the commissions aroused during the 1972–1976 period by superimposing more stringent restrictions on projects that had already been approved by local (and state) agencies. On balance, we would suggest that the 1976 act provided at least as much protection for wetlands and probably slightly greater public access to the coast than did the 1972 Initiative.[60]

NOTES

1. This view of implementation as an iterative sequence is basically consistent with those who view implementation as a learning or evolutionary process involving alternative approaches to the same set of problems; see Giadomenico Majone and Aaron Wildavsky, "Implementation as Evolution," in *Policy Studies Review Annual*, vol. 2, ed. Howard Freeman (Beverly Hills: Sage, 1978); and Leonard Goodwin and Phyllis Moen, "On the Evolution and Implementation of Welfare Policy," in *Effective Policy Implementation*, ed. Daniel Mazmanian and Paul Sabatier (Lexington, Mass.: Lexington Books, 1980).
2. Interview of Joseph Bodovitz with Stanley Scott, No. 3, Spring 1978, p. 20. The legislature's focus on previous permit actions is confirmed by our notes and published reports of legislative committee hearings; see, for example, "Coastal Plan Presented to Legislature," *State Coastal Report*, Dec. 31, 1975, pp. 4–5.
3. For example, see Eugene Bardach, *The Skill Factor in Politics* (Berkeley: University of California Press, 1972) and Bernard Asbel, *The Senate Nobody Knows* (New York: Doubleday, 1975).
4. The survey was conducted for CCEEB by Public Response Associates, a San Francisco consulting firm. It involved a personal in-home survey of 763 randomly selected respondents in the state's most populous counties (involving 93% of the state's population) during the period November 7–23, 1975 (Public Response Associates, *A Survey of Attitudes Toward the California Area*, An unpublished study conducted for the California Council on Environmental Economic Balance, San Francisco, December 1975). We are extremely grateful to Jim Carroll of CCEEB for making a copy of the survey results available to us.
5. While this represented a sensible allocation of responsibilities, it also contained some apparent inconsistencies. It is unclear, for example, how the coastal commissions could preserve scenic views if local governments retained jurisdiction over high-rise apartments and over design and architectural features. In short, additional questions clarifying the allocation of responsibilities in terms of geographical proximity to the coast would have been illuminating—but would also have substantially expanded the length and cost of the questionnaire.
6. The question was asked only of respondents who had previously indicated that at least one topical area should be a responsibility of either the state coastal commission or another state agency (Public Response Associates, *Attitudes Toward the Coastal Area*, Q. 3).
7. The detailed results of the survey were known only to CCEEB's staff and the members of its executive committee. Because of some of the ambiguities and inconsistencies in the results, it was decided not to release the poll publicly. At least the general thrust of some of the results was,

however, discussed with legislators, administration officials, and interest group lobbyists (including those from environmental groups) during the course of negotiations on specific aspects of coastal legislation (e.g., the extent of the commissions' geographical jurisdiction and the amount of money in a bond act).

8. In general, the political science literature indicates that legislators are not very knowledgeable about, or responsive to, public opinion within their district *except* on those issues of substantial importance to large numbers of their constituents. This is partially the result of the general public's overall ignorance concerning policy issues and their representatives' positions on them. See Warren Miller and Donald Stokes, "Constituency Influence in Congress," *American Political Science Review*, 57 (1963), pp. 45–56; Malcolm Jewell and Samuel Patterson, *The Legislative Process in the U.S.*, 3d ed. (New York: Random House, 1977), pp. 305–315; Donald Matthews and James Stimson, *Yeas and Nays* (New York: Wiley Interscience, 1975), pp. 26–31; and Charles Backstrom, "Congress and the Public," *American Politics Quarterly*, 5 (October 1977), pp. 411–435.

9. A recent survey of the California Legislature found that state agencies and interest groups/corporations were by far the most important sources of policy-relevant information; see Paul Sabatier, *The Sacramento Connection* (Davis, Calif.: Institute of Governmental Affairs, 1978). Also see Jewell and Patterson, *The Legislative Process*, chaps. 12–13, and Randall Ripley, *Congress: Process and Policy* (New York: W. W. Norton, 1975), chap. 8.

10. About the only publicly available sources on permit review were (a) the coastal commissions' annual reports, which listed highly aggregated figures on a straight approve–deny basis, as well as discussions of a few important issues; (b) the minutes of the various commissions; and (c) the April 1976 issue of *State Coastal Report*, which contained a summary of the data found in the annual reports and a brief article by Sabatier on permit appeals; to the best of our knowledge, the latter were the only data available on the extent of conditional approvals.

11. Economic Research Associates and Alvin Baum & Associates, *Economic Impacts of the Proposed Coastal Plan—A First Report and Further Proposals* (Sacramento: California Legislature, Joint Rules Committee, December 1975); Legislative Analyst, State of California, *Review of California Coastal Plan*, 2 vols. (Sacramento: Legislative Analyst, April 1976).

12. Following are the principal studies: (a) "Impact of Proposition 20 on Coastal Property Mixed," *State Coastal Report*, Dec. 31, 1975, pp. 13–14; (b) Real Estate Research Corporation, *Business Prospects Under Coastal Zone Management* (Chicago: Real Estate Research Corp., 1976); (c) George Goldman and David Strong, *Governmental Costs and Revenues Associated with Implementing Coastal Plan Policies in the Half Moon Bay Subregion*, Special Publication No. 3208 (Berkeley: Division of Agricultural Sciences, University of California, September 1976); (d) California Council for Environmental and Economic Balance, *An Economic Profile of the California Coastal Zone* (San Francisco: CCEEB, 1976); (e) Construction Industry Research Board, *Economic Analysis: California Coastal Zone Conservation Act* (Los Angeles: Construction Industry Research Board, 1976); (f) H. E. Freech and Ronald Lafferty, "The Economic Impact: Land Use and Land Values," in *The California Coastal Plan: A Critique*, Eugene Bardach et al. (San Francisco: Institute of Contemporary Studies, 1976), pp. 69–92; and (g) Robert Rooney, *The Economic Context of the California Coastal Plan* (San Francisco: Planning and Conservation Foundation, Fall 1975). For a review and critique of these studies, see Robert Kneisel, *Economic Impacts of Land Use Control: The California Coastal Zone Conservation Commission*, (Davis, Calif.: Institute of Governmental Affairs, February, 1979), pp. 16–33).

13. Because of our desire to include items on the 1976 legislative session and the 1976 Coastal Act, the survey questionnaire was not mailed out until the fall of 1977. Thus, there is some possibility that respondents' perceptions of impacts during the 1972–1976 period may have been contaminated by faulty memory and/or the brief experience of the commissions under the new act. Following are the number of respondents and the response rate for each of the subsamples.

	Number of respondents	Response rate
a. General mailing list of (7) commissions	188	53%
b. Outsiders active in planning process	118	59%
c. Commissioners, staff, legislators and legislative staff on two policy committees	86	51%
d. Permit applicants involved in appeals to state commission	94	48%
	486	53%

14. Respondents from categories with similar functions and views have been collapsed in the tables. These include (1) commissioners and commission staff, (2) coastal property owners and permit applicants, (3) local governments and state agency officials, (4) journalist/informed observor and interested citizens, (5) civic/community organizations and the 2 respondents from recreational organizations. In addition, the response from legislators and labor unions was so minimal that they were dropped into an "other category," which has not been reported in the tables. Finally, only 8 of the 18 impact categories in the questionnaire have been reported in Table 8-3; selections were based on our views of the principal objectives of Proposition 20 and/or politically important impacts.

15. Daniel Mazmanian and Paul Sabatier, "The Role of Attitudes and Perceptions in Policy Evaluation by Attentive Elites: The California Coastal Commissions," in *Why Policies Succeed or Fail*, ed. Helen Ingram and Dean Mann, Sage Yearbooks in Politics and Public Policy, vol. 8 (Beverly Hills: Sage, 1980), pp. 107–133.

16. Jewell and Patterson, *The Legislative Process*, chap. 12; Abraham Holtzman, *Interest Groups and Lobbying* (New York: MacMillan, 1966), chap. 4; and Jeffrey Berry, *Lobbying for the People* (Princeton: Princeton University Press, 1977), chaps. 8–9. For an excellent analysis of the significant role of campaign contributions in legislative elections within California, see John Owens and Edward Olson, "Campaign Spending and the Electoral Process in California, 1966–74," *Western Political Quarterly*, 30 (December 1977), pp. 493–512.

17. A report by the Fair Political Practices Commission revealed that various organizations spent $40 million in 1975–1976 trying to influence the legislature, with local governments and special districts spending about $6 million and various business organizations about $19 million; in contrast, environmental and public interest groups spent about $2 million ("$40.1 million spent on State Lobbying," *Los Angeles Times*, 18 August 1977). See also John Owens, Edmund Costantini, and Louis Weschler, *California Politics and Parties* (New York: MacMillan, 1970), chaps. 7–8.

18. To give some idea of the continued viability of the Coastal Alliance as an organization, a spring 1975 questionnaire sent to its 26,000 individual members netted 19,000 responses; more important, of the 300 people heavily involved in the Proposition 20 campaign invited to a December 1974 meeting, 274 attended (Sabatier interview with Janet Adams, 2 December 1975, pp. 10–13). Both the Sierra Club and the Planning and Conservation League had paid, full-time lobbyists in Sacramento, and Janet Adams served as a part-time advocate during the 1976 session. As we shall see later in this chapter, the supporters of coastal legislation had activity levels comparable to those of opponents.

19. On the other hand, to qualify for the November 1976 ballot, petition gathering would have to be completed by early June or before the legislature had an opportunity to seriously consider coastal

legislation. Thus, the most probable date would be a June or November 1977 special election, which would, of course, leave a 1-year hiatus in which there would be no commissions.

20. For example, an August 1975 survey of members of the Assembly Resources and Land Use Committee revealed that only half of the eight respondents were very familiar with provisions of the coastal plan; see "Assemblymen Divided on Coastal Plan," *State Coastal Report*, August 20, 1975, pp. 16–23. The state commission staff had, however, maintained fairly close relations with some legislative staff, notably during the drafting of the powers and funding element (see Sabatier interviews with Jack Schoop, 21 May 1975, and Joseph Petrillo, 14 August 1975).

21. There were three pieces of legislation derived from the coastal plan: (a) the major bill (SB 1579) incorporating the basic policies and suggestions for implementation, which, as introduced, was 106 pages long; (b) AB 3544 creating the Coastal Conservancy; and (c) AB 2948 for coastal acquisitions.

22. At the beginning of the 1976 legislative session, the Senate Natural Resources Committee had four members expected to vote for coastal legislation (two of whom, Rodda and Smith, had been appointed since 1972), four expected opponents, and one swing vote leaning in favor (Nejedly); in July 1976, another likely proponent (Foran) was added to the committee to fill a vacancy ("The Coastal Plan and the Legislature: Actors and Actions," *State Coastal Report*, Quarterly Materials No. 2, December 1975, pp. 2–5). Senator Mills played a critical role in these appointments as chairman of the five-member Rules Committee, which makes all committee assignments.

23. For contrasts between the rather centralized power in the Assembly and the very dispersed patterns in the Senate, see Ed Salzman, "The New Legislative Alignment," *California Journal*, 6 (February 1975), pp. 53–54.

24. This material is taken from numerous issues of the *State Coastal Report* (August 29, 1975; September 26, 1975; December 31, 1975; December 1975 special supplement on "Coastal Plan and the Legislature"), as well as interviews with Janet Adams (2 December 1975), Mel Lane (2 December 1975), and Peter Douglas (15 December 1976).

25. This assessment is based upon interviews with Janet Adams (2 December 1975), Mel Lane (2 December 1975), and Peter Douglas (25 November 1975), as well as a memo, 28 August 1975, from Bodovitz to state commissioners on Brown's remarks concerning comprehensive planning, and a brief article in *State Coastal Report*, January 30, 1976, concerning the governor's budget message.

26. Senate Bill 1579, as introduced on February 19, 1976, Sec. 30003. There were essentially three options open to framers of the principal piece of coastal legislation: (1) Incorporate the coastal plan virtually *in toto*, (2) incorporate major portions of it and additional segments by reference to specific policies in the plan, or (3) include only a few major policies in the bill itself but incorporate the remainder of the plan "by reference." The last option was favored by many members of the state commission and staff, in part because it had been used before successfully with BCDC. But the Beilenson bill wound up adopting essentially the second strategy (with Section 30003 adopting the remainder of the plan by reference), largely because of the difficulties in deciding which policies should be deleted and the insistence of Beilenson and other legislators that the legislature actually review the policies to guide the future utilization of coastal resources; as we shall see, the legislature eventually rejected *any* attempt to adopt portions of the plan by reference. (See interviews with Mel Lane, 2 December 1975; Jack Schoop, 9 January 1976; Peter Douglas, 25 November 1975; and Gail Osherenko, Beilenson's legislative assistant, 9 February 1976; as well as the *State Coastal Report*, January 30, 1976, p. 4, and November 26, 1975, pp. 1–2.)

27. For brief reviews of the substance of these two measures as introduced, see "Coastal Conservancy Measure Introduced," *State Coastal Report*, March 1976, pp. 5–6; and "Other Legislation," *State Coastal Report*, January 30, 1976, pp. 7–15.

28. This is based on interviews with Lane, Adams, Douglas, and Schoop in December 1975 and January 1976, as well as "Coastal Plan Presented to Legislature," *State Coastal Report*, December 31, 1975, p. 3; see also note 19.

29. Based largely on conversations with Joseph Petrillo, supplemented by interviews and comments from several other people.

30. Assembly Resources, Land Use, and Energy Committee, *Transcripts of Informational Hearings on Implementing the California Coastal Plan—The Role of Existing Agencies*, March 1976; Healy *et al.*, *Protecting the Golden Shore*, chap. 3; Herman Boschken, "Interorganizational Considerations in Coastal Management: The 1976 California Legislative Experience," *Coastal Zone Management Journal*, 4 (1978), pp. 47–64; and interview with Preble Stolz (Governor Brown's chief liaison on coastal legislation), 18 March 1977.

31. The Cullen-Ayala bill (SB 1920/AB 4438) was drafted by an *ad hoc* committee including representatives from the California Chamber of Commerce, CSAC, the California Realtors Association, the Sea Ranch Association, and the City of Long Beach. Numerous amendments suggested by CCEEB to make the bill workable were accepted by Senator Ayala the day of the vote in the Senate Natural Resources Committee, but CCEEB never officially endorsed the bill. The bill as originally introduced on April 2, 1976 (April 24 in the Senate) would have created two new institutions, a Coastal Planning Compliance Office in the Resources Agency and the Coastal Planning Compliance Council (composed of the lieutenant governor, the heads of five state agencies, one county supervisor, and one city councilman). Local governments within the coastal resource management area (as defined by the 1975 coastal plan) would be required to submit a coastal program (following guidelines established by the office and council) to the Planning Office, where they would be deemed approved unless denied in whole or in part by the office; such denials would then go to the council for final decision. Before or after certification, permits issued by local governments would be final unless appealed by the office (with the concurrence of the council) within 20 working days to a court. The bill also provided for the designation of sensitive coastal resource areas (including 19 major wetlands) that would be afforded greater protection; but any losses to property owners from such restrictions would have to be compensated by the state (Section 30568). The policies in the bill contained considerably more balance and were more general than those in SB 1579 (Beilenson). With respect to public access, for example, Section 30216 provided that "reasonable and adequately convenient public access to coastal zone amenities, particularly as to shoreline areas of recreational value, while providing fair and adequate protection to the privacy and security of coastal zone home and property owners, shall be provided whenever practical."

32. The Keene bill (AB 3875) passed the Assembly as a minor bill dealing with herring gill nets and then was amended in the Senate on June 25 after the defeat of the Beilenson and Cullen-Ayala bills. It was apparently drafted largely by Keene himself out of portions of the other two bills (as amended), plus his own ideas, as an alternative to the coastal commissions' bills (SB 1579 and, later, SB 1277). It was supported in Senate committee by the AFL-CIO, the building trades unions, the city councils of Los Angeles and Long Beach, the realtors, the California Chamber of Commerce, the Sea Ranch Association, and, until August, CCEEB (based upon *State Coastal Report*, August/September 1976, p. 12, as well as conversations with CCEEB officials). It would have abolished the regional commissions and restructured the state coastal commission to include three state officials, eight public members, and six local officials. It required all local governments and ports within the 1000-yard coastal zone (plus bulges for wetlands) to prepare a coastal program by July 1979 to be submitted to the coastal agency, which would be deemed approved unless denied (in whole or in part) within 90 days. The bill provided for appeals of local decisions to the coastal agency within tight time constraints, although single-family residences were generally exempt. The bill also provided for the designation of sensitive coastal resource areas (primarily wetlands) by the legislature, although requiring that property owners be com-

pensated for large losses (Section 30517). The policy sections of the bill generally contained more balance than those in the Smith bill and were considerably more detailed than the general provisions in the Cullen-Ayala bill.

33. This is based largely upon "The Coastal Plan and the Legislature: Actors and Actions," *State Coastal Report*, Quarterly Materials No. 2, December 1975, pp. 1–8, as well as an interview with Janet Adams, 2 December 1975.

34. Almost everyone involved with the negotiations on the Beilenson bill agreed that he did not know the bill as well as he should have and delegated too much responsibility to his (dedicated but inexperienced) staff; in fact, Senator Nejedly publicly complained about this on several occasions. These observers were, however, divided on the effects of Beilenson's lame duck status; while some thought it obviously hurt, others remarked that it was not Beilenson's style to really "work" other legislators and certainly not to use his powers as Finance chairman as a lever in negotiations with other members. At any rate, the constraints on his time lessened considerably when his principal opponent in the Democratic primary, Assemblyman Majority Leader Howard Berman, dropped out of the race at the urging of Governor Brown and other party leaders. (These observations are based upon interviews with six participants who requested anonymity.)

35. Senator Nejedly had indicated from the beginning of the session that, as a response to Proposition 20, he wanted a "workable" coastal bill to emerge from his committee (see, for example, *State Coastal Report*, December 31, 1975, p. 4, and May 7, 1976, p. 5). While he still had reservations about the Beilenson bill as it emerged from his committee (see his letter to Bill Press of 26 May 1976), he felt that sufficient progress had been made, and Senate rules required that bills meet a May 3 deadline for emergence from the first policy committee. Voting for the Beilenson bill were Behr, Dunalp, Nejedly, Rodda, and Smith—the latter two new members of the committee—while Ayala, Berryhill, and Dills voted against it.

36. Ascertaining the reasons for legislators' votes is admittedly full of pitfalls. Nevertheless, it seems fairly clear from talking to participants and observing the Senate hearings that Grunsky (Rep-Monterey County) and Holmdahl (Dem-Oakland) were won over by their respect for Beilenson and the concessions made to local governments and (particularly in Holmdahl's case) to the ports. Alquist (Dem-San Jose) is more difficult. While he generally supported environmental legislation (*State Coastal Report*, December 31, 1975), organized labor was very strong in his district, and he had a reputation for being rather unpredictable. Nevertheless, it seems quite clear that a major reason—publicly stated by Alquist, reiterated by his legislative assistant, and confirmed by other observers—was his anger at Beilenson for taking so much of the committee's time with SB 1579 that five bills supported by Alquist simply never got considered by the committee. In short, one could make a very plausible argument that, had SB 1579 been put in decent shape somewhat earlier, it would have taken less of the committee's time, gotten Alquist's vote, and thus been approved by Senate Finance. By far the most difficult case is Roberti. Numerous observers agreed that Roberti intensely disliked liberal environmentalists who nevertheless purchased homes in Malibu and other areas, thereby blocking both visual and scenic access for people from the interior. This was consistent with Roberti's public argument that, since the housing restrictions in the permit zone would increase densities and congestion in his own inner-city district, his constituents had to be given clear guarantees of access if he were to vote for the bill (see his testimony before the Assembly Resources Committee, June 23, 1976). Our own conversations with Roberti's legislative assistant and other observers have convinced us that his argument should be taken at face value, even if one can dispute its efficacy. Nevertheless, there were also several observers who felt that Roberti may really have been responding to pressures or enticements from organized labor. In general, the best published sources on the Senate committee's actions are found in the *State Coastal Report*, June 11 and June 23, 1976; see also the letters by Roberti and others to the *Los Angeles Times*, 24 June 1976.

37. "Coastal Commission 1977 Budget Deleted from State Budget," *State Coastal Report*, June 23,

1976, p. 4. This deletion was not contested by the commission's supporters in conference committee because they all felt uneasy about appropriating money for a nonexistant agency.
38. Senator Smith (Dem-San Mateo County) was a supporter and, as a member of the Senate Natural Resources Committee, generally knowledgeable about coastal legislation. Assemblyman Warren—one of the most powerful figures in the legislature on environmental and energy legislation—wanted to take a fresh look at coastal legislation and draft an environmentally oriented bill that was better organized than SB 1579 and met some of the objections raised against it. Thus, the bulk of the drafting was done by a small group of people including Warren, one person each from the Senate and the administration, and two staff from Warren's committee—but nobody from the commissions, Beilenson's staff, or even Smith's staff. Joe Petrillo from Smith's staff (as well as representatives of the Sierra Club and PCL) worked concurrently on another version. Toward the end of July, Warren and Smith met to hammer out the details, with the ensuing bill generally following Warren's outline. The major exception was the retention of the fairly detailed policy section in the Beilenson-Smith bill rather than the much briefer policy statement proposed by Warren. (Based on interviews with four of the participants, all of whom requested anonymity.)
39. Interview with Bob Testa, chief staff to the Senate committee.
40. While the three Assembly conferees would be appointed by the speaker, the three Senate conferees would be appointed by the five-member Rules Committee—with the two Republicans and the swing Democrat from Long Beach (Senator Kennick) expected to appoint Senator Smith and two advocates of major changes in SB 1277 ("Coastal Protection Bill Faces Do-or-Die vote," *Los Angeles Times*, 19 August 1976).
41. While the Senate Rules Committee on a 2–2 vote sent the bill to both Natural Resources and Finance, the Senate parliamentarian ruled that the second referral was discretionary—whereupon Natural Resources voted 5–3 to send it directly to the floor (interview with Bob Testa, chief committee staff, 14 December 1976). As it turns out, this parliamentary ruling may have been critical because two of the critical people on Finance—Roberti and Alquist—both subsequently voted against SB 1277 on the floor.
42. The critical amendments essentially provided (1) that urban exclusions under the conditions already specified in the bill be mandatory rather than discretionary, (2) that the regional commissions would not be activated unless deemed necessary by the state commission for work load purposes, (3) that local governments would have the option of submitting either revisions to their general plan or a coastal element, and (4) that sensitive areas would be declassified unless approved by the legislature within 2 years (*Senate Journal*, August 23, 1976, p. 16173). Almost all knowledgeable observers agreed that only the last was of any consequence. And the people we talked to who were closest to the unions and to the governor (both of whom requested anonymity) generally agreed (1) that the unions did not know the bill very well and thus had difficulty requesting meaningful amendments and (2) that John Henning, head of the California AFL-CIO, wanted to be able to negotiate something for the building trades while also not embarrassing the governor and the legislative leadership by killing the bill. There were also rumors that Brown had agreed to accede to labor's requests on subsequent appointments to the commissions, but almost everyone we talked to agreed that such deals were just not the governor's style; moreover, we could discern no "prolabor" bias in the subsequent appointments.
43. The amendments negotiated by Brown were incorporated in AB 2948, the coastal bond measure carried by Assemblyman Hart that (as we shall soon see) had been shunted aside for an alternative bond bill. After Senate approval of the amendments (25–9 on August 26), it had to return to the Assembly for concurrence. But problems developed in the manner in which the amendments had been drafted, and thus, clarifying language had to be incorporated into AB 400, carried by Speaker McCarthy, which also contained $1.5 million to fund the successor coastal agency, $23.5 million for the acquisition of coastal park sites deleted by Governor Brown from the Nejedly Bond Act, and $6.3 million for nearcoast hostels. The AB 400 amendments in the

Senate then had to be returned to Assembly committee and floor for concurrence. After apparently considerable arm-twisting by McCarthy of Assembly environmentalists (as well as opponents of coastal legislation), AB 2948 was finally approved on August 30 and AB 400 on August 31, the last day of the session. For a general review of the hectic days of August, see "Legislature Adopts Coastal Bill," *State Coastal Report*, August/September 1976, pp. 1–4; our account is also based on interviews with six participants who requested anonymity.

44. See "Coastal Conservancy Bill Survives Assembly Test," *State Coastal Report*, June 23, 1976, p. 6, as well as Sabatier interviews with Joe Petrillo of Senator Smith's staff (currently the executive director of the conservancy) and Jim Carroll of CCEEB.

45. Nejedly's bill had originally proposed a $500 million statewide park and recreation bond act to appear on the June 1976 primary ballot. Reported here is the bill as it was amended in the Senate Natural Resources Committee on January 6, 1976 (Rival Land Acquisitions Bond Measures Moving Through Legislature," *State Coastal Report*, February 17, 1976; also "Coastal Acquisition Bills continue to Move through Legislature," *State Coastal Report*, May 1976, p. 11).

46. "Coastal Acquisitions Bills Clear Committees," *State Coastal Report*, June 1976, pp. 7–8.

47. Although Californians approved 70% of bond referenda during the 1972–1976 period [Eugene Lee, "The Initiative in Perspective," *California Data Brief*, 2 (April 1978), pp. 1–4], in November 1976 a $500 million bond for low- and moderate-income housing lost 43% to 57%, and a $25 million bond issue to provide low-interest loans to homeowners to install insulation and solar energy devices lost 41% to 59%. For a brief presentation of the vote on the Nejedly-Hart Bond, see "Voters Approve Park Acquisition Bond Measure," *State Coastal Report*, November 1976, pp. 1–2.

48. "Coastal Funding Bill Enacted," *State Coastal Report*, August/September 1976, p. 4, and "Governor Signs Coastal Legislation," *State Coastal Report*, October 1976, pp. 1–2.

49. This is based upon the 298 respondents to our survey who were either on the general mailing list of the commissions or on the list of planning activists.

50. "Coastal Plan Presented to Legislature," *State Coastal Report*, December 31, 1975, p. 4. Also, "Key Coastal Bill Clears Senate Committee," *State Coastal Report*, May 7, 1976, p. 5.

51. Based on interviews with representatives of the Brown administration and CCEEB who requested anonymity.

52. This assessment of important legislators—like many of the judgments in this section—draws heavily upon interviews with a number of people who were involved with coastal legislation during the 1976 session (almost all of whom requested anonymity). These include representatives of the Brown administration, CCEEB, and most environmental groups; the principal staff of the two policy committees and Senator Beilensen; and either the legislative assistants or the members themselves of the five swing votes on the Senate Finance Committee. Senator Smith subsequently became an important member of the Democratic leadership in the upper house; see Carol Benfell, "Senator Smith, State Senate's Likely Leader of the Near Future," *California Journal*, 8 (March 1977), pp. 88–89.

53. It stands to reason that public opinion is most likely to have an effect when a legislator is in a close electoral campaign. This was apparently the case not only with Schrade but also in a Los Angeles coastal district in which Assemblyman Beverly and Senator Wedworth were competing for Wedworth's seat; both supported SB 1277, even though Wedworth had helped kill coastal legislation in 1971 and 1972 when he was a member of the Senate Natural Resources Committee. When election was not at stake, the most effective tactic was apparently not generating large volumes of mail from the district but rather arranging meetings of the legislator with a few people whom he respected; for example, Roberti was apparently most impressed by meetings he had with a small group of people from his district and with a delegation from a Chicano neighborhood in San Diego that the coastal commissions had helped protect. The most effective environmental group lobbyists were apparently Joe Edminston (the Southern California coastal coordinator for the Sierra Club), Larry Moss (the PCL's lobbyist), and John Zierold (the Sierra Club's

full-time lobbyist); in contrast, Janet Adams was not very visible and the PACE lobbyists were too inexperienced. Other examples in which environmental groups and their allies apparently did a good job of generating pressure from the districts on swing voters included senators Stevens and Marks, and to a lesser extent, senators Garcia, Greene, and Holden.

54. The thesis of declining participation by the beneficiaries of police-power regulation is an integral part of the broader theory of the cycle of decay of regulatory agencies; see Marver Bernstein, *Regulating Business by Independent Commission* (Princeton: Princeton University Press, 1955), chap. 3; and L. C. Mainzer, *Political Bureaucracy* (San Francisco: Scott, Foresman, 1973), pp. 82–83. While the data presented here suggest that supporters and opponents had comparable activity levels, they are admittedly very imprecise measures of total participation because they (1) are based on self-reports and (2) do not measure the *number* of times in which a respondent engaged in a certain activity (but only whether or not he/she did so at all).

55. Because property tax assessments are based upon the assessor's estimate of the market value of a piece of property—rather than on its present use—agricultural land on the urban fringe will be assessed as if it were residential or commercial property because that is what nearby parcels are being sold for. But estate taxes may be an even more powerful inducement for more intensive uses (particularly of agricultural land), as heirs are often forced to sell the land to pay the taxes; if the land is in an area of changing land use, it will be sold for greater than its present use value and then used for more intensive purposes.

56. For example, Article 7 of Chapter 3 indicated in considerable detail the conditions under which coastal-related and other "needed" development were acceptable even if they involved some environmental degradation and restrictions on access.

57. See California Coastal Act of 1976, Section 14, which revised Section 402.1 of the Revenue and Taxation Code. The Forest Taxation Reform Act of 1976 provided for the designation of timber reserves, which would then be assessed on the basis of timber yield rather than their full market value; for a discussion of this statute, see "Environmental Protection: Forest Taxation Reform Act of 1976," *Pacific Law Journal* 8, pp. 371–381. See also Robert Lutz, "The Inadequacy of Regulation Alone: The Metamorphosis of California's Coastal Law," in *The Government Land Developers*, ed. Neal Roberts (Lexington, Mass.: Lexington Books, 1977), pp. 202–219.

58. Following is a comparison of the scope of permissible appeals after certification of local coastal programs:

1975 plan/original 1579 (Section 30501)	1976 Coastal Act (Section 30603)
1. Energy facility or public works project	1. Energy facility or public works project
2. Residential structures of more than 4 units or any structure of more than 10,000 sq ft	2. Deleted
3. Developments within 100 ft of streams or wetlands or 300 ft of a coastal bluff, or on a beach, sand dune, floodway, or prime agricultural land	3. Development within 100 ft of streams or within 300 ft of a coastal bluff, or between the sea and the coastal road (or 300 ft from sea if no road)
4. Developments other than the principal permitted use by the zoning ordinance	4. Developments in coastal counties [but not cities] other than the principal permitted use
	5. Developments in sensitive coastal resource areas designated by the Legislature

59. 1976 Coastal Act, Section 15, amending Section 13142.5 of the Water Code. The most arduous and complex negotiations concerned the respective roles of the coastal commissions and the Energy Commission in the siting of thermal power plants. The eventual compromise established a complex process whereby the coastal commissions would first designate coastal areas where the location of new power plants would violate Coastal Act policies. This information would then be transmitted to the Energy Commission, which would be prohibited from approving power plants in these areas unless the coastal commission first found that there would be no substantial adverse environmental effect and the power plant would not be inconsistent with the principal permitted use of the land under the local coastal program. [The coastal commissions subsequently designated 46 areas—involving about 70% of the coastal zone—where power plants would be inappropriate (*State Coastal Report*, August 1977, pp. 2–3.] Outside of these special treatment areas, the coastal commission can submit a report of the Energy Commission on any proposed power plant. The coastal commission's recommendation can, however, be overridden by the Energy Commission if the latter finds the recommendations to be infeasible or to result in greater adverse environmental impact. In general, see 1976 Coastal Act, Section 30264, 30413, and Section 5 amending Section 25302 *et seq.* of the Public Resources Code.

60. This conclusion is apparently shared by many of the people in our 1977 survey who were both concerned and knowledgeable about the commissions (i.e., who responded correctly to all three questions concerning knowledge of the commissions).

	Commissioner/ staff (N = 71)	Envir.-civic group/citizen (N = 40)	Local/state official (N = 23)	Business/property owner (N = 59)	Total
The 1976 act called for *greater* protection of natural habitats than did the 1972 Initiative					
% Agree	36%	20%	42%	42%	35%
% Ambivalent	15%	27%	21%	37%	25%
% Disagree	49%	54%	37%	20%	40%
Public access to the beaches is *more* likely to be provided under the 1976 act than under the 1972 Initiative					
% Agree	49%	28%	35%	39%	39%
% Ambivalent	30%	45%	44%	34%	37%
% Disagree	21%	28%	22%	27%	24%

9

EPILOGUE
Sustaining Coastal Conservation in a Different Political Environment, 1977–1981

In the introductory chapter we argued that policy implementation should be viewed as an iterative process of formulation-implementation-reformulation. Passage of the 1976 Coastal Act and related legislation marked the end of a policy cycle that began with the 1972 Coastal Initiative and that constitutes the principal focus of this study.

While we lack the resources to provide a similarly detailed analysis of events after 1976, there are two reasons to continue the story in a more abbreviated form through mid-1981. First, from the standpoint of state involvement in coastal land use control in California, 1976 was more a midpoint than an end. Although the 1976 Coastal Act reaffirmed many of the policies pursued by the commissions during the 1973–1975 period and provided additional tools (such as the conservancy) to more effectively attain those objectives, it also embodied a fundamental policy choice to return principal responsibility for coastal land use control to local governments once they had brought their plans and zoning ordinances into conformity with Coastal Act policies. This certification of local coastal programs, initially scheduled for completion by 1980, was then extended to mid-1981.[1] In addition, by 1981 several of the major issues—e.g., Sea Ranch—that had initially spawned the commissions were finally approaching resolution. Thus, extending our chronicle for 5 years permits us to bring this unique (for both California and the United States) period of heavy state involvement in coastal land use control to something approaching a close.

Second, the coastal commissions operated in a quite different external environment in 1977–1981 from the one that existed during the Proposition 20 years. Legally, they no longer benefited from the guaranteed appropriation and obstacles to weakening legislation that the Coastal Initiative had provided during the 1973–1976 period. Like any other state agency, they now had to go through the annual anxiety of justifying their budgetary requests and trying to defeat bills weakening their legal authority. Just as the commissions were thus becoming more vulnerable to attack, the political environment in California was becoming less hospitable to aggressive land use regulation. First, the late 1970s and early 1980s witnessed a periodic housing crisis in the state, as an ever-increasing demand, dramatic increases in interest rates and the cost of building materials, and restrictions on the supply of land led to skyrocketing housing prices (particularly in the South Coast and the Bay Area) and eventually to a serious recession in the housing industry.[2] While land use restrictions were clearly only one of the causes and while criticism of the commissions did not appear to be as virulent as in 1975, the state's "housing crisis" certainly did not help implementation of the 1976 Coastal Act. Second, the 1978 and 1980 elections produced a more conservative legislature; while both houses remained Democratic, changes in the membership and, after the 1980 elections, in the Assembly and Senate leadership resulted in a legislature more skeptical of regulation, considerably more cost-conscious, and minus any powerful champions of the coastal commissions.[3] Third, while developers and opponents of land use controls were skillful about publicizing "horror stories" concerning the commissions and continued to be well represented in Sacramento, the supporters of coastal protection gradually fell by the wayside until, by 1980, they were effectively represented in the capital only by the Sierra Club.[4]

The picture was not entirely bleak. Governor Brown was reelected in 1978. While not a vocal supporter of the coastal commissions, he used his appointment and veto authority to protect the commissions from any serious assaults by opponents. A June 1980 statewide public opinion poll conducted by the respected Field organization revealed a fair degree of support for the commissions—at least among the 50% of the electorate who knew of their existence—particularly for access and many of the other principal objectives of the Coastal Act (see Table 9-1). A $285 million statewide bond issue—including $100 million for coastal acquisition and planning—was approved by 52% of the voters in November 1980. Finally, the commissions' efforts to encourage widespread public participation in the development of local coastal programs resulted in a greater understanding of coastal issues in many communities, although this did not necessarily translate into additional support in Sacramento.[5]

On the whole, however, there can be little doubt that the commissions faced a somewhat less sympathetic political environment in 1977–1981 than during the previous 4 years, and certainly that they were more vulnerable to cutbacks in budget and legal authority than previously. In addition, some new appointments to several regional commissions in 1979 made them noticeably more prodevelopment, with a consequent increase of responsibility on the state commission (whose orientation remained basically unchanged). On the other hand, while the 1976 Coastal Act called for more balance between economic

TABLE 9-1
PUBLIC OPINION TOWARD THE COASTAL COMMISSIONS [a]
JULY 1980
(N = 1039)

	Coastal counties	Noncoastal counties	Statewide total
A. Awareness of commissions [b]			
Yes	54%	45%	50%
No	46%	55%	50%
	100%	100%	100%
B. Overall rating of the commissions' performance			
Excellent or good	17%	17%	17%
Fair or poor	35%	30%	33%
Reasons [c]			
Overzealous regulations	(18%)	(18%)	(18%)
Insufficient regulation/results	(18%)	(18%)	(18%)
Don't know	48%	53%	50%
	100%	100%	100%
C. Importance to respondent of commissions' different responsibilities (% important) [d]			
1. Guaranteeing public access to beaches	77%	84%	80%
2. Protecting coastal views from being obstructed	79%	81%	80%
3. Regulating coastal development	77%	82%	79%
4. Controlling off-shore drilling	70%	75%	72%
5. Protecting wetlands	67%	72%	69%
6. Restricting subdivision of agricultural land	64%	73%	68%
7. Providing low- and moderate-income housing	51%	40%	47%

[a] Field Research Corporation, "Survey of California Public Awareness of and Attitudes Toward the California Coastal Commissions," Conducted for California Coastal Commission, July 1980. This involved personal interviews with a statewide probability sample of 1039 California adults, July 19–26, 1980.
[b] "Have you seen or heard anything about the California Coastal Commissions?"
[c] Respondents were asked to give reasons for their overall rating. Because of multiple responses, however, the sum of the reasons does not equal the subtotal. Moreover, we have not reported responses—e.g., "serves special interests"—that don't clearly fit into one of our categories. "Overzealous regulation" includes such reasons as "doesn't allow enough development," "too much red tape," "unqualified staff," "too costly." by contrast, "insufficient regulation" involves such responses as "allows too much development," "had accomplished nothing," "polluted water/beaches," etc.
[d] Sum of "extremely" and "very important" (rather than "not too important" or "not at all important").

development and environmental protection than did the Coastal Initiative, its greater specificity probably made it at least as good a resource for providing access and protecting scenic views, wetlands, and agricultural land (at least until the actual turnover of principal authority to local governments upon approval of the local coastal programs). In fact, the greater legal and financial resources provided by the conservancy and the 1976 Bond Act meant that more access should have been provided during 1977–1980 than during the previous 4 years. Given, then, that the state commission and the statutory directves remained *roughly* constant during the two periods, the more vulnerable political position in which the commissions found themselves after 1977 enables us to examine the extent to which a committed agency with a strong legal mandate can weather a rather unsympathetic political environment.

We shall first provide a brief overview of implementation during the period from January 1977 through June 1981 (when the regional commissions went out of existence). The bulk of the chapter will then examine the commissions' performance in a number of critical areas: (1) improving public access to the coast, (2) protecting wetlands, (3) mitigating the adverse effects of land development, (4) protecting low- and moderate-income housing, and (5) completing the transfer of principal authority to local governments via the development of local coastal programs consistent with Coastal Act policies. The final section will suggest some of the principal causal factors that affected the commissions' performance during this period.

I. An Overview of the Implementation Process, January 1977–June 1981

The 1976 Coastal Act gave the commissions a formidable set of responsibilities. These included the following:

1. Continuing responsibility to rule on all development permits in the coastal zone (considerably larger in size than the uniform 1000-yard permit area under Proposition 20) pending certification of the local coastal programs.
2. Developing guidelines for, assisting local governments in the preparation of, and eventually reviewing the land use plans and implementing ordinances of the 67 local governments responsible for preparing local coastal programs (LCPs). Upon certification of each program, responsibility for development review would be transferred to the local jurisdiction, subject to limited appellate review by the state coastal commission.
3. Designation, by September 1977, or sensitive coastal resource areas requiring additional protection, including continuing development re-

view by the commissions after LCP certification. Each area required an elaborate justification and was subject to legislative approval.

4. Coordination of planning and permit review with other state agencies. In several cases, the 1976 Coastal Act charged the commissions with identifying valuable coastal resources but then left principal permit responsibility to functional specialists in other agencies. For example, the commissions were given 1 year to designate (and justify) those areas along the 1100-mile coast where the location of power plants would be inconsistent with the Coastal Act. Similarly, the commissions were accorded 6 months to identify forestry areas requiring special treatment and to recommend regulations to the Board of Forestry adequate to protect the natural and scenic qualities of such special treatment areas.

Despite these extensive mandates, the coastal commissions remained rather small agencies, with about 150 full-time staff and an annual budget of around $9 million. In fact, even the commissions' critics admitted that the staff were overworked.[6]

Not surprisingly, the heavy work load and constant pressure eventually resulted in fatigue on the part of a number of key personnel. Mel Lane resigned as state commission chairman in the fall of 1977. Unfortunately for the commissions, none of his three successors—while hardworking and well-intentioned—enjoyed his enormous prestige. Joe Bodovitz resigned as state executive director in the spring of 1978, to be replaced by Michael Fischer, who had previously held that position in the North Central region. Despite the exodus of several other senior state commission staff in 1977–1978, there is little evidence that—as of June 1981—the commissions had succumbed to the oft-noted pattern whereby aggressive policy advocates in young agencies become burned out and are eventually replaced by cautious careerists more concerned with organizational maintenance and survival.[7] On the contrary, in early 1981 the commissions were apparently still perceived as an aggressive agency doing interesting and important things and thus continued to attract bright young risk-takers as staff.[8]

This perception that the commissions were still an innovative regulatory agency is supported by considerable evidence. In permit review, for example, Table 9-2 indicates that in 1977–1978 the commissions denied over a third of the permits appealed to the state commission, while imposing conditions on an additional 50%. These figures were comparable to those during the 1973–1975 period (see Table 4–9). The stringency of permit decisions did, however, decline somewhat in 1979—partially in response to the damaging bills then moving through the legislature, partially because of the state commission's decision to concentrate its efforts on the most important appeals and the development of permit guidelines.[9] The overall stringency rose again in 1980, with the increase

TABLE 9-2

ULTIMATE DECISIONS OF THE COASTAL COMMISSIONS ON PERMIT APPEALS BY YEAR, 1977–1980[a,b]

	Year			
Decision	1977 (N = 379)	1978 (N = 380)	1979 (N = 272)	1980
Approve	19%	11%	35%	9%
Approve with conditions	45%	55%	40%	70%
Deny	36%	34%	25%	21%
	100%	100%	100%	100%

[a]These figures deal only with permit applications, not with claims of exemption. By "ultimate decision" is meant the actual decision of the state commission on appeals deemed to involve "a substantial issue" and the decisions of the regional commission on cases deemed by the state commission to involve "no substantial issue."
[b]Source: For 1980, unpublished list of permit decisions provided by the California Coastal Commission; for 1977–1979, personal review of commission minutes. Because we do not know precisely the decision rules the commission staff used in tabulating their list, one should use a little caution in comparing the 1980 figures with those for 1977–1979.

in conditional approvals almost certainly reflecting the application of packages of conditions resulting from the development of guidelines.

In addition, as we shall see in the next section, the commissions often used their discretion to pursue more stringent policies than were legally required or than political prudence might have dictated. The clearest example was low- and moderate-income housing, where the commissions transformed a brief and rather ambiguous legislative mandate into one of the most stringent programs anywhere in the country. This involved requiring not only one-for-one replacement housing but also that 25% of new housing in developments of five or more units be reserved (at reduced rates) for people of low and moderate incomes. In several cases in Southern California, the commissions worked hard to protect the area's remaining—and often only marginally productive—wetlands against developments involving millions of dollars. Finally, in the Santa Monica-Malibu area the commissions have attempted, with some success, to restrict development on hundreds of parcels in mountainous terrain through what are called "transferable development credits"—an instrument long advocated by land use lawyers that has proven to be extremely difficult to translate into practice.

Like any aggressive regulatory agency, the commissions have been subjected to considerable criticism by those who must bear the costs. Given the increasingly conservative political climate in the state, the result has been continuous turmoil. Every legislative session since 1979 has witnessed the introduction of dozens of bills seeking to weaken the authority of the commissions. Proposals have been made almost yearly to change the voting rules, to require compensation for any access easements required as a permit condition, to delete specific developments or issues (e.g., housing) from the commissions' jurisdiction, all the way to the outright abolition of the agency (see Table 9-3 for major bills). There have also been repeated attacks on the commissions' budget, of which the most serious was the 1980 Senate Republican caucus vote to seek deletion of the commissions' appropriation—a not altogether idle threat, given the two-thirds vote required in each house to pass the budget bill. While virtually all of these efforts have been defeated or substantially weakened because of the efforts of Speaker McCarthy, the Assembly Resources Committee, and, to a lesser extent, the Senate Natural Resources Committee, the commissions have been continually on the defensive. Virtually the only area of active support has been public access, where the legislature in 1979 substantially strengthened the authority of the commissions and the Coastal Conservancy. Finally, the changes in legislative leadership following the 1980 elections resulted in several changes in the Assembly and Senate appointees to the state and regional commissions, although not in any major reorientation of the state commissions' basic policy (at least as of June 1981).[10]

The commissions have continued to receive strong support from the courts. The only possible exception was the 1980 Chula Vista case, which, at the very least, clearly indicated that the commissions could not legally require local governments to accept their proposed changes in LCPs—thereby forcing additional negotiations and further delaying certification. In addition, the election of a conservative Republican as attorney general in 1978 risked seriously compromising the commissions' activities, as he threatened to reduce their legal staff and to refuse to represent them on cases in which he disagreed with their decisions. After a period of intense controversy—including an effort by Governor Brown to reduce the attorney general's budget—the issue was eventually resolved with only minor damage.[11] But it certainly illustrated the enormous potential leverage that any attorney general has to frustrate a stringent regulatory program.

In sum, there is not much evidence that, as of mid-1981, the coastal commissions as a whole had begun to succumb to Bernstein's "cycle of decay" of regulatory agencies.[12] They continued to aggressively pursue the limits of their statutory mandate and, as a consequence, to be the target of substantial criticism and attempts to weaken their authority. This does not mean, however, that their efforts have always been successful.

II. The Commissions' Performance on Selected Major Issues

This section will examine the performance of the coastal commissions and, to a lesser extent, the conservancy in a number of areas covering most of the important objectives of the 1976 Coastal Act: (1) improving public access to the

Table 9-3
Selected[a] Major Bills Involving the Coastal Commissions, 1977–1981

Bill (author)	Final provisions	Outcome
1977 session		
SB 1081 (Alquist)	Transferred final permit authority over siting of LNG terminal from commissions to PUC	Passed
1978 session		
SB 1873 (Smith)	Omnibus bill making a number of minor changes in Coastal Act	Passed
AB 3478 (Nestande)	Delays expiration of regional commissions from 6/79 to 6/81	Passed
1979 session		
SB 175 (Cusonovich)	Excludes single-family residences (SFR) from commissions' permit authority and prohibits permit conditions for public use (e.g., accessways) without compensation	Killed in Senate Resources Committee
SB 569 (Beverly)	Loosens conditions under which urban areas are excluded from commissions' permit authority	Killed in Senate Resources Committee
SB 779 (Dills)	Exempts SFRs from commissions' permit review under certain conditions and mandates access program	Folded into AB 643 in Assembly Resources Committee after passing Senate
SB 643 (Calvo)	Exempts SFRs from commissions' permit view under limited conditions; changes voting requirements on permit decisions; reimburses local government for carrying out certified LCPs	Passed
AB 13 (Pappan)	Excludes large development in Santa Monica mountains (Headlands Properties) from commissions' permit jurisdiction; first of many important "leverage bills"	After passing Assembly and Senate, killed in last days of session
AB 462 (Mello)	Made a number of adjustments in commissions' geographical jurisdiction, including fairly important deletions in Cannery Row, Ferndale, and Carlsbad	Passed
AB 989 (Kapiloff)	Transfers principal authority for development of accessways from State Parks Dept. to commissions and conservancy; mandates development of access inventory	Passed
AB 988 (Kapiloff)	Extends liability waiver from undeveloped public property to include accessways	Passed
SB 751 (Keene)	Requires commissions to prepare guide to coastal accessways	Passed

1980 session

Bill (Author)	Description	Outcome
AB 2973 (Vasconcellos)	Authorizes funds from tideland oil revenues to coastal accessways, among other purposes	Passed
AB 1971 (Mello)	Makes minor adjustments in coastal boundary and requires the commission to approve an LCP for the Aliso Viejo Development (Orange County) consistent with specified affordable housing requirements	Passed
SB 1581 (Sieroty)	Omnibus bill modifying the Coastal Act, including changes in voting rules and restrictions on application of moderate-cost housing regulations to motels	Passed
SB 778 (Keene)	Reduces commissions' jurisdiction in the siting of energy facilities	Killed in Assembly committee after passing Senate
SB 1922 (Keene)	Exempts a new sewage treatment plant in Eureka from certain Coastal Act policies subject to an approved wetland restoration plan	Passed
AB 2706 (Bane)	Exempts Sea Ranch from commission review in return for specified accessways, tree trimming, etc.	Passed
AB 2081 (Bergeson)	Deletes commissions' authority to require affordable housing	Killed in Assembly Resources Committee

1981 session

Bill (Author)	Description	Outcome
SB 260 (Ellis)	Repeals the 1976 Coastal Act	Killed in Senate Natural Resources Committee
AB 937 (Tucker)	Exempts two hotels in Marina del Rey (Los Angeles) from provisions of Coastal Act	Killed in Assembly Resources Committee
SB 493 (Carpenter)	Exempts Bolsa Chica Gap (Orange County) from commission review upon completion of county plan	Held in Assembly Resources Committee
AB 626 (Mello)	Removes affordable housing from the jurisdiction of the commissions in return for a modest strengthening of the state mandate to local governments for the provision of low- and moderate-income housing	Killed in Assembly committee (?)
AB 385 (Hannigan)	Revises the LCP certification process to provide for local government assumption of permit review after approval of its land use plan (but prior to approval of its zoning ordinances), subject to appeals to commission	Passed
AB 321 (Hannigan)	Makes minor adjustments in definition of prime agricultural land and in development controls in wetlands	Passed

[a]This includes all of the important bills that passed and many—but certainly not all—of those that were not approved by the legislature.

beach, (2) protecting wetlands, (3) mitigating the adverse effects of land development, (4) protecting low- and moderate-income housing along the coast, and (5) completing the local coastal programs. We make no pretense of an exhaustive analysis of any of these topics. Instead, our intent is to sketch a general portrait of the commissions' performance in these areas during the 1977–1981 period.

A. *Improving Public Access to the Coast*

Maximizing public access to the coast (consistent with habitat constraints and respect for the legal right of property owners) was one of—if not the—major substantive objective(s) of the 1976 Coastal Act. It was explicitly listed in Section 30001.5 as one of the four basic goals. It was the *only* substantive policy mentioned in the section on local coastal programs, which specifically required that "each LCP . . . shall contain a specific public access component to assure that maximum public access to the coast and public recreation areas is provided" (Section 30500a). Access was included in the limited scope of postcertification appeals to the state commission.[13] And the 1976 legislature approved a bond act providing for substantial park acquisition (and thus access) along the coast.

There are essentially two means of providing physical access to the coast: (1) The approval of development permits can be conditioned on the dedication of accessways or, in large developments, miniparks. (2) The state can acquire land and then turn it over to some public (or private) agency for development and management of a park.

Turning first to the provision of access through the regulatory process, the actual opening of an accessway is contingent upon (a) a permit condition requiring the dedication of an access easement; (b) the actual recording of that deed restriction by the property owner; (c) the assumption by some public (or private) agency of responsibility for (1) liability in the case of accidents, (2) development of the accessway (e.g., construction of stairs), and (3) management of the accessway (e.g., litter collection).

While recording deed restrictions may seem a trivial task, experience has shown that property owners often neglect for one reason or another to do so. For example, of the 1026 permits with access conditions in the 1973–1980 period, by early 1981 only 540 (or 53%) of the restrictions had been legally recorded.[14] Unfortunately, the commissions did not become aware of the seriousness of the problem until 1978, and not until 1980 did they assign sufficient staff to begin clearing the accumulated backlog (which they hope to reduce to a year's permits by the end of 1981).

Monitoring compliance with deed restrictions is a minor problem compared to the difficulty of finding someone to actually develop and manage deeded accessways. For example, data provided in Table 9-4 indicate that, of the 439 lateral accessways (i.e., those traversing the property along the wet sand beach)

TABLE 9-4

NUMBER OF ACCESSWAYS WHERE MANAGEMENT RESPONSIBILITY HAS BEEN ACCEPTED BY A PUBLIC OR PRIVATE AGENCY AS OF JANUARY 1981[a]

| | Acceptance of management responsibility | | | | |
| | Accepted by | | | | |
Type of access	Public agency or private assoc.	Owner (deed restriction)	Subtotal	Not yet accepted	Total
Lateral accessway (miles along coast)	17 (2.5 mi)	233 (4.0 mi)	250 (6.5 mi)	189 (13.6 mi)	439 (20.1 mi)
Vertical accessway	19	20	39	47	86
Trail or path (miles)	11 (3.2 mi)	1 (0.0 mi)	12 (3.2 mi)	21 (2.1 mi)	33 (5.3 mi)
Viewpoint	2	3	5	2	7
Others	20	9	29	6	35

[a]Source: California Coastal Commission, Coastal Access Inventory, January 15, 1981.

whose deeds had been recorded, only 250 (involving 6.5 miles) had been accepted by a public or private agency for management by early 1981, while 189 (involving 13.6 miles) had not yet been; and many of those accepted are on isolated parcels and thus largely unreachable. On the even more critical vertical accessways (i.e., those from the coastal road to the beach), 39 had been accepted for management by a public or private agency (and thus opened for public use), while 47 had not yet been.

While one might assume that the management of these accessways would be the natural responsibility of the state parks department, that agency viewed its task as developing and managing *parks*, not scattered accessways; in addition, it regarded accessways (e.g., in Malibu) as a political hot potato best left alone.[15] The other logical candidate would be local parks departments, but they too were often reluctant to get involved in a controversial issue and, moreover, faced significant budgetary restrictions following the passage of Proposition 13 in June 1978. Recognizing these difficulties, staff from the commissions, the conservancy, and Assemblyman Kapiloff's office drafted AB 988 and AB 989—subsequently approved by the 1979 legislature—which (1) exempted public agencies from liability requirements in the case of accessways and (2) transferred principal management responsibility from the state parks department to the coastal commission and the conservancy, chiefly through the vehicle of conservancy grants to local public and private agencies. AB 2973 (Vasconcellos), approved a year later, authorized the utilization of tideland oil and gas revenues for the acquisition and development of accessways. As a result of these efforts, the commission and the conservancy developed a comprehensive access program, and the conservancy during 1979–1980 made 54 grants totaling $800,000 to local agencies for the development of access facilities; these do not, however, cover management costs, estimated at $200–$2000 annually for each accessway.[16]

Despite these advances, actually opening accessways remains problematic. In Malibu, for example, of the 193 lateral accessways mandated by commission permit decisions, in only about 45 cases had management authority actually been assumed and the area fully open to public use by the beginning of 1981. Of the 10 vertical access conditions required by the commissions since 1973, by early 1981 only 2 accessways had actually been opened. In 2 more cases, the county parks department had agreed to accept management responsibility and a conservancy grant had been made to develop access facilities, but these accessways—as well as three recent park acquisitions that also had received conservancy development grants—were vetoed at the last moment by the Los Angeles Board of Supervisors. Allegedly because of fiscal constraints, the board refused to allow the use of county funds to maintain these new accessways and parks.[17] While Malibu is admittedly one of the areas of the state where access arouses the most intense opposition by local residents, its proximity to Los Angeles's 5 million residents makes it one of the most important. And it certainly illustrates the multitude of clearance points involved in the implementation of any access program.

A related aspect of the access issue concerns the commissions' efforts to require park dedications and visitor-serving facilities in large new development projects. The most important example is probably the Avco case in southern Orange County.[18] The developer originally proposed an immense project involving 8000 housing units (2000 of them in multistory dwellings seaward of the coast highway) and practically no public facilities. After the company's application for a vested right exemption from the Coastal Initiative was eventually rejected by the state supreme court in 1977, it began to replan the entire project. Finally, in September 1979 the state commission approved a greatly revised proposal: The number of units was reduced from 8000 to 3000, while those west of the highway went from 2000 to 400 carefully clustered ones. In addition, the company agreed to add (1) a 7½-acre public park along the shoreline (valued at perhaps $16 million), (2) a 300-unit hotel, (3) 9 miles of public trails, and (4) 900 units of moderate-income housing, as well as agreeing to greatly expand an inland park and to open the previously private golf course and recreation center to public use approximately half the time. In short, what would have been an immense private development on 600 acres of coastal shoreline and hillsides was transformed into a smaller project with substantially greater public access.

Turning now to park acquisitions, the 1976 Parks Bond Act was discussed in Chapter 6. The process was repeated 4 years later, when 52% of the voters in November 1980 approved a similarly structured $285 million bond measure (including $60 million for coastal parks and $10 million for conservancy grants). Between 1976 and 1981 the state acquired approximately 20,000 acres of coastal parks and wetlands, of which about 15,000 had been recommended by the coastal commissions in their 1975 acquisitions list (although not necessarily as Priority I). In addition, the federal government had rather extensive acquisition programs in the Santa Monica Mountains (north of Los Angeles), Pt. Reyes National Seashore (north of San Francisco), and the Redwood National Park. Altogether, the amount of shoreline publicly owned for recreation almost doubled between 1973 and 1981 from 243 miles to approximately 447 miles.[19]

While these figures are impressive, they can mask a variety of problems. Sometimes, as in Malibu, park development encountered vetoes from local elected officials. At other times, opposition from local landowners created real difficulties. In the case of Crystal Cove State Park in southern Orange County, the state purchased 1898 acres of coastal land (included 3.25 miles of beachfront) for $32.6 million in May 1979. But development of park facilities has been delayed by owners of 70 mobile homes and 45 houses holding leases from the previous owner (the Irvine Company). Most have indicated a desire to accept 20-

year leases rather than relocation benefits. And one mobile home owner has refused to move from a critical pathway to the beach that is also the most desirable location for bathroom facilities.[20]

Finally, any discussion of access along the coast would be incomplete without an update of the Sea Ranch saga. As will be recalled from Chapter 5, in 1974 the state commission imposed overall conditions on the development involving additional accessways, tree removal, septic tank controls, and site and bulk restrictions on individual houses. At the same time, it refused to set a figure for eventual buildout of the subdivision and allowed individual lot owners to build after posting a $1500 "environmental deposit." The decision pleased virtually no one and was challenged in several court suits.

Following passage of the 1976 Coastal Act—with its more detailed and, at least with respect to public access, stringent permit review criteria—the North Central Regional Commission decided to reexamine the situation. After almost 2 years of negotiations involving the coastal commissions, the Sea Ranch Association, and the developer, the state commission in March 1979 adopted a new set of overall conditions that were similar to the earlier ones except for the deletion of the $1500 deposit, an agreement to permit 2029 single-family dwellings, and requirement of additional access: Instead of the two accessways and pass system required under the 1974 conditions (in addition to the park and two accessways previously deeded by the developer), the commissions now required six accessways (with limited parking) plus a 3-mile blufftop trail at the northern end of the development. While these conditions were acceptable to the developer, the lot owners' association was outraged by the additional access and the revival of the "hostage strategy"; it thus decided to continue the fight in both the legislature and the courts.[21]

The legislature attempted to fashion a solution in 1980 when it approved AB 2706 (Bane). This exempted single-family residences at Sea Ranch from commission review essentially in return for the 1979 overall conditions, return of the "environmental deposits," and payment of $500,000 to the Sea Ranch Association in settlement of litigation. Although the association opposed the Bane bill, it decided in May 1981 to accept its provisions after a three-judge federal court ruled that the commissions' requirements concerning access, tree-trimming, etc., were constitutional. A few issues—including the assumption by the county or state of management responsibility for the accessways, and a suit by COAAST challenging the constitutionality of the litigation settlement—remain to be resolved.[22] Nevertheless, actual conveyance of the deed restrictions to the conservancy in June 1981 appeared to mark the beginning of the end of the long and bitter dispute over Sea Ranch.

In sum, there was almost certainly more access actually provided in 1977–1980 than in the previous 4 years after Proposition 20. By 1978 the commissions had become aware of the serious difficulties in recording deed restrictions and in finding someone to assume management responsibilities. With strong legislative support, they began to resolve those difficulties in many areas. Moreover, support from both the legislature and the Federal courts for public access at Sea Ranch appeared to have resolved that conflict. Finally, the repeated willingness of both the legislature and the state's voters to provide funds for park acquisitions meant that an increasing percentage of the coast would be available for public recreation, albeit often after innumerable delays. But access clearly illustrates the incredible difficulties involved in translating clear legislative intent into on-the-ground performance. It also raises the question of why the commissions were rather slow to recognize and propose solutions for many of those difficulties, given that this was probably their program area with the greatest legislative and public support. This is an issue to which we shall return in the final section of this chapter.

B. Protecting and, Where Feasible, Restoring Coastal Wetlands

In addition to promoting public access to the shore, the 1976 Coastal Act placed a high priority on protecting and, where feasible, increasing the biological productivity of coastal wetlands. These were defined rather broadly to include "lands . . . which may be covered periodically or permanently with shallow water," including freshwater and saltwater marshes, swamps, mudflats, and fens. Within wetlands, the act placed severe restrictions on the diking, filling, and dredging operations that are the prerequisite to any development. In addition to requiring mitigation measures to minimize adverse environmental effects, Section 30233 limited such operations to the following purposes: (1) port, energy, and coastal-dependent industrial facilities; (2) maintaining existing navigation channels; (3) incidental public services, e.g., cables, piers, intake lines; (4) mineral extractions, except in environmentally sensitive areas; (5) nature study, aquaculture, or other resource-dependent activities; (6) new or expanded boating facilities, but only in degraded wetlands and only if such facilities occupied no more than 25% of the area, with the remaining 75% restored to greater biological productivity. Given that these uses clearly precluded normal residential and commercial development, it is not surprising that many of the major battles during the 1977–1981 period focused on defining precisely what constituted a "wetland." Even in those cases where development was permitted, the act clearly mandated a policy of "no net fill," i.e., that the developer had to acquire and restore to tidal action an area of equivalent biological productivity to that which had been lost.[23] Finally, the commission's post-LCP appellate authority included all developments in tidelands, submerged lands, or within 100 feet of any wetland or stream.

Given this strong legislative mandate and the importance attached to protecting wetlands by most staff and commissioners (particularly at the state level), it is hardly surprising that—with one possible minor exception—the commissions pursued this mandate aggressively and apparently with considerable success during the 1977–1981 period.

The exception concerned the state commission's decision in both 1977 and 1978 to refrain from proposing to the legislature the designation of "sensitive coastal resource areas"—presumably largely wetlands and surrounding areas. The advantage of such designation was that any development in them, even after certification of LCPs, would have been subject to appeal to the state commission. But the commission decided that the uncertainties involved in obtaining legislative approval of recommended areas would so delay the LCP process as not to be worth the additional protection, especially since the normal planning and permit processes seemed to be working reasonably well.[24]

In terms of the coastal commissions' regulatory activities,[25] as of mid-1980 there had been approximately 30 permit appeals involving development within or adjacent to a wetland. Of these, 12 were denied and the remaining 18 were approved with conditions. If the commissions actions involving Humboldt Bay (to be discussed shortly) were at all indicative of statewide trends, they actually managed to *increase* wetland acreage during this period through requiring extensive restoration as a condition for allowing some development. On the other hand, even commission staff admit that they had no estimate of wetland acreage lost through sedimentation and other indirect impacts of development. Part of the reason, of course, is that such development often occurred outside the coastal zone; on developments within the commissions' jurisdiction they tried quite hard to minimize sedimentation through erosion controls or outright denials.

Probably the best indicator of the commissions' general policy toward wetlands protection can be found in the 2-year effort to develop an "interpretive guideline for wetlands" that would be used to guide permit decisions and the certification of local coastal programs. From the beginning, the state staff developed three basic lines of argument:[26]

1. Wetlands are environmentally sensitive habitat areas that are strongly protected by the 1976 Coastal Act. In particular, development should be allowed only if (a) the proposed project involves one of the permitted uses discussed previously, (b) there are no feasible less environmentally damaging alternatives, (c) damages are mitigated as much as possible, and (d) the net biological productivity of the wetland is not impaired.

2. In defining wetlands, the state staff initially sought to expand the Coastal Act's definition (focusing on land periodically or permanently covered with water) with a more biologically sophisticated classification system developed by the U.S. Fish and Wildlife Service. Based upon soil and plant types found in water-saturated systems, the U.S. Fish and Wildlife Service scheme provided the commissions with a means of identifying

degraded (former) wetlands that were only infrequently covered with standing water. This position was roundly criticized by local governments and developers in Southern California. The commission was eventually forced to retreat to a position in which the U.S. Fish and Wildlife Service scheme based on hydric soils would provide "supplementary guidance" to the basic Coastal Act definition.

3. Finally, the staff argued that degraded wetlands should be subject to essentially the same development restrictions as nondegraded wetlands, and that the commissions should actively seek to restore them. This position raised howls of protest in the south, where most of the historic wetlands have been substantially degraded through diking and filling over the past century. In response, the state staff argued that the Coastal Act required decisions on specific parcels to be deferred pending analyses by the California Department of Fish and Game concerning the extent of degradation and the feasibility of restoration.

As one would expect, the final guidelines—approved by the state commission on February 4, 1981—were generally applauded by environmental groups and criticized by most local governments and land developers as being more restrictive of development than the act intended and as unduly limiting the discretion of local governments in the formulation of LCPs.

The policy disputes at the core of the guidelines were neatly illustrated in a number of major controversies surrounding the development of LCPs in Southern California. These included Chula Vista (to be discussed in a later section), the highly degraded Ballona wetlands near Los Angeles airport, and the Bolsa Chica Gap in Orange County,

Bolsa Chica involved approximately 1600 acres served by a couple of small creeks just behind a state beach. [27] Over a third of it was involved in a tidelands ownership dispute between the state and the principal property owner, although it had not been subject to tidal action since the area was diked in the 1890s. The 1973 settlement gave 300 acres of the best wetland habitat to the California Department of Fish and Game for an ecological reserve and an additional 230 acres to the state if and when a second outlet to the ocean were opened. Since that time, the county and the property owner (a subsidiary of Signal Oil) have been preparing a plan for the area involving a pleasure boat harbor, about 5700 homes, a commercial area, and 350–500 acres of restored marshland. The state commission staff—supported by the relevant state and federal agencies—argued that about 1200 acres of the lowland area were (degraded) wetlands and thus that development planning should be put into abeyance until, consistent with the Coastal Act, the Department of Fish and Game had made a study of the extent of degradation and the feasibility of increasing its biological productivity. This position was contested by the landowner and the county, who argued (1) that the designation of "wetlands" relied too heavily on the USFWS scheme rather than susceptibility to tidal action or the periodically submerged water test of the Coastal Act (2) that even some of the periodically submerged areas were man-made rather than "natural;" and, in any event, (3) that the county should be allowed to envisage other (e.g., residential and commercial) uses for the area beyond the highly restricted ones permitted by the Coastal Act. The state com-mission, on a 9–2 vote, accepted the staff's recommendation on March 19, 1980. Subsequently, Bolsa Signal Corp. filed several court suits and persuaded a local state senator to introduce legislation (SB 493) that would exempt the area from the LCP process. On the other hand, a local environmental group brought litigation to declare the *entire* area a wetland. As of the summer of 1981, the entire issue was still in limbo (although SB 493 had, at least temporarily, been killed by the Assembly Resources Committee). It seems likely, however, that the commissions will eventually accept a scaled-down development proposal.

Wetlands protection was considerably less controversial along the central and northern coasts, in part because the areas were not so degraded as in the south, in part because they were not subject to as intense development pressures. The 1976 and 1980 bond acts funded some major acquisitions, including 2000 acres at Lake Earl (North Coast) and 1510 acres at Elkhorn Slough on Monterey Bay. And the federal government established a marine sanctuary among a group of islands in the Santa Barbara Channel. Even acquisition, however, does not always guarantee protection. For example, there is some evidence that Elkhorn Slough is gradually being filled because of runoff from upstream agricultural land beyond the jurisdiction of the coastal commissions.[28]

Despite these problems, there is little doubt that the commissions—working in conjunction with the conservancy—have demonstrated substantial capacity to preserve, and even to increase, the biological productivity of wetlands. This is best illustrated by Humboldt Bay.[29] In the period from 1977 through 1980, approximately 35 acres of wetland habitat were lost through a half dozen dredg-ing or filling operations. In each case, however, the commissions required the developer to restore to tidal action an area of at least comparable biological productivity. These actions, plus a variety of acquisition efforts, resulted in the restoration of approximately 170 acres during this period—or a net increase of about 135 acres in wetland habitat in and around the bay. One of the most interesting cases involved the Eureka Pocket Marshes, a series of extremely degraded, small (less than 1 acre) wetlands in an industrial waterfront area. Given that the area was beyond restoration, the coastal commissions and the conservancy established a program whereby developers pay in-lieu fees to the conservancy, which then uses the funds to acquire and restore to tidal action a larger and more productive area elsewhere on the bay. Another case involves the new sewage treatment plant for the City of Eureka on a 90-acre site of seasonally flooded wetland diked since the turn of the century. Given that this is not one of the permissible uses for fill under the Coastal Act, the sewage district got together with the North Coast staff and the Department of Fish and Game to draft special legislation (SB 1922) approved by the 1980 legislature. Under the agreement, 10 acres will be filled and used for the treatment plant, an additional 60 acres will be treated with secondary effluent (free of toxic metals), and the remaining 20 acres will be returned to full tidal action. Thus, the project involves the loss of 10 acres in return for much greater productivity on 80 acres. Finally, as will be recalled from Chapter 5, one of the major controversies during the 1973–1976 period concerned a proposed sewer interceptor—and increased development—on the lowlands between Eureka and Arcata. According to the draft LCP (which is expected to be approved by county and coastal commission officials in the fall of 1981), the entire lowlands will be zoned exclusively for agriculture—with, of course, no sewer extension.

In sum, the record suggests that, during the 1977–1981 period, the coastal commissions—often in conjunction with the conservancy and other state agencies—did a good job of minimizing additional fills and of improving the productivity of degraded wetlands. This positive evaluation should, however, be cautioned by uncertainties concerning (1) the adverse effects of upstream sedimentation and (2) several large development projects in degraded wetlands in Southern California that, as of mid-1981, were still in limbo. On the whole, though, the combination of a protective statutory mandate, sympathetic interpretation by the commissions, and strong support from other state agencies and local environmental groups appears to have halted the century-long reduction in the acreage and biological productivity of California's coastal wetlands.

C. Mitigating the Adverse Consequences of Land Development

In addition to promoting public access and protecting wetlands, the 1976 Coastal Act sought to direct development along the coast so as to, among other things, (1) give priority to coastal-dependent development, (2) protect scenic views, particularly seaward of the coastal highway, (3) preserve prime agricultural land, (4) concentrate development in existing urbanized areas, and (5) restrict development in geologically hazardous zones.[30] A systematic overview of the commissions' performance in directing development during 1977–1981 would be beyond the scope of this chapter. But three significant trends are worth noting.

First, even though the Coastal Act accorded very low priority to residential development along the coast, the commissions had a great deal of difficulty dealing with the hardships a conscientious application of those policies imposed on applicants seeking to build single-family residences (SFRs). The issue grew in importance because legislators tended to sympathize with the plight of individual property owners fighting (what was perceived to be) a rigid, insensitive bureaucracy.[31] For example, the fall 1978 hearings of the Senate Natural Resources Committee revealed that several key members—including senators John Nejedly and Jerry Smith—were quite receptive to the pleas of individual lot owners at Sea Ranch. Another celebrated case involved the efforts of Victoria Consiglio to build a house on the Big Sur coast between the highway and the ocean. The commissions denied her a permit on the grounds that the house could not be hidden from the view of motorists—even though there were several more "exposed" houses in the area that antedated Proposition 20 and even though the commissions had granted a permit to another applicant whose house was hidden from view but whose access road visibly crossed the Consiglio property. Finally, the 1979 session of the legislature—after rejecting a number of bills that would have drastically curtailed the commissions' authority over single-family residences—approved a compromise measure (AB 643, Calvo) that gave local governments the authority to assume sole jurisdiction over SFRs in (commission-designated) areas with adequate water, road, and waste-disposal facilities that were not in geologic hazard zones and not between the coastal highway and the sea.

A second issue concerned the commissions' efforts to deal with several large subdivisions approved before 1972 in which buildout would adversely affect coastal resources. It was here the Coastal Conservancy proved invaluable. In the case of Whiskey Shoals, a 72-lot subdivision on the Mendocino coast between the highway and the ocean, the conservancy purchased all the lots, redesigned the subdivision to cluster development in relatively hidden areas and provide greater public access, and then resold the lots on the open market—at apparently no net cost to the state.

A more complicated case involves the Santa Monica Mountains (just inland from Malibu), where there are over 10,000 unbuilt lots, many of them small in size and in hazardous terrain. Full buildout would seriously impair public access to Malibu (by exacerbating already crowded roads), as well as creating significant erosion and water-quality problems in many areas. In response, the commissions (in consultation with Los Angeles County) developed a Transfer of Development Credit (TDC) program whereby everyone wishing to build along the coast would first be required to purchase and retire the development credits from one or more hazardous lots in the mountains. Like most TDC schemes, however, the program did not initially function very well, in part because the responsibility for locating lot owners willing to retire their development credits lay entirely with applicants. But in 1980 the conservancy began to actively purchase development credits, thereby serving as a "bank" for prospective purchasers. In addition, it initiated several projects similar to Whiskey Shoals in which it purchases lots, redesigns the subdivision, and then resells the lots—ideally at little net cost to the state. While there are still problems to be worked out in this very complicated scheme, the prospects for success are reasonably bright.[32]

Finally, we'd like to say a few words about the siting of power plants and other major energy facilities. This was a major point of controversy during the 1976 session, resulting in an elaborate compromise whereby the final decision on the siting of electrical generating plants was given to the California Energy Commission but in which the coastal commissions were charged with mapping coastal areas where a power plant would have adverse environmental or recreational consequences. The whole issue turned out to be moot—at least for the 1977–1981 period—because the leveling off of electrical demand within the state meant that the utilities found no reason to propose additional generating facilities along the coast.[33]

There was a major siting decision, but it involved a liquified natural gas (LNG) terminal. In response to the utilities' contention that Southern California would face a major natural gas shortage in the early 1980s, the 1977 legislature approved a bill (SB 1081) placing sole authority for the siting of such a facility in the hands of the Public Utilities Commission (PUC). The role of the coastal commissions was restricted to evaluating prospective sites based upon Coastal Act criteria. After a massive study, the commissions in May 1978 released their rankings: (1) Horno Canyon on the Camp Pendleton Marine Base (San Diego County), (2) Rattlesnake Canyon (San Luis Obispo County), (3) Point Conception (Santa Barbara County), the site clearly favored by the utilities and mandated by the legislature for inclusion. While the rankings were consistent with the Coastal Act's emphasis on protection of natural resources, they were greeted with some bemusement in Sacramento—as obtaining the defense department's approval for the Camp Pendleton site was deemed *highly* unlikely; in addition, the Rattlesnake Canyon site would require the construction of a $175 million breakwater and was located within 5 miles of the Diablo Canyon nuclear power plant. Not surprisingly, the PUC 2 months later unanimously approved a provisional permit for the Point Conception site as the only one able to meet the statutorily imposed construction deadline. Although the political repercussions of the coastal commissions' recommendation are difficult to measure, the preference accorded the almost certainly infeasible marine base provided additional ammunition to the commissions' detractors and probably strengthened the perception of even moderate legislators that the commissions were often incapable of striking a reasonable balance between environmental and socioeconomic goals.[34]

D. *Protecting Low- and Moderate-Income Housing and Visitor Facilities*

Given the high demand, limited supply, and resultant high prices of land along the coast, the natural operation of the market will eventually produce a situation in which only the wealthy can afford to live there and in which even moderately priced hotels will gradually disappear. Although the coastal commissions wrestled with this problem during the 1973–1976 period, it remained a rather minor issue in both permit review and the coastal plan.[35] The relevant provision of the 1976 coastal legislation was strengthened slightly—largely at the insistence of Senator David Roberti—but the final result in Section 30213 was still rather ambiguous and certainly a minor part of the entire Coastal Act: "Lower cost visitor and recreational facilities and housing opportunities for persons of low and moderate income shall be protected, encouraged, and, where feasible, provided. Developments providing public recreational opportunities are preferred."

Out of this modest cloth, the state commission over the next 3 years developed one of the most aggressive affordable housing programs in the entire country. It began incrementally. The housing guidelines approved by the state commission in the fall of 1977 were quite similar to those previously outlined in the coastal plan.[36] In particular, they provided that (1) in new residential developments, priority should be given to proposals that include housing opportunities for persons of low and moderate income, particularly where governmental funds are available to help finance or subsidize such units; (2) the demolition of existing housing should be discouraged unless necessary for health or safety reasons; replacement housing of comparable cost should be provided; (3) when reviewing proposals to convert moderate-income apartments to condominiums, the commissions should require that tenants be given the right of first refusal and that units of comparable rent be available in the same general coastal area. In applying these policies to specific permit cases, however, the state commission gradually became more specific and more stringent in its demands.

By January 1980 it had evolved a set of policies that essentially required the following:

1. Twenty-five percent of all housing in developments of five or more units were to be reserved for persons with 80–120% of the median household income of the region in question and prices charged for affordable units were to be scaled to the buyer's income level, i.e., subsidized by the developer. Alternatively, the applicant was usually given the option of dedicating a certain amount of land to the local housing authority or of paying in-lieu fees that would then be used to construct affordable units. In order to compensate for providing such units, the developer was sometimes given a density bonus. And if he provided more than the minimum of affordable units, he was allowed to use these "credits" on other projects.

2. In the demolition of existing low and moderate income units, the applicant was usually required to provide replacement housing for displaced tenants at comparable rents.

3. The conversion of apartments to condominiums was discouraged and,

in those cases where it was permitted, the commissions usually required that at least one-third of the project's units be reserved for low- and moderate-income persons. For example, as a condition for permitting a 103-unit conversion project in Del Mar, the commissions required that 28 one- and two-bedroom units be sold for $20,000–$47,000 (in contrast to the normal price of $49,000–$70,000).

4. In the case of new hotels and motels, the commissions usually stipulated that some of the units be rented at reduced rates to people of moderate income. In a celebrated case involving two 300-unit hotels at Marine del Rey (Los Angeles), the commission accepted the owner's offer to reserve 45 rooms during weekends for low- and moderate-income people at no more than 50% of normal rates (with eligibility to be determined by guests' zip codes) in lieu of the commissions' more ambitious (and workable) proposal.

In explaining this rather remarkable policy escalation, it appears that the principal factors were (1) the strong commitment of a few state staff and commissioners to affordable housing, in part to counter charges that they were "environmental elitists"; (2) the ability of housing advocacy groups to bring appeals, mobilize support at the hearings, and pose a credible litigation threat; and (3) gradually accumulated evidence that the high prices that developers were able to charge along the coast enabled them to absorb the losses from controlled units and still receive a fair return on investment. [37]

These very innovative policies had at least two types of impacts. On the one hand, they met the statutory directive to "protect and, where feasible, provide" lower-cost housing. As of March 1981, the state and regional commissions had approved approximately 152 permit applications involving about 4828 units of affordable housing; of these, 404 were already occupied and 405 were in construction. In addition, the commission estimated that approximately 500 affordable units had been protected from demolition or condominium conversion. [38] On the other hand, the housing policies aroused substantial opposition from developers resentful of the additional delays and costs and, perhaps more importantly, from local government officials who opposed the intrusion of a state agency into what had traditionally been a local domain and who argued that affordable housing was not a *coastal* issue but rather one involving the entire community and even region. In the words of one local government representative: "Coastal boundaries don't make much sense when you look at housing. You have boundaries in urban areas of only a few blocks wide. It's hard to say that's a basis upon which to make meaningful housing policies. . . . A city's housing policies critically affect [its] very nature and must, to a certain measure, reflect the values . . . that are perhaps the very reasons that people have chosen to live in particular communities." [39] Although the commissions did not apply the housing guidelines as strictly in the review of local coastal programs as in permit review, by mid-1980 it was clear that the additional conflict between local and coastal officials over this issue was significantly exacerbating the formulation and certification of local coastal programs. Even state commission staff admitted that affordable housing was one of the major stumbling blocks in about half of the uncertified LCPs, particularly in the urbanized communities of Southern California.[40]

Given the dissatisfaction of developers and local government officials and the resultant delays in LCP certification, the legislature predictably became involved in the housing issue. The signals coming out of the 1980 session were, however, somewhat mixed. On the one hand, the Assembly Resources Committee killed a measure (AB 2081, Bergeson) sponsored by conservative Republicans that would have simply deleted housing from the commissions' authority. And the legislature ratified a negotiated settlement involving Aliso Viejo (a large Orange County development) that, in part, provided for 1000 units of affordable housing (AB 1971, Mello). While some commission staff interpreted these bills as supporting their program, the legislature also approved a measure by the commissions' strongest supporter, Senator Alan Sieroty (SB 1581), that among other things, expressly prohibited the commissions from fixing overnight room rentals at visitor-serving facilities or from approving any method for the identification of low- or moderate-income persons for the purpose of determining eligibility for overnight room rentals. More important than these coastal-specific measures, however, the 1980 legislature approved a series of bills (most noticeably AB 2853, Roos) that gave much clearer guidelines to local governments concerning the need to provide adequate housing opportunities for all income classes on a regional (metropolitan) basis but that also left the ultimate decisions in the hands of local authorities. Given that there was now a middle ground between the commission and pure local control, it became politically feasible for the 1981 legislature to remove the commissions' jurisdiction over affordable housing and instead require that local governments conform to the provisions of the Roos bill (while also somewhat strengthening that statute's provisions concerning the replacement and construction of low- and moderate-cost housing within the coastal zone.)[41]

In the final section of this chapter we shall discuss the repercusions of the commissions' very innovative housing program on its ability to faithfully implement the Coastal Act as a whole.

E. Formulating and Certifying Local Coastal Programs (LCPs)

As will be recalled from Chapter 7, one of the crucial compromises incorporated in the 1975 coastal plan involved the recommendation to return principal responsibility for regulating coastal development to local governments once they had brought their general plans and zoning ordinances into conformity with state coastal policies. Likewise, one of the principal issues during the 1976 legislative session concerned the precise procedures of LCP certification and the eventual allocation of responsibilities between state and local authorities—with the result that the chapter on LCPs was one of the most detailed in the 1976 Coastal Act.[42]

The legislation required local governments in the coastal zone to revise their land use plans and zoning ordinances to be consistent with the substantive policies of the Coastal Act and submit their proposed local coastal programs to the coastal commissions by January 1, 1980. The regional and state commissions were then given about 6 months to review local proposals. In the event the LCP was disapproved in whole or in part, the state commission was required to provide the affected local government with an explanation of its action and to suggest ways in which its concerns could be met. The entire process was expected to be completed by January 1, 1981 (subsequently revised to June 30, 1981). In any jurisdiction without a certified LCP (including zoning ordinances) by that date, the state commission was authorized to assume sole authority for issuing development permits. Upon certification of a local coastal program, principal authority for reviewing development was to be transferred back to the local government—subject only to limited appellate review by the state commission (to be discussed shortly). Finally, the 1976 Coastal Act authorized a local government to request the coastal commissions to prepare its LCP, but this was clearly viewed as an alternative to be used only under exceptional circumstances.

The importance attached by many legislators—including senators Nejedly and Smith—to completing the LCP process within the statutory timetable was underscored by the prominent role given this topic in the 1977 and 1978 oversight hearings by the Senate and Assembly resources committees. When it became obvious in 1978 that the process was behind schedule, the legislature requested a study of causes of delay from the Assembly Office of Research; at the same time, it extended the life of the regional commissions (which obviously would play a critical role in the review process) from mid-1979 to June 30, 1981.[43]

Nevertheless, by July 1, 1981, the vast majority of LCPs had not been certified. Because many of the 67 local governments decided to divide their planning efforts into geographical segments, there were actually 106 LCPs. In addition, most jurisdictions chose to submit their zoning ordinances after gaining approval of their land use plans (as the former had to be consistent with the latter). As indicated in Table 9-5, only 6 LCPs (including San Mateo County, half of Marin County, Long Beach, and three small cities) had been fully certified by the statutory deadline. In addition, the city of Redondo Beach had had its land use plan certified and was preparing its zoning ordinances. In 40 other cases, the coastal commission approved the land use plan and/or zoning ordinances, but the conditions that they imposed had not yet been accepted by

TABLE 9-5
STATUS OF LOCAL COASTAL PROGRAMS AS OF JULY 1, 1981[a]

Status of LCP	Number of LCPs
A. Land use plan and zoning ordinances "Approved" by coastal commissions	
1. Commissions' conditions accepted by local government, i.e., LCP fully certified	6 (plus all 4 port master plans)
2. Commissions' conditions *not yet* accepted by local government	15
B. Only land use plan "approved" by coastal commissions	
1. Commissions' conditions accepted by local government	1
2. Commissions conditions *not yet* accepted by local government	25
C. Land use plans not yet submitted by local government, and/or not yet "approved" by coastal commissions	59
Total	106

[a]Source: California Coastal Commission, *LCP Status Report*, as updated by planning staff in a personal communication, on 20 July 1981.

local government; thus, local authorities were still deciding whether to accept the commissions' suggestions or to submit a new (revised) LCP and have the process start all over again. (This illustrates the importance, at least symbolic, of the Chula Vista court decision, which rejected the commissions' contention that they had the authority to conditionally approve LCPs; had their argument been accepted, they could have legitimately claimed that 21 LCPs and an additional 26 land use plans had been approved by July 1981.)[44] Finally, in the remaining 59 cases the commissions had not even acted on the land use plans by the deadline (in most instances because they had not yet been officially submitted by local authorities).

To point to this shortfall is not, however, to denigrate some significant accomplishments. As one official for the League of California Cities observed[45]:

> Local coastal programs—the plans prepared by the [local] jurisdictions . . . are excellent. They are, almost without exception, some of the best land use planning efforts that local jurisdictions have ever carried out. . . .

> We can't say that these would have been done on their own. . . . The Coastal
> Act and Proposition 20 have worked. They have induced and encouraged and en-
> forced local governments to do some coastal management that I don't think would
> have happened if we didn't have this law. And if we were to get rid of the [coastal]
> commission I'm not sure how long those commitments, in a number of jurisdictions,
> would be expected to continue.

Moreover, the list of certified LCPs contained some real surprises—most notably, the city of Long Beach, which had been one of the most vociferous opponents of the coastal commissions throughout the 1973–1976 period.[46]

Nevertheless, the LCP process clearly fell substantially behind schedule.
What were some of the principal reasons?

First, it is important to note that collaborative planning between state and
local authorities has traditionally been a problematic enterprise.[47] For example,
as of 1979 about two-thirds of local jurisdictions in California were still in
violation of state planning laws mandating the development of various elements
(e.g., transportation, housing, open space), some of which had been statutorily
required since 1955. Moreover, conformance with Coastal Act policies required
changes in local planning ordinances that were unprecedented in both breadth
and specificity.

Second, all parties acknowledge that the amount and/or timing of LCP
funding was a problem.[48] Although by late 1980 the commissions had distributed $7 million—much of it from federal OCZM grants—to local govern-
ments for LCP preparation, local governments often had to pay the initial costs
of identifying coastal issues and developing (quite elaborate) work programs.
While they were supposed to be reimbursed for these start-up expenses, problems
sometimes developed. By late 1978, about 25% of local governments still did not
have approved work plans (and thus state–federal funding for their program).
Given the substantial constraints that Proposition 13 placed on local governments, they were very reluctant to allocate staff time to LCP preparation unless
they could be ensured of reimbursement from state funds. Even state commission officials admitted in late 1980 that they had probably overestimated the
ability of local governments to accomplish a Herculean task with limited funds.
Finally, quite beyond the funding of the LCP process itself, the cutbacks in state
and federal funding for a wide variety of programs—e.g., mass transit, sewer and
water systems, farm subsidies—undoubtedly complicated the task of preparing
LCPs and thus resulted in some delay. (Supposedly the $30 million provided to
local governments by the 1980 Bond Act to fund projects designated in certified
LCPs would ameliorate this situation in the post-1981 period.)

Third, as in any new program involving so many actors, a number of
administrative problems arose in the management of time and information.
Local governments sometimes had difficulty obtaining basic planning data from
other state agencies. They also complained that the two-tier coastal commission

structure left them uncertain about whether elaborate compromises negotiated at the regional level would be upheld by the state commission. The extensive public participation efforts—normally involving citizen advisory committees, extensive review of documents, and numerous public hearings—probably delayed some LCPs, particularly in areas not familiar with such programs. Finally, several legislators questioned whether the approximately 33% of commission staff time through 1978 devoted to providing guidance and assistance to local LCP staff was really sufficient.[49] (Given the magnitude and tight deadlines in the permit review process, however, any reduction in permit staff probably would have resulted in sloppy work and/or howls of protest from applicants.)

Fourth, there were persistent disagreements over the appropriate extent and detail of coastal commission review of local coastal programs. On the one hand, local governments often sought refuge in the platitudes so characteristic of local plans as a means of minimizing change in the *status quo* and/or of "papering over" local disagreements. This was judged unacceptable by coastal commission officials anxious to induce local officials to make precise commitments. But this in turn led to persistent complaints from local officials that commission staff were failing to distinguish critical statewide issues—where detailed review was appropriate—from purely local or less important ones, where locals should be left more discretion. As one of their representatives put it,[50] "We think the Commission has become a coastal zone planning [and] land use commission within the coastal zone which does not differentiate between what are the real priorities in the Coastal Act, and those things [e.g., housing] which, while arguably covered in the Act, are not critical. The Commission is striving in LCPs for the same level of control that it now has in permits." While this issue was never fully resolved, the state executive director did issue a directive in 1978 cautioning staff to focus on critical statewide issues.[51]

Fifth, and perhaps most important, delays in developing and certifying LCPs were often the result of disagreements over specific policies. Interestingly, surveys of local officials in both 1978 and 1980 revealed that by far the most frequently mentioned stumbling block was the commissions' guidelines on affordable housing.[52] For example, 14 of the 26 cities responding to a survey by the League of California Cities in late 1980 indicated that housing was the major point of controversy; no other issue received more than five responses. On the other hand, access—which had been a significant point of contention in 1978— was no longer a major issue, as the actions of the 1979 legislature in strengthening the access program apparently convinced local authorities that this was, in fact, a legitimate statewide concern.[53]

In addition to housing, the major points of substantive disagreement were protection of agricultural lands and wetlands. The former was particularly an issue in San Mateo and in northern San Diego County, while the latter was a major point of controversy in several areas of Southern California. In Chula

Vista (San Diego County), for example, the city proposed to fill a wetland in order to carry out its waterfront improvement plan, which would create jobs, increase tax revenues, and enhance the city's image. The commissions rejected (or, more precisely, rewrote) the relevant portions of the LCP because of the adverse effects on wildlife habitat—whereupon the process became ensnarled in litigation and, as of May 1981, was still far from resolution.[54]

Finally, it should be noted that a few local governments simply rejected the legitimacy of state involvement in "their" land use decisions.[55] This was, for example, the attitude of a city in San Mateo County until the 1978 elections produced a change in the city council. Likewise, the hostility of the chairman of the Los Angeles City Council's planning committee produced an impasse until he was replaced in early 1980. Finally, most of the six cases in which local jurisdictions (and/or the legislature) asked the coastal commissions to prepare the LCP involved a fundamental unwillingness by local officials to participate in the enterprise (with Carlsbad in San Diego County being the most notable example).

In sum, the reasons for delays in LCP preparation and certification were varied and complex. In effect, both sides had things to gain and to lose from meeting the LCP schedule. Local governments presumably had an incentive to complete LCPs as soon as possible, thus regaining permit authority from the commissions and gaining some control over the development decisions of other state agencies. On the other hand, overeagerness to meet the statutory deadlines would have conferred considerable bargaining advantages on the commissions, and some local officials probably preferred to let the commissions bear the continued responsibility for controversial permit decisions. Local governments also had some incentive to delay in the hope that the legislature would dilute the policies mandated by the 1976 act (as eventually happened with respect to housing). Even the 1976 act's principal sanction against delay—the possibility that the commissions would have exclusive permit jurisdiction in those areas without certified LCPs after June 1981—proved not to be credible. As for the coastal commissions, they had strong incentives to push for detailed LCPs representing strict adherence to Coastal Act policies—even at the cost of some delay in LCP certification—because this was their singular opportunity to have those policies written into local ordinances; this was particularly crucial in jurisdictions where those policies had little local support. Another factor militating against a willingness to compromise on the part of the commissions in order to meet the statutory deadlines was that certification of LCPs would ultimately entail a reduction in their ability to control coastal development. On the other hand, the commissions had to appear to be making a good-faith effort at negotiating acceptable solutions with local governments if they were not to incur the wrath of the legislature.

The 1981 legislature attempted in several ways to expedite the certification of LCPs. First, it simply removed one of the major points of controversy—low-

and moderate-income housing—from the commissions' jurisdiction. Second, in lieu of the 1976 act's provision that gave the commissions exclusive permit jurisdiction after June 1981 in areas without LCPs, the legislature approved a compromise measure (AB 385, Hannigan) that would turn over permit review authority to local governments upon certification of their land use plans but prior to commission action on their zoning ordinances; in order to protect statewide interests during the interim period, however, the measure gave the state commission appellate review over any local permit decision within the coastal zone. Third, it retained funding for regional ("district") offices and most regional staff positions even after the demise of the regional commission on June 30, 1981. Finally, the 1980 legislature approved a bond issue authorizing $30 million in additional funds for the implementation of projects in LCPs. Only time will tell if these changes will significantly speed up the LCP process.

But the story will not end after certification of all local coastal programs. There will still be a need to periodically revise and update them. Moreover, recall that the state commission's postcertification appellate authority will be limited to developments seaward of the coastal highway, within 300 feet of a beach or coastal bluff, or within 100 feet of any wetland or stream; projects that were not the principal permitted use under a county's (but not a city's) certified zoning ordinances; or major public works and energy facilities.[56] This rather restricted scope of review—and the fact that the standard of review in any litigation would be conformance with the certified LCP—was one of the critical factors behind the commissions' efforts to make LCPs as detailed as possible. But the commissions also realized that if they pushed policies that were not willingly agreed upon by local agencies, those policies would probably not be implemented in their intended manner. These limitations are particularly critical in cases requiring local governments to expend money (rather than simply to review private proposals). For example, the Long Beach LCP contained a proposal to build a bicycle path along the beach as a means of providing access in areas of limited automobile parking. But opposition from local residents—concerned that it might attract thieves and muggers on roller skates—made it impossible for the city council to muster the two-thirds vote needed to approve construction of the path, even though state funds were available to build the facility.[57] In short, certification of LCPs is simply the beginning of another implementation saga.

III. Conclusions

This chapter has examined the performance of the coastal commissions (and the conservancy) in addressing a number of the goals of the 1976 Coastal Act during the period from January 1977 through June 1981. Within the constraints of the data available from governmental and published sources, it appears that the coastal agencies have certainly fulfilled the statutory mandate to preserve and, where feasible, to restore the state's wetlands (although proposed legislation could certainly change that in Southern California). Likewise, the commissions did everything that could reasonably have been expected to protect low- and moderate-income housing along the coast (although we have some reservations about the side effects of their policies in this area). The two agencies launched an innovative program to deal with the problem of old, largely unbuilt subdivisions that shows some real promise of success, although the jury is still out on the TDC scheme in Malibu. On the other hand, the record is more mixed when it comes to providing public access to the ocean. While the commissions pushed the regulatory process as far as it could reasonably go (e.g., in the case of Sea Ranch) and made a major effort to increase access in Orange County via conditions on the large Irvine and Avco developments, it was not until 1979 or even 1980 that they began to adequately address the obstacles to actually providing vertical and lateral access in Malibu and elsewhere. With the aid of the conservancy and various acquisition programs, however, there is some hope that the promise of greater access made to the voters in 1972 will finally be realized. Finally, the record of persuading local governments to revise their land use ordinances to conform to Coastal Act policies can only be described as frustrating and disappointing—although perhaps no more so than previous efforts at state–local collaborative planning.

In seeking to understand these outcomes in the context of the factors discussed in Chapter 1, it is clear that the 1976 Coastal Act did a rather good job of structuring the implementation process. Certainly the clear directives to maximize access in new developments seaward of the coastal highway and to require a policy of "no net fill" in wetlands developments were an important resource to the commissions. On the other hand, the equally precise deadlines for completion of LCPs went unfulfilled, and the commissions' innovative affordable housing policies and TDC schemes could only partially be traced to statutory policies. As for the adequacy of the causal theory and the incentives for compliance incorporated in the Coastal Act, these were clearly inadequate only in the case of LCPs. The 1976 legislature never did find a set of mechanisms to force local governments and the commissions to bargain seriously with each other; this was particularly lacking in the case of the commissions, although even the threat of commission continuation of permit review did not prove to be an important factor in the case of many local governments. With respect to wetlands, the legal authority of the two coastal agencies was sufficient except in the case of controlling sedimentation originating outside the coastal zone. And their ability to actually provide access was greatly enhanced by the 1979 amendments, although the statutes still placed critical vetoes in the hands of local governments and, in some cases, the state parks department. Finally, while the financial resources provided the commissions and local governments were probably adequate prior to mid-1978, the passage of Proposition 13 adversely affected the access program and LCP preparation.

Turning to the nonlegal factors affecting implementation, the personal commitment of agency officials was clearly an important element in the evolution of the very stringent policies dealing with wetlands, affordable housing, and the TDC scheme. On the other hand, differences in political philosophy between commission and local government officials were probably the most important proximate cause of many of the LCP stalemates. Active interest-group support almost surely played an important role in the stringent policies pursued with respect to wetlands protection, affordable housing, and access at Sea Ranch, while the absence of such support (particularly in contrast to the vociferous opposition of local property owners) probably lay behind many of the difficulties in providing access to Malibu. Finally, the general political and economic climate of the late 1970s was, at best, a mixed blessing in terms of the effective implementation of the 1976 Coastal Act. The growing skepticism of public officials concerning the efficacy of regulation and the increasingly tight fiscal constraints under which they operated, particularly after 1978, certainly did not help the commissions' efforts to provide access, protect wetlands, or develop LCPs. On the other hand, the 1980 Field poll suggests that there was still broad public support for many Coastal Act objectives; high interest rates and the fiscal impacts of Proposition 13 probably helped restrain demand for new construction, particularly outside existing urbanized areas;[58] and the leveling off of the state's consumption of electricity and natural gas reduced demand for new energy facilities along the coast.

As we noted at the beginning of this chapter, however, the commissions no longer benefited from the protective umbrella of the Coastal Initiative after 1976. They thus became much more vulnerable to changes in interest-group and public support, particularly as these were reflected in the legislature. And one of the most salient features of the 1977–1981 period, particularly after the 1978 elections, was a gradual chipping away of the commissions' legal authority—with the notable exception of public access. In 1977 the commissions' role in the siting of LNG facilities was substantially reduced. The 1979 session witnessed some restrictions on the commissions' ability to regulate the construction of single-family residences and numerous changes, a few of them significant, in the boundaries of the coastal zone. The 1980 legislature made some modest changes in the commissions' voting rules and in their regulation of hotel room rates. And in 1981 the legislature removed the commissions' authority to require affordable housing, as well as somewhat restricting their ability to regulate development after the certification of the land use portion of the local coastal programs. Moreover, throughout this period there were numerous bills to delete specific developments from the commissions' jurisdiction; while few were approved, the threat of passage sometimes gave developers important leverage in their negotiations with coastal officials.[59]

There were numerous reasons behind this decline in the commissions' legal authority. They included (1) the election of more conservative legislators in 1978 and 1980, (2) the general receptivity of legislators to constituent complaints about administrative agency decisions, (3) the leverage that increasing campaign costs gave to development groups,[60] (4) the decline of environmental interest-group representation in Sacramento, with the exception of the Sierra Club and a few local wetlands groups, and (5) the lack of a real "fixer," i.e., someone capable of providing guidance as well as protection to the agency.[61]

But we would suggest that the coastal commissions were partly to blame for this gradual erosion of their authority. Of course, any effort to faithfully implement the 1976 Coastal Act would have aroused substantial opposition from coastal property owners. Nevertheless, the commissions seem to have generated more opposition, and less support, than was necessary. Their permit review activities were periodically plagued by what came to be known in Sacramento as "horror stories," i.e., instances of apparent insensitivity to the rights of property owners and/or of attention to petty details[62]: After fires wrecked a number of Malibu beach-front homes in the winter of 1977, some commission staff suggested that vertical access might be required as condition of rebuilding, after which Governor Brown accused the commissions of being "bureaucratic thugs." And in 1981 the commissions acceded to the request of one of their legislative supporters to expedite the processing of a sewage treatment plant so that it would not lose a federal matching grant, but then required that the facility be painted in "earth tones"—at a cost of $150,000. In like vein, there were some actions— such as the decision to recommend a clearly infeasible LNG site a few months after the legislature had voted massively to expedite siting—which certainly undermined the agency's basic credibility. Even though such horror stories constituted a tiny fraction of the 50,000 permits processed by the commissions and even though many of them turned out, upon close inspection, to have involved a rather selective presentation of the total picture, the mere recitation nurtured a perception of the commissions as unreasonable zealots.[63]

In addition to generating needless opposition, the commissions appear to have been rather ineffective in developing legislative and constituency support. Very few of the commissioners had any "clout" in Sacramento, and it was apparently not until 1980 that even the staff made a major effort to systematically cultivate the support of key members of the Assembly and Senate resources committees.[64] Nor did they undertake any serious efforts to actively present their side of the picture to the mass media in an effort to counter the media's proclivity for horror stories.[65] And it is still unclear why the commissions failed until 1978–1980 to accord higher priority to the development of a systematic access program when that was the issue that had "sold" Proposition 20, which was a major priority in the 1976 Coastal Act, and which constituted one of the few tangible, positive benefits that the commissions could provide.[66] Even the one clear case of the commissions' efforts to expand their base of constituency support—affordable housing—was continued after it had become fairly obvious that the policy was arousing intense opposition and, in turn, having an adverse effect on issues such as LCP completion and (thus) access that were clearly more important to the 1976 (and subsequent) legislature(s).

In sum, while commission officials enjoyed an excellent reputation for hard work, sound fiscal management, and commitment to statutory objectives, they appear to have neglected one of the cardinal rules of American administrative politics: In a system with strong interest groups, weak political parties, and strong legislative committees, the long-term viability of an agency demands assiduous cultivation of key legislators and conscious efforts to maximize the balance of constituency-group support.[67] To demean this truism is to endanger the commissions' ability to faithfully implement the 1976 Coastal Act.

NOTES

1. Although the 1976 Coastal Act did not set a precise deadline for the certification of local coastal programs prepared by local governments, it did require that they be submitted to the regional commissions no later than January 1, 1980, and that LCPs actually prepared by the commissions be certified by the end of that year. This schedule was revised slightly by 1978 legislation (AB 3478) that delayed the expiration date for the regional commissions from mid-1979 to June 30, 1981—by which time almost all of the LCPs were expected to be certified.

2. Marty Gellen, *The Housing Crisis in California*, Working Paper No. 340, Institute of Urban and Regional Development, University of California-Berkeley, February 1981.

3. The legislature went from 55–25 and 25–15 Democratic majorities in the Assembly and Senate, respectively, during the 1976 session (57–23 and 26–14 following the 1976 elections) to 50–30 and 25–15 after November 1978 and 47–33 and 23–17 following the 1980 elections. It was the unanimous view of the legislative staff we interviewed (Tom Willoughby of the Assembly Energy and Natural Resources Committee, Jeff Arthur of the Senate Natural Resources Committee, and Bill Kier of the Senate Office of Research) that the 1979–1981 legislatures were less sympathetic to regulation, including coastal land use, than were their 1976 counterparts. This was accentuated following the 1980 elections when two strong supporters of the coastal commissions, Speaker Leo McCarthy and Senate Pro-Tem James Mills, were replaced by less supportive Democrats.

4. Of the environmental (and other) organizations that were the core of the commissions' support during 1976, only the Sierra Club continued to have full-time paid lobbyists by 1980–1981. The Coastal Alliance and PACE were dead, the PCL virtually moribund, and the League of Women Voters in less evidence than before. In contrast, CCEEB and the traditional development, labor, and local government organizations were still well represented and had been joined by the California Coastal Council, a membership-based organization composed of vociferous opponents of the commissions who were adept at gaining media coverage although less skillful at lobbying legislators. [See Anne Jackson, "Is the Environmental Movement an Endangered Species?" *California Journal* (March 1981).]

5. The coastal commissions made a major effort to encourage public participation in local governments' LCP efforts, and most communities developed an LCP Advisory Committee as well as extensive public hearings and notification efforts. In many locales—including Big Sur, Humboldt, Long Beach, Santa Barbara, San Mateo—there was extensive participation. Apparently, much of it was from people previously uninvolved in the coastal permit process who were using the LCPs as a forum to debate local planning issues and who were not active in Sacramento. Interviews with Judy Rosener, state commissioner, 3 July 1981, and Bob Brown, the commissions' chief planner, 6 June 1981.

6. Senate Committee on Natural Resources, *Implementation of the 1976 Coastal Act, Hearings,* December 8–10, 1977, *passim.* The budgetary figures are taken from the commission's *Biennial Reports.*

7. See, for example, Anthony Downs, *Inside Bureaucracy* (Boston: Little, Brown, 1967), Chap. 2, and Marver Bernstein, *Regulating Business by Independent Commission* (Princeton: Princeton University Press, 1955), chap. 3.

8. Interview with Peter Douglas, assistant director of the commissions, San Francisco, 5 June 1981.

9. Among the more serious threats in the 1979 session were (1) SB 175 (Cusonovich), which would have excluded single-family residences from the commissions' jurisdiction and required compensation for all public access conditions; (2) SB 779 (Dills), a more moderate attempt to exclude single-family residences; (3) SB 569 (Beverly), which would have substantially weakened the commissions' authority in urbanized areas; and (4) numerous bills seeking to exclude specific projects. These issues were eventually resolved by AB 643 (Calvo) and AB 462 (Mello), which represented reasonable compromises between the commissions and their critics. As admittedly indirect evidence that the commissions may have been sensitive to these weakening bills, the percentage of outright approvals on permit appeals declined from 38% during the legislative session (January through August) to 27% after it had ended. See also Rosener interview, July 1981.

10. In 1980–1981 there were at least two changes among state appointees on the regional commissions and three such changes on the state commission; of the latter, only one marked a substantial change in orientation less favorable to the Coastal Act.

11. For discussions of this controversy, see (1) Robert Jones, "Deukmejian Style Triggers Growing Political Storm," *Los Angeles Times,* 26 December 1979; (2) Larry Stamer, "Panel Restores $1 Million to Deukmejian's Budget," *Los Angeles Times,* 8 April 1980; and (3) Douglas interview.

12. Bernstein, *Regulating Business by Independent Commission,* chap. 3. For critiques, see Paul Sabatier, "Social Movements and Regulatory Agencies," *Policy Sciences* (Fall 1975), pp. 301–342; and Kenneth Meier and John Plumlee, "Regulatory Administration and Organizational Rigidity," *Western Political Quarterly* 31 (March 1978), pp. 80–95. While the commissions (particularly the state) have not noticeably changed, in the opinion of some observers the South Coast and San Diego regional commissions became less sympathetic to coastal protection after 1977.

13. These provisions are found in the California Coastal Act (Statutes of 1976, Chapter 1330), Sections 30001.5, 30210-30212.5, 30500, and 30603. After certification of LCPs, appeals from local governments are limited to (1) developments between the sea and the first coastal road or within 300 ft of the beach; (2) developments on any public trust lands, within 100 ft of any estuary or stream, or within 300 ft of the top of the coastal bluff; (3) developments in legislatively approved sensitive coastal resource areas; (4) developments in coastal counties (not cities) that were not the principal permitted use; and (5) major public works or energy facilities.

14. These figures are taken from the commissions' Coastal Access Inventory, January 15, 1981, as well as the Howe interview.

15. Interviews with (1) Peter Douglas; (2) Joseph Petrillo, executive director of the conservancy, 30

June 1981; and (3) Bill Fishman, legislative assistant to Assemblyman Kapiloff, 23 June 1981. These were the three principal staff on the 1979 Kapiloff bill (AB 989), which was originally intended to prod Parks on access but was subsequently amended to give the conservancy and commissions principal authority after Parks had made it clear that it was not really interested in the program.

16. California Coastal Commission, *Coastal News*, October 1980 (the joint Commission/Conservancy Recommendations for Access) and January 1981 (the annual progress report on the coastal access program).

17. These figures are based on the coastal access inventory maps for Malibu and on an interview with Steve Howe of the commissions' access program. It is reasonably certain, however, that the board's action was motivated at least as much by respect for the intense hostility of most Malibu residents to granting additional access, particularly given the recent election of a conservative Republican to represent the coastal supervisorial district. (See Richard O'Reilly, "State, Landowners Pitted on Beach Access," *Los Angeles Times*, 12 May 1981.)

18. These figures are taken from Steve Howe and from Richard O'Reilly, "Coast Panel: What It Did and Didn't," *Los Angeles Times*, 10 May 1981.

19. Richard O'Reilly, "Coast Panel's OK a Matter of Public, Private Benefit," *Los Angeles Times*, 11 May 1981. Also an interview with Ronnie Rogers of Avco, 12 August 1981.

20. Based on articles in the *Los Angeles Times*, 5 December 1979, 24 July 1980, and 25 June 1981.

21. See California Coastal Commission, "Sea Ranch Overall Conditions and Findings," Adopted March 7, 1979; and Senate Committee on Natural Resources and Wildlife, *Hearings on the Coastal Zone in Sonoma County*, December 14, 1978, Santa Rosa.

22. In a February 1981 referendum, a clear majority of the members of the Sea Ranch Association voted to continue the court suit rather than accept the compromise contained in the Bane bill. Following the April 7, 1981, court decision (*Sea Ranch Association* v. *California Coastal Commission*, Opinion No. C-74 1320), a group of moderates were elected to the association's board of directors. (See articles in the *Los Angeles Times*, 8 April 1981, and the *Independent Coast Observer*, 29 May 1981, as well as interviews with George Wickstead of the Sea Ranch Association, 6 November 1980, and Steven Scholl of the North Central Regional Commission, 5 June 1981.) The COAAST suit alleged that the $500,000 actually constituted a payment for the accessways. Given the approval of the Bane bill by both parties and the legislature, the chances of its success are probably quite slim. Were it to be accepted by a court, however, it would nullify the entire settlement. In the meantime, everything has been proceeding on schedule.

23. According to Section 30607.1 of the 1976 Coastal Act, "where any dike and fill development is permitted in a wetland . . . mitigation measures shall include, at a minimum, either acquisition of equivalent areas of equal or greater biological productivity or opening up equivalent areas to tidal action; provided, however, that if no appropriate restoration site is available, an in-lieu fee sufficient to provide an area of equivalent productive value or surface areas shall be dedicated to an appropriate public agency." The general policy of "no net fill" had also been developed by BCDC [Robert Feinbaum, "The Growing Use of Tradeoffs to Solve Development Problems," *California Journal* (1978), pp. 159–161].

24. As will be recalled from the previous chapter, the requirement for legislative approval of recommended "sensitive coastal resource areas" was one of the concessions made in the dying days of the 1976 session to win labor's support for the act (Section 30502.5 Statutes of 1976, Chapter 1331). For the commission's reasons, see the staff recommendations of April 28, 1977 and July 12, 1978, as well as the State Commission *Minutes* of August 15, 1978.

25. Letter from William Travis (state commission staff) to Rosella Sussman of the U.S. Office of Coastal Zone Management, 15 February 1980, pp. 7–11; Eric Metz and Michael DeLapa, "California's Wetland Regulatory Program: Developing an Interpretive Guideline for Protecting

Significant Natural Resources" (Paper presented at Coastal Zone 80 Conference, November 17–20, 1980, Hollywood, Florida), p. 6.

26. Metz and DeLapa, "California's Wetland Program," pp. 7–19; California Coastal Commission, "Statewide Interpretive Guideline for Wetlands and Other Wet Environmentally Sensitive Habitat Areas," Adopted February 4, 1981; and Coastal Commission *Minutes,* July 19, 1979; December 28, 1979; July 17, 1980; and February 4, 1981.

27. This discussion of Bolsa Chica is based on (1) Richard O'Reilly, "State, Developers Battle Over Wetlands Definition," *Los Angeles Times,* 15 May 1981; (2) the March 11, 1980, Coastal Commission Staff Report, "Preliminary Wetlands Determination for the Bolsa Chica Local Coastal Plan"; (3) the State Commission *Minutes,* March 19–20, 1980, pp. 11–16; and (4) Joe Anguiano, "Bolsa Chica Land Agreement in Danger of Collapse," *Los Angeles Times,* 29 April 1979.

28. O'Reilly, "Wetlands"; Travis letter to Sussman.

29. This discussion of Humboldt Bay is based largely upon an interview with Dan Ray of the North Coast staff, 6 July 1981. On the Eureka pocket marshes, see State Coastal Conservancy, staff recommendation, December 11, 1980, File No. 77-14, and *1981 Activities Report,* p. 15.

30. California Coastal Act, Statutes of 1976, Chapter 1330, Chapter 3, especially Article 6.

31. For specific cases, see Senate Natural Resource Committee, *1978 Hearings on the Coastal Zone in Sonoma County,* pp. 156 ff; and Richard O'Reilly, "Coast Panel: What It Did and Didn't," *Los Angeles Times,* 10 May 1981. On the more general issue of legislators' receptivity to complaints of individual hardships, see Schultze, *The Public Use of Private Interest,* pp. 70–83; and Morris Fiorina, *Congress: Keystone of the Washington Establishment* (New Haven: Yale University Press, 1977), chap. 5.

32. On Whiskey Shoals and the TDC program, see Conservancy, *1981 Activities Report,* pp. 20–22; Richard O'Reilly, "Mountain Development Plan Okd," *Los Angeles Times,* 3 November 1980; State Commission *Minutes,* July 22, 1980, p. 35; the conservancy's staff reports of November 13, 1980 (El Nido Subdivision), and December 11, 1980 (Cold Creek Watershed); and memo from Michael Fischer (state commission executive director) on Revisions to the South Coast Region Interpretive Guidelines for the Malibu-Santa Monica Mountain Area, March 11, 1981. On TDCs in general, see John Costonis, *Space Adrift Landmark Preservation and the Marketplace,* (Urbana: University of Illinois Press, 1974); and Steven Woodbury, "Transfer of Development Rights: A New Tool for Planners," *Journal of the American Institute of Planners* 41 (January 1975), pp. 3–14.

33. On this topic, see (1) the 1976 Coastal Act, Chapter 3, Article 7; (2) California Coastal Commission, "Designation of Coastal Zone Areas Where Construction of an Electric Power Plant Would Present Achievement of the Objectives of the 1976 Coastal Act," September 1978; and (3) the forecasts of electrical supply and demand found in the 1977, 1979, and 1981 Biennial Reports of the California Energy Commission.

34. The infeasibility of Camp Pendleton stemmed from at least two factors. First, the military was strongly opposed because of the risks posed by the terminal to the 40,000 people on the base and to the viability of training for amphibious landings. Second, it clearly had the capacity to veto the project—absent an override by the president—because state authorities could not exercise their powers of eminent domain on federal land. While the 1977 LNG Terminal Act left the ultimate decision of feasibility to the PUC, for the coastal commissions to have treated the issue so lightly did not help their credibility among legislators convinced of the need to accelerate the siting process. It should be noted, however, that for a variety of reasons—including uncertainties over geologic hazards, the opposition of Indians, and the leveling off of demand for natural gas—construction at the Pt. Conception facility is very much in limbo (as of July 1981). In addition, the commissions may have regained some of their credibility through an apparently excellent study of the feasibility of an *offshore* LNG facility. This discussion relies heavily on the following

sources: (1) Lawrence Susskind and Stephen Cassella, "The Dangers of Preemptive Legislation: The Case of LNG Facility Siting in California," *Environmental Impact Assessment Review*, 1 (March 1980), pp. 9–26; (2) California Coastal Commission, *Final Report Evaluating and Ranking LNG Terminal Sites*, May 24, 1978; (3) an interview with Tom Willoughby, chief staff to the Assembly Energy and Natural Resources Committee, 12 July 1981; (4) Herb Fox, "California's LNG Project: Afflicted with a Terminal Illness," *California Journal* 12 (May 1981), pp. 185–187; and (5) an interview with Tim Davis, chief staff to the Senate Energy and Public Utilities Committee, 15 July 1981.

35. For example, the relevant policies (nos. 125–126) in the 1975 coastal plan occupied about 1 of the 150 pages of substantive policies in that document. Likewise, in only about 6% of appeals to the state commission during 1973–1975 was affordable housing an issue (Sabatier, "State Review of Local Land Use Decisions," p. 267). For a discussion of selected housing cases, see Douglas and Petrillo, "California's Coast," pp. 230–231.

36. California Coastal Commission, "Interpretive Guidelines on Housing in the Coastal Zone," September 28, 1977.

37. This discussion is based primarily on Drew Liebert, "The Essence of Administrative Discretion: A Case-Study Approach to the Use of Bureaucratic Discretionary Powers" (Unpublished master's thesis, Claremont Graduate School, March 1981), chap. 2. See also Robert Jones, "Coastal Housing Plan for Needy Sparks Controversy," *Los Angeles Times*, 28 October 1979; Richard O'Reilly, "Affordable Housing Puts Poor into Seaside Homes," *Los Angeles Times*, 13 May 1981; and State Commission *Minutes*, July 17, 1979, pp. 21–27, and August 14, 1979, pp. 19–21.

38. California Coastal Commission, Memorandum from Michael Fischer (executive director) Concerning Revision of the Housing Guidelines, March 2, 1981, p. 5.

39. Testimony of Russell Selix, League of California Cities, before the Senate Natural Resources Committee, Concord, November 14, 1980, pp. 4–5.

40. Interview with Bob Brown, chief planner of the coastal commissions, 5 June 1981. This position was confirmed by Peter Douglas (deputy director of the commissions, interview of 5 June 1981) and by a survey of coastal cities conducted by the League of California Cities (Senate Natural Resources Committee, *Hearings on Problems of Local Governments with Compliance with the Coastal Act*, Concord, November 14, 1980, p. 112).

41. We are indebted to Jeff Arthur, chief of staff to the Senate Natural Resources Committee, for this interpretation of the evolution of housing bills (interview, 12 June 1981). It should be noted that, in direct response to requests from the South Coast and San Diego Regional Commissions, the state commission in May 1981 made some minor revisions in the housing guidelines, chiefly by raising the "floor" for exempted developments from 5 to 10 units (State Commission *Minutes*, May 5, 1981). But this was definitely "too little, too late" to save the program in the legislature.

42. California Coastal Act, Statutes of 1976, Chapter 1330, Sections 30500–30603.

43. Assembly Office of Research, *An Assessment of the California Coastal Planning Process*, February 1979, pp. 1–2 [hereafter referred to as AOR, *LCP Study*]. See also Senate Natural Resources Committee, *1977 Hearings on the Implementation of the Coastal Act*, pp. 27–34, 46–55, 145–155.

44. In a suit brought by the City of Chula Vista, a state superior (district) court in December 1980 ruled that the 1976 Coastal Act did not give the commissions authority to attach conditions to LCPs, but only to approve or reject specific portions and, in the latter case, to recommend to local governments how the commissions' concerns might be addressed. Somewhat surprisingly, the commissions decided not to appeal the ruling. Jack Jones, "Judge Rules Against State Coastal Panel," *Los Angeles Times*, 9 December 1980.

45. Testimony of Russell Selix in Senate Natural Resources Committee, *1980 Hearings on Progress of LCPs*, pp. 108–109.

46. Apparently the preparation of the Long Beach LCP was due to (1) a political scandal that led to the hiring of a new, and very skillful, planning director and (2) the city's ability to skirt two of the major controversies plaguing LCP programs, housing and wetlands. With respect to the former, the city could legitimately argue that it already had an adequate overall program in this area. While there was a potentially controversial wetland within the city's jurisdiction, it was placed in a separate LCP. See Richard O'Reilly, "Two Plans Highlight Coastal Dilemmas," *Los Angeles Times*, 17 May 1981.

47. AOR, *LCP Study*, pp. 28, 46. On efforts in other states, see Jens Sorenson, *State–Local Collaborative Planning*, (Washington, D.C.: Office of Coastal Zone Management, 1979); and Frank Popper, *The Politics for Land Use Reform*, (Madison: University of Wisconsin Press, 1981).

48. See, for example, AOR, *LCP Study*, pp. 24, 30–33; and Senate Natural Resources Committee, *1980 Hearings on the Progress of LCPs*, pp. 6–16, 157.

49. AOR, *LCP Study*, pp. 23–24, 39–44; also testimony of Leonard Grote, chairman of the state commission, before the Assembly Committee on Energy and Natural Resources, February 10, 1981, pp. 13–15.

50. Testimony of Russell Selix, League of California Cities, in Senate Natural Resources Committee, *1980 Hearings on Progress of LCPs*, p. 113.

51. Stanley Scott, "Coastal Planning in California," in *Coastal Conservation: Essays on Experiments in Governance*, ed. Stanley Scott (Berkeley: Institute of Governmental Studies, 1981), pp. 29–32. On the more general issue of the appropriate level of specificity, see the testimony by Norbert Dall (Sierra Club) and Naida West (CCEEB) in the 1980 Senate hearings, as well as that of Leonard Grote, chairman of the state commission, before the Assembly Committee on Energy and Natural Resources, February 10, 1981, p. 11.

52. In October 1980, Russell Selix of the League of California Cities surveyed 49 cities concerning the status, expectations, and points of controversy in their LCPs. Thirty-six cities (or 73% of the total) responded, of which 26 reported serious substantive difficulties with the coastal commissions. Fourteen of these (or 54%) indicated that housing was a serious point of controversy. The next most important (with 5 responses) were (1) land use conflicts, (2) specificity/detail, (3) wetlands, and (4) agricultural lands. These results, which were briefly reported in the 1980 Senate hearings (p. 112), are based upon personal communication with Mr. Selix. Housing was an issue raised throughout those hearings, with only the Sierra Club defending the commissions' policies. Affordable housing was likewise by far the most frequently mentioned substantive issue by 17 local government planners in an April 1978 workshop sponsored by the Sea Grant Program (see Stanley Scott, "Coastal Planning Issues," in *Coastal Conservation*, pp. 23–26). Finally, recall that the state commission's chief planner concurred in 1981 that housing was a substantial roadblock in about 50% of LCPs (*supra*, note 40).

53. While access was the second most important issue in the April 1978 Sea Grant Workshop (17 weighted points, compared to 28 for affordable housing and 9 for the third most important issue), it was mentioned by only 2 of the 26 respondents reporting problems in the October 1980 League of Cities survey.

54. On the Chula Vista controversy, see (1) O'Reilly, "Two Plans Highlight Coastal Dilemmas"; (2) Grote's 1981 testimony before the Assembly Resource Committee, pp. 12–13; and (3) the state staff memo on Chula Vista's LCP, April 16, 1981.

55. See, for example, Senate Natural Resources Committee, *1980 Hearings on the Progress of LCPs*, pp. 168–170; AOR, *LCP Study*, pp. 47–48; and a personal interview with Russell Selix, 16 July 1981.

56. 1976 Coastal Act, Section 30603.

57. On Long Beach, see O'Reilly, "Two Plans Highlight Coastal Dilemmas." On the more general issue of local compliance with certified LCPs, see Senate Natural Resources Committee, *1980 Hearings on the Progress of LCPs*, p. 111; and Leonard Grote, "Coastal Conservation and

Development: Balancing Local and Statewide Interests," in *Coastal Conservation*, ed. Scott, pp. 16–22.

58. Office of Planning and Research, Governor's Office, New Housing: *Paying Its Way?*, May 1979.

59. Perhaps the best-known case involves a proposed 2100-unit subdivision in the Santa Monica Mountains by Headlands Properties, which the state commission subsequently reduced to 180 units on a portion of the land while under the threat of a boundary change bill (AB 13, Pappan) that almost passed the legislature. Another case involves Bolsa Chica, in which the Bolsa Signal Corp. has hired the former Assembly Pro-Tem to serve as its lobbyist. [See Mark and Miriam Rafferty, "Santa Monica Mountains Controversy Continues," *Environs* 4 (July 1979), pp. 6–8.]

60. On the escalating costs of legislative campaigns and their effects on voting patterns, see Douglas Shutt, "Huge Campaign Spending Continues, Survey Finds," *Los Angeles Times*, 19 August 1981; and John Owens and Edward Olson, "Campaign Spending and the Electoral Process in California, *Western Political Quarterly*, (December 30, 1977), pp. 493–512.

61. While Speaker Leo McCarthy and Assembly Resources Committee chairman Victor Calvo protected the commissions from gutting legislation and carried some important compromise bills during the 1977–1980 period, they apparently did not provide affirmative guidance to the commissions nor draft legislation that would have helped the agency carry out its more important mandates. This opinion is based on interviews with Peter Douglas, Tom Willoughby, Jeff Arthur, Bill Kier, Bill Yeates (agency lobbyist, 18 June 1981), and Judy Rosener (state commissioner, 3 July 1981).

62. Based on interviews with Willoughby, Arthur, and Yeates.

63. For example, while one might question the commissions' concern with the visual impacts of a sewage treatment plant (particularly one near the Moss Landing power plant), it is nevertheless true that the water district did not object to the "earth tones" condition during commission review but only afterwards. (State Commission, *Minutes*, June 23, 1981, as well as interview with Peter Douglas, 27 August 1981).

64. For example, Rosener tried unsuccessfully for several years to persuade the commission to make a serious effort to provide key legislators with information on the positive aspects of the commissions' performance (to counteract the horror stories) and to inform them of commission decisions involving their districts. In a similar vein, Mel Lane, former chairman of the state commission, argued: "What the commissioners and staff haven't adequately appreciated is that in a sense the single most important constituency group they have never comes to their meetings. It is the California State Legislature" (quoted in Liebert, "Administrative Discretion," pp. 2–32).

65. According to Judy Rosener, none of the major papers of the state (with the occasional exception of the *Los Angeles Times*) assigned a reporter to continuously monitor the commissions and thus none were capable of providing an overall assessment of the commissions' performance (with the exception of O'Reilly's superb series in May 1981, which may have come too late to balance the horror stories).

66. Interviews with Douglas, Arthur, and Willoughby. Bill Yeates and Judy Rosener thought it may have been the result of understaffing and the difficulty of making the transition from a regulatory to a mangement agency. But the commissions surely could have learned from BCDC's extensive experience with identical access problems and its publication of an access guide.

67. For the classic statement, see Norton Long, "Power and Administration," *Public Administration Review* (Fall 1949), pp. 257–264.

THE IMPLICATIONS OF THE CALIFORNIA COASTAL COMMISSIONS FOR THE EFFECTIVE IMPLEMENTATION OF REGULATORY POLICY

The 20th century has witnessed a vast expansion of governmental activity designed to correct market deficiencies in the areas of environmental and consumer protection, worker safety, and racial discrimination through the creation of agencies charged with regulating A's behavior in order to minimize his adverse effects on B. This growth of governmental regulation in the United States has been repeatedly criticized for its adverse effects on the freedom of regulated groups, the efficient allocation of resources, and the purchasing power of lower- and moderate-income people (through raising consumer prices).

Nevertheless, in many ways the most damaging criticism raised against regulatory programs is that they are often unable to achieve their stated objectives. It is, after all, extremely difficult to justify the restrictions on individual liberty and the utilization of scarce resources if the program is not at least moderately successful in attaining statutory goals. This criticism has been raised against not only traditional regulatory programs in air and water pollution control, consumer product safety, and civil rights that directly proscribe (or prescribe) certain activities but also against programs in these and other policy areas that employ conditional (restricted) grants as inducements to behavioral change. In fact, there is at present a rather pervasive pessimism about the ability of governmental regulation to significantly alter behavior.[1]

On the other hand, there have been some success stories—chiefly at the state and local level—in the fields of civil rights, land use control, and pollution control.[2]

On the basis of a review of previous studies of both successes and failures, we developed a general model of the variables affecting the implementation of regulatory policy and then extracted from that model a set of six conditions under which a program that seeks a substantial departure from the *status quo* can achieve its objectives[3]:

1. The enabling legislation or other legal directive mandates policy objectives that are clear and consistent (or at least provides substantive criteria for resolving goal conflicts).
2. The enabling legislation incorporates a sound theory identifying the principal factors and causal linkages affecting policy objectives, and gives implementing officials sufficient jurisdiction over target groups and other points of leverage to attain, at least potentially, the desired goals.
3. The enabling legislation structures the implementation process so as to maximize the probability that implementing officials and target groups will perform as desired. This involves assignment to sympathetic agencies with adequate hierarchical integration, supportive decision rules, sufficient financial resources, and adequate access to supporters.
4. The leaders of the implementing agencies possess substantial managerial and political skill and are committed to statutory goals.
5. The program is actively supported by organized constituency groups and by a few key legislators (or a chief executive) throughout the implementation process, with the courts being neutral or supportive.
6. The relative priority of statutory objectives is not significantly undermined over time by the emergence of conflicting public policies or by changes in relevant socioeconomic conditions that undermine the statute's causal theory or political support.

These conditions will always be sufficient and generally necessary to achieve statutory objectives.

In assessing the probability that a statute will be effectively implemented, however, one must consider what might be termed the "tractability" of the problem(s) being addressed.[4] The likelihood of changing target group behavior is obviously a function of (1) the amount of change required and (2) the orientation of target groups towards the mandated change. It is also dependent (3) on the diversity in proscribed activities, as diversity makes it more difficult for implementing agencies to frame precise and understandable regulations. Thus, we argued in Chapter 1 that the strength or degree of bias in the six conditions necessary to achieve statutory objectives would be a function of these three aspects of the tractability of the problem: In other words, the greater the amount of mandated change, target group resistance, and diversity in proscribed activites, the clearer the statutory objectives, the more structured the implementation process, the greater the need for constituency group support, etc., if the objectives were to be achieved. But the tractability of the problem is also a function of (4) the availability of adequate knowledge concerning the extent of, and factors affecting, the problem(s) and (5) the percentage of the population being regulated. Obviously, the statute cannot incorporate a valid causal theory if none exists. Moreover, the degree of net political support for a regulatory program is likely to be inversely related to the size of the target group as a percentage of the total population: It is easier to develop political support for changing the behavior of 10% of the population than for regulating 90%.

Within the context of this general analytical framework, we focus in this study on the implementation of the 1972 California Coastal Zone Conservation Act both because it mandated substantial behavioral change (and thus presented a rather difficult problem of implementation) and because it appeared to meet most of the six conditions for effective implementation. Our basic hypothesis, then, is that the policy decisions of the coastal commissions and the actual impacts of those decisions should have been in substantial conformance with statutory objectives.

In this concluding chapter we shall first review the general argument concerning the implementation of Proposition 20 and then explore some of the more general implications of this case for the validity of our conceptual framework and for the effectiveness of government regulation.

I. Application of the Framework to the California Case

The California Coastal Initiative mandated the commissions to "preserve, protect, and where possible, to restore the resources of the coastal zone for the enjoyment of current and succeeding generations." More specifically, the commissions were charged with maximizing public access to the beaches, preserving the scenic beauty of the coast and particularly views of the ocean from the coastal highway, and protecting wetlands and other wildlife habitat. They were to accomplish these objectives both during the interim (1973–1976) permit review process and on a continuing basis by proposing, and then persuading the legislature to adopt, an enforceable coastal plan consistent with these goals.

That was no mean task. While precise estimates are extremely difficult to obtain, it is nevertheless quite clear from changes in permit applications made by the commissions during the 1973–1976 period that the principal target groups (i.e., people wishing to develop property in the 1000-yard permit area) often had to change their behavior substantially to comply with statutory objectives. Moreover, evidence from the permit hearings and the results of our survey of businessmen and property owners revealed that they generally opposed such changes. It was also quite clear from the beginning that there was an enormous variety of target group activities along the 1100-mile coast that were potentially inconsistent with statutory objectives and thus that no relatively simple set of regulations would suffice. Finally, the repeated failure of the legislature to pass the precursors of Proposition 20 indicated that the commissions would face a very difficult task in persuading the 1976 legislature to approve an enforceable coastal plan consistent with the Coastal Initiative's objectives.

The commissions were thus confronted with a difficult task in bringing target group behavior into conformity with statutory objectives even within the 1973–1976 period, let alone on a long-term basis. Nevertheless, there were two aspects of the problem that made it more tractable. First, the commissions were regulating only a small percentage of the population of the state (or even of coastal counties), while a much greater percentage of both populations were potential beneficiaries. Thus, there was certainly an excellent opportunity—as demonstrated by the vote on Proposition 20—for obtaining a net balance of political support for regulatory objectives. Second, the knowledge base of the factors affecting—and the means to obtain—access, scenic protection, and wetlands preservation was relatively well established (although not always incorporated into the statute). At any rate, the commissions were not confronted with the very substantial problems of a lack of knowledge and an adequate causal theory that have afflicted, for example, the regulation of both air pollution and nuclear power plants.

On the whole, then, one could conclude that the commissions were confronted with a difficult task in bringing about target group compliance but one that was nevertheless potentially tractable from a political and scientific perspective. While the overall task was more difficult than that facing most land use regulatory agencies, it was nevertheless not as intractable as the one confronting, for example, the U.S. Environmental Protection Agency in achieving ambient air standards for automotive pollutants.[5]

Within this context, Table 10-1 summarizes the conclusions of our study concerning the extent to which the implementation of the 1972 Coastal Initiative met the hypothesized conditions of effective implementation. The table deals primarily with the conditions affecting the permit and planning decisions of the commissions during the 1973–1976 period and the extent to which those decisions—and the attendant impacts—could be expected to conform to statutory objectives to protect wetlands, scenic views, and public access to the wet sand beaches.

The table indicates a number of things: (1) The 1972 Coastal Conservation Act gave clear priority to wetland protection and to physical and scenic access over economic development, although these objectives were not stated in very precise terms. (2) In relying entirely on police power regulation during the interim period, the statute incorporated a causal theory that was quite adequate to protect scenic views (through controlling the size and location of development), but whose limited geographic jurisdiction would occasionally create problems in protecting wetlands from perturbations upstream beyond the 1000-yard permit boundary. More important, the commissions' lack of authority to acquire and manage accessways during the 1973–1976 period promised substantial problems in actually *providing* access, as the commissions would be dependent upon other units of government over which they had absolutely no control for these services. On the other hand, (3) the statute generally did an excellent job of structuring the implementation process so as to maximize the probability that implementing officials and target groups would act consistently with statutory objectives. Of particular note were the largely successful efforts of statutory framers to assign implementation to a generally sympathetic agency (the coastal commissions) that was highly integrated, that had ultimate responsibility for plan development, and that placed responsibility for the most important permit decisions in the hands of the state commission. In addition, the statute contained penalties adequate to discourage noncompliance by target groups; incorporated highly supportive decision rules; and provided excellent access to statutory supporters, while precluding weakening amendments by the legislature during the 1973–1976 interim period. The only potential problem areas were funding, where the guaranteed appropriation contained in the Initiative should have been sufficient but turned out to be only marginally so, and in enforcement, where defense of the statute in the courts would be dependent upon the willingness of the commissions and the attorney general to devote scarce resources to this task.

Turning to the nonstatutory conditions, the combination of statutory structuring, popular support for Proposition 20, and the prestige of the commissions as an important innovation in environmental planning resulted in (4) the selection of implementing officials who were generally supportive of statutory objectives and relatively skillful in using available resources. While not the case in regions such as the North Coast, where local government representatives were strongly opposed to Proposition 20, it was certainly true at the state commission, where such support, combined with the hierarchically integrated decision process, could be expected to result in final decisions on important planning and permit matters generally consistent with statutory objectives. In addition, (5) the implementation effort benefited from the continued participation of environmental and other constituency groups that had originally supported it, as well as strong support during the 1973–1976 period from the state appellate courts, the Federal Office of Coastal Zone Management, and the governor. Finally, (6) the implementation effort suffered from a substantial downturn in the state economy that hit one of the principal target groups (the construction industry) particularly hard, but the adverse effects of this change in socioeconomic conditions were minimized by the insulation that the Initiative provided the commissions from weaking amendments by the legislature.

In sum, the implementation of the Coastal Initiative clearly met four of our six conditions of effective implementation at least moderately well during 1973–1976: Condition 1 (relatively clear and consistent objectives), Condition 3 (a structured process to maximize agency and target group compliance), Condition 4 (supportive and skillful top implementing officials), and Condition 5

TABLE 10-1

EXTENT TO WHICH THE IMPLEMENTATION OF THE 1972 CALIFORNIA COASTAL INITIATIVE MET THE HYPOTHESIZED CONDITIONS OF EFFECTIVE IMPLEMENTATION

Conditions of effective implementation	Overall assessment	Discussion
Cond. 1. Statute contains clear and consistent policy directives	Moderate	Statute had a consistent tilt in favor of protecting wetlands and increasing public's scenic and physical access to the coast; no explicit need to balance with economic development; while priorities clear, objectives were not very precise, and no guidelines provided for resolving conflict among 3 major goals
Cond. 2. Statute incorporates sound causal theory identifying sufficient factors and target groups to attain statutory objectives	Mixed: High for scenic access, moderate for wetlands, low for public access	Factors affecting statutory objectives were fairly well understood, but adequate causal theory not always incorporated into statute; commissions' authority to regulate all development within the permit zone sufficient to protect scenic access; lack of control over upstream development potentially a problem in protecting wetlands; land of management and acquisition authority a major problem in providing physical access; requirement for interim permit review and preparation of a plan in the context of considerable administrative discretion within a highly structured biased system a good way to deal with general uncertainty and extensive variation in target group behavior
Cond. 3. Statute not only provides jurisdiction over target groups but also structures implementation to maximize probability of compliance from implementing officials and target groups	High/moderate	
a. Assignment to a sympathetic agency	Moderate/high	Statute assigned implementation solely to newly created coastal commissions, thus guaranteeing high priority and—given their environmental mandate and genesis in initiative process—a highly supportive professional staff; attempt to structure regional commissions through state vs. local appointees worked reasonably well, though dependent on local support for Proposition 20; structuring of state commission quite successful

b. Hierarchically integrated implementing agencies with few veto points and adequate incentives for compliance	High except for public access	Commissions had final control over all development during interim period, with state commission making final decisions on coastal plan and most important permit decisions; *provision* of physical access required cooperation of legislature and parks departments, i.e., loosely integrated; adequate fines, although enforcement required cooperation of attorney general; sunset clause meant, however, that fate of commissions and coastal plan ultimately dependent upon 1976 legislature
c. Supportive decision rules	High	Burden of proof on permit applicants; approval of most potentially harmful developments required ⅔ vote of commission membership
d. Financial resources	Moderate to high	Guaranteed appropriation in Initiative should have been sufficient—especially when combined with federal grants—but inflation and commissions' initiatives in energy conservation and low-cost housing strained resources
e. Formal access to supporters	High	Strong requirements for public hearings in permit review and planning vigorously implemented by commissions; liberal rules of standing to appeal permit decisions to state commission and to courts; Initiative specifically precluded weakening amendments by legislature prior to 1976 (although everything then subject to review)
Cond. 4. Commitment and skill of top implementing officials	Mixed, but generally high	Although regional commissioners showed wide variation in support for statutory objectives (based largely on degree of local support for Proposition 20), regional staff were uniformly sympathetic; more important, state commissioners and staff were very supportive; skill difficult to measure, but certainly high at the state level and moderate to high in the 3 regions studied (especially the North Central)
Cond. 5. Continuing support from constituency groups and sovereigns	Generally high, although mixed in 1976 legislature	Environmental groups and other supporters of Proposition 20 continued to participate actively in both planning and permit review, as well as the 1976 session; courts and federal OCZM very supportive; governor and legislature supportive during 1973–76 interim (e.g., supplemental appropriations), though eventually demanded compromise in coastal plan
Cond. 6. Changing socioeconomic conditions (and thus political support) over time	Poor, but commissions insulated until 1976	Recession of 1974–75, which hit the construction industry particularly hard, affected public and elite support for Proposition 20, but statute essentially insulated commissions from political repercussions until 1976 legislative session; effect at that point uncertain, although legislature clearly sensitive to economic impacts of regulation

(continuing support from constituency groups and sovereigns). The only one that was not met (Condition 6, regarding changing socio-economic conditions) was essentially neutralized during the interim period by the statute. Finally, Condition 2, regarding the incorporation of an adequate causal theory, was met much better for some objectives (protection of scenic resources and, to a lesser extent, wetlands) than for others, most notably the maximization of public access to the wet sand beaches.

In the opening chapter, we suggested that implementation should be conceptualized as a sequential process, beginning with the passage of the basic statute and followed by the policy outputs (decisions) of the implementing agencies, the compliance of target groups with those decisions, the actual and perceived impacts of those decisions, and, finally, revisions (or attempted revisions) in the basic statute. We further argued that—assuming internally consistent statutory objectives—a statute would achieve its desired (actual) impacts if (1) the policy decisions of the implementing agencies were consistent with those objectives, (2) target groups complied with those decisions, and (3) the statute incorporated an adequate causal theory in which target group complicance would be sufficient to attain statutory objectives.

Combining this sequential view of the stages of implementation with our previous conclusions concerning the extent to which the implementation of the 1972 Coastal Initiative met the overall conditions of effective implementation, one arrives at the following propositions:

Proposition 1: The final permit and planning decisions of the coastal commissions during the 1973–1976 period should have been consistent with all three major statutory objectives to (a) protect scenic views from the coastal highway to the ocean, (b) protect wetlands, and (c) provide (maximize?) public access to the wet sand beaches. This proposition flows directly from the four conditions that were met, as they include the principal factors affecting agency decisions. Within this general hypothesis of overall consistency, we would also suggest two subsidiary propositions:

Proposition 1.1: The consistency of regional commission decisions with statutory objectives should have been strongly correlated with the basic policy orientation of regional commissioners, with nonconforming decisions being overturned upon appeal to the state commission.

Proposition 1.2: In cases of conflict between major statutory objectives— principally, protection of fragile coastal wildlife habitat versus maximization of public access—the commissions should have attempted to give them more or less equal weight, as the two objectives probably had roughly equal support among commissioners, staff, and supportive constituency groups.

Proposition 2: Target group compliance with commission permit decisions should have been quite high because of the ability of commission staff and supportive constituency groups to monitor compliance and the substantial penalties for noncompliance incorporated in the Initiative. Within this overall hypothesis of general compliance, we would suggest a subsidiary proposition:

Proposition 2.1: Target group compliance would be reduced in the case where (a) the costs of monitoring noncompliance substantially increased (as commission staff resources were somewhat marginal) and/or (b) in areas (e.g., the North Coast region) where local opinion was hostile to Proposition 20.

Proposition 3: Given the general conformity of commission decisions and target group behavior with statutory objectives, the determinative factor in actually attaining statutory objectives would be the extent to which the statute incorporated a valid causal theory for different objectives. Hence, one would expect attainment to be high in the case of scenic views, moderately high with respect to wetland protection, and rather low with respect to physical access to the beach.

Proposition 4: Given the support of most of the commissioners for the statutory objectives and their frustration with potential limitations in their interim permit authority to actually protect wetlands and to provide physical access, the coastal plan should have contained provisions to rectify the major deficiencies by (a) expanding their geographical jurisdiction over upstream areas affecting wetlands and by (b) expanding their authority to acquire and manage access areas.

Given these "predictions" on the basis of the general legal and political context in which the commissions operated (taken largely from Chapters 2–4), to what extent did the commissions actually realize the statutory objectives of Proposition 20?

Our conclusions on this topic are summarized in Table 10-2. For each of the three major objectives (scenic access, wetlands protection, and physical access), part A indicates the extent to which each was met during the period of interim permit review (1973–1976) by tracing out the temporal sequence from commission decisions through target group compliance and causal theory to the end result, i.e., the actual impacts. Part B indicates the extent to which the long-term goal of Proposition 20—namely, the passage by the 1976 legislature of an enforceable coastal plan consistent with these three objectives—was realized. Admittedly, these summary judgments are somewhat rough, first, because the Coastal Initiative's objectives were not stated in very precise terms and, second, because of the extreme difficulty of determining the actual impacts of the thousands of permit decisions made by the commissions and the very complex provisions of the 1976 coastal legislation. Nevertheless, these represent our best judgments based upon the information presented in Chapters 5–8.

Turning first to the interim permit period, the table suggests that the commissions had an excellent record with respect to protecting the scenic resources

TABLE 10-2

EXTENT TO WHICH THE PRINCIPAL SUBSTANTIVE OBJECTIVES OF THE 1972 COASTAL
INITIATIVE WERE MET (1) DURING THE PERIOD OF INTERIM PERMIT REVIEW
(1973–1976) AND (2) IN THE FUTURE POLICIES GOVERNING THE UTILIZATION OF
COASTAL RESOURCES

	Objective of Coastal Initiative		
	Protect scenic resources, especially views of the coast from the highway	Protect wetlands and other valuable habitat	Increase the physical access of the public to wet sand beaches
Part A: Interim permit review			
1. Consistency of commissions' final permit decisions with statutory objectives	High	High	Moderate/high
2. Compliance of target groups with those decisions	High	High	Moderate/low
3. Incorporation of a valid causal theory in the Coastal Initiative	High	Moderate	Moderate/low
4. End result: Extent to which actual impacts were consistent with statutory objectives	High	High	Rather low
Part B: Future policies governing the coast			
1. Consistency of coastal plan proposed by the commissions with objectives of Proposition 20[a]	High	High	High
2. Consistency of 1976 coastal legislation with objectives of Proposition 20[b]	Moderate/high	Moderate/high	Moderate

[a]Includes both the plan's substantive policies and its implementation proposals.
[b]Includes the 1976 Coastal Act (both policies and implementation scheme), as well as the conservancy legislation and the Nejedly-Hart Bond Act as approved and subsequently implemented by the legislature.

of the coast. Data presented in Chapter 5 indicated that this issue was addressed by the state commission (either by imposing relevant conditions or by denying permits) 91% of the time it was raised, and our review of the situation at Sea Ranch indicated that the commissions tried very hard to protect viewsheds. Moreover, target groups compliance was not a problem with this objectives because noncompliance was easily detected. Finally, the underlying causal theory was quite adequate. Much the same situation existed in the case of protecting wetlands and other valuable wildlife habitat. Our case study of Humboldt Bay indicated that even in this hostile region the commissions ultimately had an excellent record, although the success was heavily dependent upon the hierarchical integration provided by the appeals process. Wetlands protection was the objective upon which the respondents in our survey of the "attentive public" perceived the commissions to have had the greatest impact. Moreover, given the efforts of environmental groups, the Department of Fish and Game, and most water quality control boards to monitor compliance in this area, it is unlikely that noncompliance was a serious problem. Finally, the commission officials we interviewed did not feel that upstream perturbations had had any noticeable effect on coastal wetlands, although this was a topic that unfortunately was not examined in any detail. On the whole, the evidence suggests that the potential inadequacy in geographic jurisdiction did not prove to be a significant problem in practice and thus the commissions appear to have had an excellent record in actually protecting wetlands.

Their record, however, indicates that the commissions were deficient with respect to providing physical access to the ocean. While the aggregate data on permit decisions and the case studies of Sea Ranch and the South Coast region indicated that the commissions' permit decisions were generally consistent with this objective—with the probable exception of the first year at Malibu, when the appeals process failed to work properly—this good permit record was subsequently undermined by serious problems with compliance and with the underlying causal theory. As indicated in Chapter 6, the commissions often failed to make sure that accessways supposedly provided as a condition of permit approval were legally recorded as deed restrictions by applicants. More important, the statute did not give the commissions adequate authority to provide access. The commissions could only require access conditions or deny a permit to protect the option of future acquisition. The actual opening of new accessways and public beaches was generally contingent upon the willingness of state and local parks departments to provide the management services (i.e., liability insurance, routine maintenance), which in turn was contingent upon the willingness and ability of other units of government to acquire the land and provide maintenance funds. During the interim permit period, the needed intergovernmental cooperation was seldom forthcoming.

Part B of Table 10-2 gives our summary assessment of the extent to which the commissions achieved the *long-term* goal of Proposition 20 of persuading the 1976 legislature to approve an enforceable plan consistent with the three substantive objectives of protecting scenic views and wetlands and providing physical access. To reiterate, one must recognize that—unlike permit review—the commissions did not have the legal authority to achieve this goal. Instead, they could only propose a coastal plan consistent with Proposition 20 and try to develop sufficient political support to persuade the legislature not to emasculate it. The commissions certainly did an excellent job on the first task. Our analysis of the coastal plan in Chapter 7 indicated that its substantive policies and implementation structure would have been quite adequate to protect scenic views, wetlands, and access. Moreover, the plan would have corrected most of the inadequacies in Proposition 20 in terms of upstream jurisdiction and authority to acquire and manage access points. In terms of developing political support, we saw in Chapter 7 that the commissions' extensive effort in "participatory planning" apparently did not win much new support for the plan: Those who had opposed Proposition 20 opposed the plan, while those who had supported Proposition 20 thought the commissions had done a good job. The plan itself and the planning process were, however, sufficiently in accord with the expectations of the interest groups and legislators who had originally helped pass Proposition 20 to persuade them to fight hard for the plan in the 1976 legislature. Moreover, we saw in Chapter 8 that the state commission staff played an important role in negotiating the amendments necessary to win additional support. Largely (but not entirely) as a result of what might broadly be construed as the implementation of Proposition 20, the legislation approved in 1976—while not as "strong" as the coastal plan proposed by the commissions—nevertheless was generally consistent with the Initiative's objectives of protecting scenic resources, wetlands, and public access and at least as likely as Proposition 20 to actually attain them.[6]

This summary has confirmed the four principal propositions advanced on the basis of our framework (*supra*, pp. 358–359 of this chapter). First, as would be "predicted" by the framework, the commissions' permit decisions were generally consistent with statutory objectives. Even the major exception uncovered— namely, decisions regarding public access at Malibu in 1973—was directly attributable to the absence of one of our overall conditions, i.e., the failure of environmental groups in the South Coast region to appeal decisions concerning access at Malibu. Second, as suggested by the framework, target group compliance was generally high. And the major exception to this trend—i.e., poor compliance in recording deed restrictions concerning access—was attributable to the relatively high costs of monitoring compliance in this matter (under the system of issuing permits in use during the 1973–1976 period), which, in turn, could be attributed to the marginally sufficient staff resources available to the commissions and/or to a failure on the part of the commissions' leadership to give sufficient priority to this critical, but unglamorous, clerical task. Third, consistent with the framework, the critical factor affecting the extent to which each of the three objectives was actually attained during the 1973–1976 period was the adequacy of the causal theory incorporated into the statute and, particularly, the commissions' inadequate authority to provide physical access to the wet sand beaches. Finally, the commissions' experience with 3 years of permit review and their support for Proposition 20's objectives led them to propose specific remedies for the Initiative's deficiencies in the coastal plan.

II. Implications of the Case Study for the Adequacy of the Conceptual Framework and the Possibility of Effective Regulation

In the introductory chapter we proposed a set of six sufficient and generally necessary conditions for the achievement of regulatory objectives requiring a substantial departure from the *status quo*. At the same time, we selected the implementation of the 1972 California Coastal Zone Conservation Act for detailed analysis because it seemed to satisfy most of the conditions and thus provided an ideal test case of the framework in identifying the conditions crucial to successful implementation.

A. Adequacy of the Conceptual Framework

Through the detailed analysis of this test case, we feel that our basic argument has been confirmed: The commissions did achieve two of their three major short-term objectives, as well as substantially attaining their long-term goal of persuading the legislature to renew substantially the mandate of Proposition 20. Moreover, the principal exception to this history of successful implementation—namely, the commissions' failure to substantially increase physical access to the coast—was predictable within the context of the framework and, specifically, by the deficient causal theory incorporated into the Initiative with respect to this objective. Our set of six conditions thus seems to represent a parsimonious integration of the principal factors affecting the effective implementation of statutory (or other legal) objectives.

This is not to imply that we view the framework as the final word on the topic. In the first place, several of the terms are not as precise as we would like: What exactly constitutes, for example, "substantial managerial and political skill," "active support by constituency groups and a few key legislators," or "the significant undermining of statutory objectives over time by the emergence of conflicting public policies?" If the framework is to adequately serve its predictive function, further refinements in this regard will certainly be necessary. Second, even in this detailed study we had difficulty operationalizing some of the concepts, most notably "managerial and political skill." In effect, we found such skill to be relatively easy to recognize in a *posthoc* fashion but incredibly difficult to develop quantitative indicators for or to postulate empirical predictors of.

Third, we frankly have some reservations about the form of our framework. Instead of the traditional form of a scientific theory as a system of (preferably mathematical) propositions directly stipulating causal relationships, our framework has been presented as a set of sufficient conditions for the accomplishment of a state of affairs. Nevertheless, we have presented a simplified version of the causal model underlying our framework in Chapter 5 and have tested it empirically. Moreover, we have repeatedly indicated the manner in which critical variables in the conditions actually affect the implementation process. For example, organized constituency groups affect (or are at least highly correlated with) permit decisions, compliance of target groups, and reformulation of the statutory mandate. In general, however, our previous efforts to develop a causal model in terms of a flow diagram specifying the variables affecting different stages of the implementation process proved to be so complex that reviewers kept demanding a short list of critical factors.[7] The set of sufficient (and generally necessary) conditions was our response to that request.

These *caveats* notwithstanding, the framework does seem to be an important advance in the analysis of policy implementation. In particular, we would cite (1) our efforts to show how statutory (and other legal) variables can structure behavior; (2) the critical role played by supportive constituency groups; (3) the attempt to integrate socioeconomic, legal, organizational, and participatory variables into a modified version of Richard Hofferbert's "general developmental model" and then to test it empirically[8]; (4) the specification of the important stages of the implementation of regulatory programs; and (5) our efforts to address what we have termed "the tractability of the problem." In sum, while admitting that the framework (and the general causal model upon which it rests) is not the last word in implementation analysis, we nevertheless feel that it is an important advance over previous conceptual frameworks.[9]

B. Implications for Effective Regulation

Unfortunately, our conclusions concerning the prospects for the effective implementation of regulatory programs (i.e., attempts to change the behavior of private or public target groups through legal directives or conditional grants) are not quite so sanguine.

On the one hand, the California Coastal Initiative was, by and large, effectively implemented. Two of its three major objectives were attained in the short term, and its long-term objective of legislative approval of policies and implementing institutions capable of protecting scenic views and wetlands and providing public access was substantially achieved. In short, this case certainly serves as a counter example to the prevailing pessimism concerning the ability of regulatory statutes seeking substantial behavioral change. Moreover, the analytical framework indicates the conditions that need to be fulfilled if other programs are to be at least as successful.

On the other hand, this study has confirmed the existence of significant obstacles confronting effective regulation. It has illustrated what might be termed "the weakest link in the chain argument," i.e., that objectives will be attained *only if* the decisions of implementing agencies are consistent with statutory objectives *and* if target groups comply with those decision *and* if the statute incorporates a valid causal theory. Failure to meet any one of these requirements will result in a suboptimal achievement of objectives. For example, the substantial efforts of the coastal commissions in imposing access easements and in denying permits to preserve acquisition options—both of which engendered vociferous opposition from target groups—were wasted in many cases because the commissions did not adequately monitor compliance or did not have the authority to acquire and/or manage accessways.

The access issue also illustrates the importance of a hierarchically integrated implementation system, i.e., of minimizing the number of veto/clearance points over which statutory supporters do not possess sufficient incentives to bring about compliance. For example, to open a beach to the public for recreational purposes generally required the following steps: (1) denial by the commissions of permit applications to develop the property until the remaining steps could be fulfilled; (2) legislative approval of a bond act authorizing sufficient funds to purchase the property; (3) voter approval of the bond act; (4) legislative authorization to purchase that specific piece of property at market value; (5) approval by the State Public Works Review Board of purchase of the property; (6) agreement by a state or local parks department to manage the property (e.g., provide liability insurance and maintenance functions); (7) actual performance of those responsibilities, which generally involve the letting of contracts for construction of facilities, as well as annual appropriations for maintenance—both of which entail several additional clearance points. The coastal commissions, of course, had practically no control over any of these steps. Therefore, it was hardly surprising that, as of the summer of 1979 (or 4 years after the commissions' proposed list of acquisitions was presented to the legislature), very few of the properties on that list had been acquired and practically none had been officially opened to the public. [10]

On a related and, from the standpoint of effective regulation, more ominous point, there is very likely to be a trade-off between the degree of hierarchical integration and assignment to sympathetic agencies, on the one hand, and the incorporation of a valid causal theory, on the other. This results directly from the fact that the achievement of almost any important policy objective involves causal factors under the jurisdiction of a wide variety of agencies at the same level of government and often at different levels of government. Thus, statutory framers are confronted with the dilemma of either (1) trying to deal with sufficient factors to at least theoretically attain the objective—which will almost certainly result in a loosely integrated system involving numerous semiautonomous agencies with widely varying commitments to the achievement of the (new) objective—or (2) delegating authority to one or more highly committed agencies with authority to overrule some—but not all—of the agencies potentially affecting the problem.

This dilemma was perfectly illustrated by our case study of the California coast. The framers of the Coastal Initiative—who were, after all, subject only to the constraint of obtaining the approval of a majority of the state's voters— essentially adopted the second strategy, with the mixed results that we have noted. In contrast, the coastal plan represented a fairly coherent effort to combine *both* hierarchical integration and jurisdiction over sufficient causal factors. But this proposal was rejected by the 1976 legislature, which adopted the first strategy of addressing all the principal factors but establishing a system that was only moderately well integrated (i.e., in which the commissions had only moderate control over the relevant decisions of other state and local agencies affecting critical coastal resources).

The problem of developing a hierarchically integrated system with sufficient jurisdiction to attain statutory objectives is, however, likely to be particularly acute in societies, like the United States, in which political power is widely dispersed within and among levels of government. This can be contrasted, for example, to France and other European countries that have more centralized political systems and public agencies with the requisite authority to acquire vast tracts of land for the purpose of developing a coherent plan, set aside part of it for parks and other public purposes, and then resell the remainder to private developers. Even after being brought into the picture, the private developer is usually bound by fairly detailed zoning provisions.[11]

Within the United States, we would hypothesize that the incidence of successful implementation is likely to be higher for state programs than for those of the federal government or perhaps even local governments. Almost any federal program is confronted with having to frame regulations to cover enormous numbers of people in widely disparate situations. When, for example, is Texas similar to New York? Rather than risk arousing intense resentment from regions subject to unreasonably uniform restrictions, most federal programs rely on general regulations that are then implemented by state (and local) governments subject to only rather loose federal review (because of the constitutional autonomy of states and the political autonomy of most federal legislators).[12] In contrast, state programs generally affect fewer numbers of people (and thus involve less bureaucratic coordination) and deal with more homogeneous problems, thereby permitting more precise regulations, which are more suited to most applicable situations. In addition, while state programs often involve local governments in the implementation process, most local governments have considerably less constitutional autonomy vis-à-vis their state superiors than do the states vis-à-vis the federal government.[13] Finally, while local governments usually confront the most homogeneous problems and offer the greatest possibilities for hierarchical integration, they are also often subject to social and economic forces and to state and federal legal constraints essentially beyond their control. In short, it would seem that the states would offer the best possibility for avoiding the trade-off between adequate jurisdiction and hierarchical integration.

Whatever the actual advantages of state versus local programs, it is almost certain that federal domestic programs generally have the least probability of being effectively implemented. Unfortunately, such programs have dominated most academic policy research. Thus, it is entirely possible that the dominant pessimism concerning the efficacy of governmental regulation may be partially a function of this selection bias.

Another potentially optimistic note concerns the prospects for avoiding what Marver Bernstein has termed the "cycle of decay" of regulatory agencies, whereby such programs eventually become dominated by—or at least reach a modus vivendi with—target groups.[14] Although the limited time period of this study has not been sufficient to evaluate the appropriateness of the Bernstein thesis in this case, the permit data presented in Chapter 5 and the content of the coastal plan certainly indicate that the commissions were not subject to such a cycle during the 1973–1976 period. Moreover, the important role of environmental groups in the permit hearings, the planning process, and the 1976 legislative session certainly suggests that the active participation of a supportive constituency is a crucial factor in at least forestalling such "decay."

But the California case also illustrates the difficulty that any political system has in maintaining over an extended period of time the tension inherent in seeking substantial involuntary change in target group behavior. In fact, we would suggest that one of the keys to avoiding "the cycle of decay"—and thus a distinctly suboptimal attainment of statutory objectives—is to reduce the costs (real and perceived) of regulation on target groups and thus their incentive to undermine the program through continuous appeals to judicial and legislative sovereigns. The California case suggests a number of ways in which such costs can be reduced without undermining the achievement of statutory objectives. First, direct financial assistance can be provided in cases of extreme hardship. This was one of the principal functions of both the Coastal Conservancy and the Nejedly-Hart Bond Act that emerged out of the initial 2 years of experience under Proposition 20; much the same principle is illustrated by programs in most states to provide low-interest loans to companies who find pollution control regulations to be particularly burdensome. Second, the change in state commission decisions from a heavy reliance on denials in 1973 to a much greater utilization of conditional approvals thereafter illustrates the potential for regulatory agencies to develop more sophisticated mechanisms that deal precisely with the adverse aspects of target group behavior and thus minimize the overall costs to the regulated. Third, it seems reasonable to assume that opposition from target groups to regulatory programs is a function not simply of costs imposed but also of altered expectations. Thus, to the extent that regulatory objectives can be incorporated very early into the general expectations or way of doing business of target groups, they are more likely to be accepted as "given." For example, at least part of the vociferous opposition of coastal property owners to the commissions was probably a result of the fact that the commissions imposed an entirely new set of constraints *after* developers had already obtained the approval of local (and state) agencies. Thus, one of the potentially major advantages of the local certification process incorporated into the 1976 Coastal Act was that commission regulations would become incorporated into local plans and zoning ordinances and thus become part of the general expectations of developers from a very early date. A fourth mechanism for potentially reducing the opposition of target groups to regulatory programs without adversely affecting the attainment of statutory objectives is through a greater reliance on economic incentives (such as emission fees and liability insurance) rather than on legal directives. While not applicable to the California case, such mechanisms do offer at least the potential for reducing the adverse effects of legal directives in many regulatory areas on the efficient allocation of resources within a firm and on technological innovation. [15]

C. A Concluding Note on Generalizability

In concluding this study, we would simply like to reiterate that the implementation of the 1972 California Coastal Initiative was obviously an unusual case. It was selected in the hope that an example of relatively successful implementation could shed light on (the probably much more numerous examples of) problematic implementation. And there are two important aspects of the California case—both arising from its genesis on the initiative process—that potentially affect the generalizability of our conclusions. First, the November 1972 vote on the Initiative provided a very clear record of public support for the general objectives of Proposition 20, which, in turn, affected the appointments to the commissions and probably the 1976 legislature's reception of the coastal plan. Most statutes do not, of course, enjoy such clear indications of public support. Nevertheless, such clear indications are clearly not precluded in the case of legislative statutes through, for example, public opinion surveys or even long campaigns in which the general tenor of public opinion becomes known to legislators.

Second, the protection that the Initiative provided the commissions from normal legislative oversight during the 1973–1976 period probably affected their permit decisions and particularly their ability to withstand the serious 1974–1975 recession without having their program (or statute) emasculated because of the (temporary) change in public and legislative priorities. On the other hand, the complete legislative review that the commissions underwent in 1976—when statewide unemployment was still over 9%—would suggest that the commitment to statutory objectives on the part of most agency officials and their political support from constituency groups and state officials was sufficiently great that the protective umbrella provided by the Initiative probably did not have that great an effect.

The possible uniqueness of this implementation effort, however, should not be confused with the generalizability of the conceptual framework developed for analyzing it, which, we believe, is universally applicable. The six necessary and generally sufficient conditions of successful implementation of statutory objectives identified—from clear and consistent objectives through no significant undermining over time by changing social concerns or technological developments—equally pertain to all implementation efforts. These conditions are especially pertinent to the success of public policies that have as their goal a significant divergence from the *status quo*. Also important is an appreciation of the five stages of the implementation process—from the output of the implementing agency through feedback and reformulation—and how the turn of events at each stage plays such an important role in the overall success of the implementation effort.

NOTES

1. See, for example, Charles Jones, *Clear Air* (Pittsburgh: University of Pittsburgh Press, 1972); Paul Downey and Gordon Brady, "Implementing the Clean Air Act: A Case Study of Oxidant Control in Los Angeles," *Natural Resources Journal* 18 (April 1978), pp. 237–283; Allen Kneese and Charles Schultze, *Pollution, Prices, and Public Policy* (Washington, D.C.: Brookings Institution, 1975), chaps. 3, 5; James Q. Wilson, "The Dead Hand of Regulation," *Public Interest*, 25 (Fall 1971), pp. 39–58; Marver Bernstein, *Regulating Business By Independent Commission* (Princeton: Princeton University Press, 1954); Helen Ingram, "Implementation Through Bargaining: The Case of Federal Grants in Aid," *Public Policy* 25 (Fall 1977), pp. 498–526; Jeffrey Pressman and Aaron Wildowsky, *Implementation* (Berkeley: University of California Press, 1973); and Jerome Murphy, "Title I of ESEA," *Harvard Educational Review* 41 (1971), pp. 35–63.

2. For example, see Steven Kelman, "Regulation That Works," *The New Republic*, (November 28, 1978), pp 16–20; Harrell Rodgers and Charles Bullock, *Coercion to Compliance* (Lexington, Mass.: D. C. Heath, 1976); Gerald Swanson, "Coastal Zone Management from an Administrative Perspective: A Case Study of the San Francisco Bay Conservation and Development Commissions," *Coastal Zone Management Journal* 2 (1975), pp. 81–102; and Paul Sabatier, "Social Movements and Regulatory Agencies," *Policy Sciences* 6 (Fall 1975), pp. 301–342.

3. The most thorough version of our framework in terms of a set of variables and hypotheses—complete with an incredibly complex flow diagram—is found in Daniel Mazmanian and Paul Sabatier, *Implementation and Public Policy* (Glenview: Scott, Foresman, & Co., 1983), chap. 2. For another version organized around five conditions of effective implementation, see Paul Sabatier and Daniel Mazmanian, "The Conditions of Effective Implementation," *Policy Analysis* 5 (Fall 1979): 481–504.

4. The variables affecting the "tractability" of a problem are discussed more fully in Sabatier and Mazmanian, *The Implementation of Regulatory Policy*, Part II.

5. Most land use statutes mandate less behavioral change or involve less diverse target group behavior than the Coastal Act; for example, the San Francisco Bay Conservation and Development Commission deals only with dredging, filling and access within a 100-foot band around San Francisco Bay. On the other hand, achievement of the air quality standards affected by auto emissions would have required a 90% reduction in emissions per vehicle mile over the 50,000-

mile lifetime of autos and massive changes in the driving habits of urban residents. Moreover, the 1970 Clean Air Act had incredible distributional consequences, imposing costs on all car buyers, while benefiting only urban residents. In general, see Henry Jacoby and John Steinbruner, *Clearing the Air* (Cambridge, Mass: Ballinger, 1973).

6. As we saw in chapter 9, the coastal legislation retained much of the basic policy thrust and implementation structure of the coastal plan. But the commissions' postcertification review of local decisions (especially in urban areas) was somewhat less; the policies regarding energy (and other industrial) facilities risked creating some adverse impacts on scenic resources and perhaps wetlands; the policies regarding marinas provided less protection to wetlands than those in the plan; and the substantial cuts made by the legislature (in the 1976 Bond Act and 1977 implementing legislation) in the commissions' list of recommended acquisitions would almost certainly reduce the commissions' subsequent ability to provide access.

7. For the underlying model, see Mazmanian and Sabatier, *Implementation and Public Policy*, chap. 2.

8. In addition to chapter 6 in this volume, see Daniel Mazmanian and Paul Sabatier, "A Multi-stage Approach to Public Policy Making," *American Journal of Political Science* 24 (August 1980): 439–468.

9. See Sabatier and Mazmanian, *The Implementation of Regulatory Policy*, for a discussion of the limits of the conceptual frameworks of Bardach, Lutz, Berman, and Van Meter and Van Horn.

10. Illustrative of the point is the acquisition of the 5000-acre state park, including 6 miles of beach front at lakes Earl and Talawa, just outside of Crescent City in the northernmost reaches of the state. The acquisition effort, which is just now reaching conclusion, was given official impetus as early as 1971 with recognition of the importance of the wetlands and dunes at the site by the state Department of Parks and Recreation, subsequent support in 1974 in an important cataloguing of remaining wetlands by the State Department of Fish and Game and the U.S. Bureau of Sports Fisheries and Wildlife, inclusion of the lakes in the coastal commissions' 1976 acquisition list, and finally, the authorization for the acquisition by the 1977 legislature under the 1976 Bond Act. After 3 years of processing the proposal through the state-required checkpoints and negotiating with the principal landholder in the area, acquisition is today reasonably assured. Yet to come, of course, is the planning by the park department for the use and preservation of the dunes and wetlands, development of visitor facilities, and, only then, opening of the park to the public, which is likely to be several more years away. [See, "A New Park", *Coastal News* (September, 1979), pp. 4–5.]

11. A particularly strong example is the activities of the French government over the past decade in Languedoc-Rousillion, in which it purchased several hundred kilometers of coast, drained many of the marshes, retained substantial portions of land in open space and for public recreation, and then sold the remainder to private developers on the basis of detailed zoning regulations. In general, see Peter Harrison and W. R. Derrick Sewell, "Shorelines Management: The French Approach," *Coastal Zone Management Journal*, 5(3):161–180, and *The Government Land Developers*, ed., Neal Roberts, (Lexington, Mass.: Lexington Books, 1977).

12. Typical of federal programs that rely heavily on state and local officials for actual implementation are Title I of the 1965 Elementary and Secondary Education Act and the stationary source provisions of the 1970 Clean Air Amendments. Atypical are the uniform nationwide emission standards, which have been vigorously opposed by legislators from rural areas and which have been relaxed by Congress in both 1974 and 1977.

13. Herbert Kaufman, *Politics and Policies in State and Local Government* (Englewood Cliffs N.J.: Prentice-Hall, 1963).

14. Marver Bernstein, *Regulating Business by Independent Commission*, chap. 3. For a review and critique of the "cycle of decay" hypothesis, see Paul Sabatier, "Social Movements and Regulatory Agencies: Toward a More Adequate View of 'Clientele Capture,'" *Policy Sciences* 6

(Fall 1975), pp. 301–342; and Kenneth Meier and John Plumlee, "Regulatory Administration and Organizational Rigidity," *Western Political Quarterly* 31 (March 1978), pp. 80–95.

15. For the argument in favor of economic incentives, see Charles Schultze, *The Public Use of Private Interest* (Washington, D.C.: Brookings Institution, 1977).

16. For applications of the framework by other scholars in a variety of policy areas, see *Effective Policy Implementation*, ed., Mazmanian and Sabatier, (Lexington, Mass.: Lexington Books,, 1980).

Appendix A
Principal Data Bases

Following are the principal data bases for the study of the California Coastal Commissions:

1. A round of semistructured interviews with most commissioners, staff members, and interest group leaders were conducted in 1973–1974 and were repeated on a more limited basis in 1978. These dealt with a wide variety of topics, including the methods of commissioner and staff appointment, informal relationships among the three groups of actors, perceptions of relative influence, general attitudes toward the Coastal Initiative, and methods of information acquisition.

2. At the conclusion of the initial semistructured interviews, structured questionnaires were distributed to commissioners and staff members concerning attitudes toward the Initiative and other political issues, perceptions of the major issues confronting their commission, conceptions of the proper role of staff vis-à-vis commissioners, and perceptions of the quality of public participation in planning and permit review. The overall response rate was 79%, or 65 individuals.

3. Information in the permit files of the three regional commissions was coded on a random sample of 413 applications for development permits in the 1973–1975 period. Information concerned the type and location of the development, the staff recommendation and commission decisions, the major issues discussed in the staff report and at the public hearing, the voting record of commissioners, and the information presented by staff, applicants, and environmental groups on a wide variety of potential impacts of the development.

4. A similar analysis was made of a random sample of 166 permits appealed to the state commission from all six regions in the 1973–1975 period. It concerned the type and location of development, the staff recommendation and commission decision at both the regional and state levels, the issues discussed at the state hearing, and the voting record of state commissioners.

5. With the assistance of Judy Rosener of the University of California, Irvine, the entire population of hearing calendar permits considered by the three regional commissions in the 1973–1975 period was examined. But the data here were based solely on the commissions' *Minutes* and thus were limited to the type and location of development, the staff recommendation and commission decision, and the commissioners' voting record.

6. The analysis of the development of the 1976 coastal plan included a detailed examination of policy changes made in three of the nine plan elements as the proposals made their way from the original state staff recommendations, through the staff and commissioners in the North Coast and North Central Regions, and ultimately back to the state commission. The analysis involved coding the type of change in the approximately 30 policies per element and then attempting to ascertain the reasons for those changes through interviews with the principal participants and examination of relevant documents.

7. The final major data source involves an evaluation survey mailed in the fall of 1977 to random samples of people on the state commission mailing list, a list of people who actively commented on at least one of the plan elements, and permit applicants who appealed a regional commission decision to the state commission in the 1973–1976 period. Questionnaires were also mailed to all staff and commissioners serving during that period. Responses were received from 486 people, or about 55% of the sample. The questionnaire involved items concerning perceptions of the objectives of the Coastal Initiative and the extent to which each was achieved, attitudes toward the Initiative and the 1976 coastal plan, activities during the 1976 legislative session concerning the plan, and a comparison of the 1972 Coastal Initiative and the 1976 Coastal Act eventually approved by the legislature.

These major efforts at systematic data acquisition were supplemented with library research, interviews, attendance at commission meetings, and generally following the activities of the commissions.

The data were assembled with the anticipation that they would be useful in examining each of the variables in our framework. We feel particularly confident about data regarding statutory variables, constituency groups, commitment of agency officials, the planning and permit review decisions of the commissions, perceived impacts, and revisions in the statute during the 1976 session. The data on public opinion are good for the state as a whole, although reliable regional breakdowns are available only for November 1972. While our information concerning the attitudes and activities of sovereigns is certainly not systematic, this is probably not a serious defect because, by all accounts, legislative and executive sovereigns played a very small role in the 1973–1975 period—in part because the Initiative really isolated the commissions from normal oversight process, except for the courts.

The only serious data deficiencies concern target group compliance and, particularly, the actual impacts of commission decisions. With respect to the first, we had hoped to have our interviews with commission staff and other observers supplemented by a more full-blown analysis of compliance with coastal commission permits by the Urban Institute under a separate NSF-supported study. Unfortunately, the project did not materialize and it was the opinion of a number of observers that noncompliance was probably not a serious enough problem to warrant a major analysis on our part. As for the actual impacts of commission decisions, a comprehensive analysis is essentially impossible, in part because so many of the economic impacts will not be fully known for another 5–10 years. Thus, we shall rely upon existing studies of economic impacts— particularly Robert Kneisel's work on housing costs—to be supplemented by our own admittedly limited work on shoreline access, park acquisitions, and wetland protection. On the whole, however, we feel we have marshaled a substantial amount of information in analyzing the coastal commissions' implementation effort.

THE CALIFORNIA COASTAL ZONE CONSERVATION ACT OF 1972 ('Proposition 20')*

The people of the State of California do enact as follows:

SECTION 1. Division 18 (commencing with Section 27000) is added to the Public Resources Code, to read:

DIVISION 18. CALIFORNIA COASTAL ZONE CONSERVATION COMMISSION

CHAPTER 1. GENERAL PROVISIONS AND FINDINGS AND DECLARATIONS OF POLICY

27000. This division may be cited as the California Coastal Zone Conservation Act of 1972.

27001. The people of the State of California hereby find and declare that the California coastal zone is a distinct and valuable natural resource belonging to all the people and existing as a delicately balanced ecosystem; that the permanent protection of the remaining natural and scenic resources of the coastal zone is a paramount concern to present and future residents of the state and nation; that in order to promote the public safety, health, and welfare, and to protect public and private property, wildlife, marine fisheries, and other ocean resources, and the natural environment, it is necessary to preserve the ecological balance of the coastal zone and prevent its further deterioration and destruction; that it is the policy of the state to preserve, protect, and, where possible, to restore the resources of the coastal zone for the enjoyment of the current and succeeding generations; and that to protect the coastal zone it is necessary:

(a) To study the coastal zone to determine the ecological planning principles and assumptions needed to ensure conservation of coastal zone resources.

(b) To prepare, based upon such study and in full consultation with all affected governmental agencies, private interests, and the general public, a comprehensive, coordinated, enforceable plan for the orderly, long-range conservation and management of the natural resources of the coastal zone, to be known as the California Coastal Zone Conservation Plan.

(c) To ensure that any development which occurs in the permit area during the study and planning period will be consistent with the objectives of this division.

(d) To create the California Coastal Zone Conservation Commission, and six regional coastal zone conservation commissions, to implement the provisions of this division.

CHAPTER 2. DEFINITIONS

27100. "Coastal zone" means that land and water area of the State of California from the border of the State of Oregon to the border of the Republic of Mexico, extending seaward to the outer limit of state jurisdiction, including all islands within the jurisdiction of the state, and extending inland to the highest elevation of the nearest coastal mountain range, except that in Los Angeles, Orange, and San Diego Counties, the inland boundary of the coastal zone shall be the highest elevation of the nearest coastal mountain range or five miles from the mean high tide line, whichever is the shorter distance.

27101. "Coastal zone plan" means the California Coastal Zone Conservation Plan.

27102. (a) "Commission" means the California Coastal Zone Conservation Commission.

(b) "Regional commission" means any regional coastal zone conservation commission.

27103. "Development" means, on land, in or under water, the placement or erection of any solid material or structure; discharge or disposal of any dredged material or of any gaseous, liquid, solid, or thermal waste; grading, removing, dredging, mining, or extraction of any materials; change in the density or intensity of use of land, including, but not limited to, subdivision of land pursuant to the Subdivision Map Act and any other division of land, including lot

*As amended by Chapters 28 and 1014 (1973), amendments printed in bold type.

splits; change in the intensity of use of water, ecology related thereto, or of access thereto; construction, reconstruction, demolition, or alteration of the size of any structure, including any facility of any private, public, or municipal utility, and the removal or logging of major vegetation. As used in this section, "structure" includes, but is not limited to, any building, road, pipe, flume, conduit, siphon, aqueduct, telephone line, and electrical power transmission and distribution line.

27104. "Permit area" means that portion of the coastal zone lying between the seaward limit of the jurisdiction of the state and 1,000 yards landward from the mean high tide line of the sea subject to the following provisions:

(a) The area of jurisdiction of the San Francisco Bay Conservation and Development Commission, **together with all contiguous areas 2,900 feet landward thereof, and any river, stream, tributary, creek, or flood control or drainage channel which flows into such area,** is excluded.

(b) If any portion of any body of water which is not subject to tidal action lies within the permit area, the body of water together with a strip of land 1,000-feet wide surrounding it shall be included; **provided, however, that this subdivision does not apply to any river, stream, tributary, creek, or flood control or drainage channel when a portion of it lies within the permit area.**

(c) Any urban land area which is (1) a residential area zoned, stabilized and developed to a density of four or more dwelling units per acre on or before January 1, 1972; or (2) a commercial or industrial area zoned, developed, and stabilized for such use on or before January 1, 1972, may, after public hearing, be excluded by the regional commission at the request of a city or county within which such area is located. An urban land area is "stabilized" if 80 per cent of the lots are built upon to the maximum density or intensity of use permitted by the applicable zoning regulations existing on January 1, 1972.

Tidal and submerged lands, beaches, and lots immediately adjacent to the inland extent of any beach or of the mean high tide line where there is no beach shall not be excluded.

Orders granting such exclusion shall be subject to conditions which shall assure that no significant change in density, height, or nature of uses occurs.

An order granting exclusion may be revoked at any time by the regional commission, after public hearing.

(d) Each regional commission shall adopt a map delineating the precise boundaries of the permit area within 60 days after its first meeting and file a copy of such map in the office of the county clerk of each county within its region. **In delineating any inland boundary of the permit area, the regional commission may adjust such boundary by moving it seaward by**

not more than 50 yards. Such adjustments may only be made to avoid bisecting any lot or parcel owned by the same person or to conform to identifiable physical natural or manmade features such as streets, highways, or any structures, in order to more efficiently carry out the provisions of Chapter 5 (commencing with Section 27400) of this division.

27105. "Person" includes any individual, organization, partnership, and corporation, including any utility and any agency of federal, state, and local government.

27106. "Sea" means the Pacific Ocean and all the harbors, bays, channels, estuaries, salt marshes, sloughs, and other areas subject to tidal action through a connection with the Pacific Ocean, excluding non-estuarine rivers, **streams, tributaries, creeks and flood control and drainage channels.**

CHAPTER 3. CREATION, MEMBERSHIP, AND POWERS OF COMMISSION AND REGIONAL COMMISSIONS

Article 1. Creation and Membership of Commissions and Regional Commissions

27200. The California Coastal Zone Conservation Commission is hereby created and shall consist of the following members:

(a) Six representatives from the regional commissions, selected by each regional commission from among its members.

(b) Six representatives of the public who shall not be members of a regional commission.

27201. The following six regional commissions are hereby created:

(a) The North Coast Regional Commission for Del Norte, Humboldt, and Mendocino Counties shall consist of the following members:

    (1) One supervisor and one city councilman from each county.
    (2) Six representatives of the public.

(b) The North Central Coast Regional Commission for Sonoma, Marin, and San Francisco Counties shall consist of the following members:

    (1) One supervisor and one city councilman from Sonoma County and Marin County.
    (2) Two supervisors of the City and County of San Francisco.
    (3) One delegate to the Association of Bay Area Governments.
    (4) Seven representatives of the public.

(c) The Central Coast Regional Commission for San Mateo, Santa Cruz, and Monterey Counties shall consist of the following members:

    (1) One supervisor and one city councilman from each county.

(2) One delegate to the Association of Bay Area Governments.

(3) One delegate to the Association of Monterey Bay Area Governments.

(4) Eight representatives of the public.

(d) The South Central Coast Regional Commission for San Luis Obispo, Santa Barbara, and Ventura Counties shall consist of the following members:

(1) One supervisor and one city councilman from each county.

(2) Six representatives of the public.

(e) The South Coast Regional Commission for Los Angeles and Orange Counties shall consist of the following members:

(1) One supervisor from each county.

(2) One city councilman from the City of Los Angeles selected by the president of such city council.

(3) One city councilman from Los Angeles County from a city other than Los Angeles.

(4) One city councilman from Orange County.

(5) One delegate to the Southern California Association of Governments.

(6) Six representatives of the public.

(f) The San Diego Coast Regional Commission for San Diego County, shall consist of the following members:

(1) Two supervisors from San Diego County and two city councilmen from San Diego County, at least one of whom shall be from a city which lies within the permit area.

(2) One city councilman from the City of San Diego, selected by the city council of such city.

(3) One member of the San Diego Comprehensive Planning Organization.

(4) Six representatives of the public.

27202. All members of the regional commissions and public members of the commission shall be selected or appointed as follows:

(a) All supervisors, by the board of supervisors on which they sit;

(b) All city councilmen except under subsections (e) (2) and (f) (2), by the city selection committee of their respective counties;

(c) All delegates of regional agencies, by their respective agency;

(d) All public representatives, equally by the Governor, the Senate Rules Committee and the Speaker of the Assembly, provided that the extra member under (b) (4) and the extra member under (c) (4) shall be appointed by the Governor, the Senate Rules Committee and the Speaker of the Assembly respectively.

27203. A member of a regional commission who is also a supervisor from a county or city and county with a population greater than 650,000 may, subject to confirmation by his appointing power, appoint an alternate member to represent him at any regional commission meeting. The alternate member shall serve at the pleasure of the member who appointed him. The alternate member shall have the same qualifications as a public member pursuant to Section 27220. An alternate member shall have all of the powers and duties as a member of the regional commission, except that the alternate member shall only participate and vote in meetings in the absence of the member who appointed him.

An alternate member shall be entitled to a payment and reimbursement for the necessary expenses, pursuant to Section 27223, incurred in participating in regional commission meetings. Either the member of the regional commission or his alternate member shall receive such payment and reimbursement for the necessary expenses pursuant to this division. If both the member of the regional commission and his alternate member attend and participate in any portion of a regional commission meeting, only the alternate member shall be entitled to such a payment and reimbursement for attending that particular meeting.

An alternate member shall not be eligible for appointment to the commission.

Article 2. Organization

27220. Each public member of the commission or of a regional commission shall be a person who, as a result of his training, experience, and attainments, is exceptionally well qualified to analyze and interpret environmental trends and information, to appraise resource uses in light of the policies set forth in this division, to be responsive to the scientific, social, aesthetic, recreational, and cultural needs of the state. Expertise in conservation, recreation, ecological and physical sciences, planning, and education shall be represented on the commission and regional commissions.

27221. Each member of the commission and each regional commission shall be appointed or selected not later than December 31, 1972.

Each appointee of the Governor shall be subject to confirmation by the Senate.

27222. In the case of persons qualified for membership because they hold a specified office, such membership ceases when their term of office ceases. Vacancies which occur shall be filled in the same manner in which the original member was selected or appointed.

27223. **Except as provided in this section**, members shall serve without compensation but shall be reimbursed for the actual and necessary expenses incurred in the performance of their duties to the extent that reimbursement is not otherwise provided by another public agency. All members shall receive fifty dollars ($50) for each full day of attending meetings of the commission or of any regional commission.

27224. The commission and regional commissions shall meet no less than once a month at a place convenient to the public. Unless otherwise provided in this division, no decision on permit applications or on the adoption of the coastal zone plan or any part thereof shall be made without a prior public hearing. All meetings of the commission and each regional commission shall be open to the public. A majority affirmative vote of the total authorized membership shall be necessary to approve any action required or permitted by this division, unless otherwise provided.

27225. The first meeting of the commission shall be no later than February 15, 1973. The first meeting of the regional commissions shall be no later than February 1, 1973.

27226. The headquarters of the commission shall be in a city, county, or city and county which lies, in whole or in part, within the coastal zone.

Article 2.5. Conflicts of Interest

27230. Except as hereinafter provided none of the following persons shall appear or act, in any capacity whatsoever except as a representative of the state, or political subdivision thereof, in connection with any proceeding, hearing, application, request for ruling or other official determination, judicial or otherwise, in which the coastal zone plan, or the commission or any regional commission is involved in an official capacity:

(a) Any member or employee of the commission or regional commission;

(b) Any former member or employee of the commission or regional commission during the year following termination of such membership or employment;

(c) Any partner, employer, an employee of a member or employee of the commission or any regional commission, when the matter in issue is one which is under the official responsibility of such member or employee, or in connection with which such member or employee has acted or is scheduled to act, in any official capacity whatsoever.

27231. No member or employee of the commission or any regional commission shall participate, in any official capacity whatsoever, in any proceeding, hearing, application, request for ruling or other official determination, judicial or otherwise, in which any of the following has a financial interest: the member or employee himself; his spouse; his child; his partner; any organization in which he is then serving or has, within two years prior to his selection or appointment to or employment by such commission or regional commission, served, in the capacity of officer, director, trustee, partner, employer or employee; any organization within which he is negotiating for or has any arrangement or understanding concerning prospective partnership or employment.

27232. In any case within the coverage of Section 27230, or Section 27231 when the commission determines that in any case within the coverage of such section the financial interest involved is not substan-

tial, the prohibitions therein contained shall not apply if the person concerned advises the commission in advance of the nature and circumstances thereof, including full public disclosure of the facts which may potentially give rise to a violation of this article, and obtains from the commission a written determination that the contemplated action will not adversely affect the integrity of the commission or any regional commission. Any determination made pursuant to this section shall require the affirmative vote of two-thirds of the members of the commission.

27233. Nothing in this article shall preclude or prevent any member of the commission or any regional commission, or any employee thereof, who is also an employee of another public agency, a county supervisor, city councilman, member of the Association of Bay Area Governments, member of the Association of Monterey Bay Associated Governments, delegate to the Southern California Association of Governments, or member of the San Diego Comprehensive Planning Organization, and who has in such designated capacity voted or acted upon a particular matter, from voting or otherwise acting upon such matter as a member of the commission or any regional commission, or employee thereof, as the case may be. Nothing in this section shall be construed to exempt any such member of the commission or any regional commission, or any employee thereof, from any other provision of this article.

27234. Any person who violates any provision of this article shall, upon conviction, and for each such offense, be subject to a fine of not more than ten thousand dollars ($10,000) or imprisonment in the state prison for not more than two years, or both.

Article 3. Powers and Duties

27240. The commission and each regional commission, may:

(a) Accept grants, contributions, and appropriations;

(b) Contract for any professional services if such work or services cannot satisfactorily be performed by its employees;

(c) Be sued and sue to obtain any remedy to restrain violations of this division. Upon request of the commission or any regional commission, the State Attorney General shall provide necessary legal representation.

(d) Adopt any regulations or take any action it deems reasonable and necessary to carry out the provisions of this division, but no regulations shall be adopted without a prior public hearing.

27241. The commission and regional commissions may request and utilize the advice and services of all federal, state, and local agencies. Upon request of a regional commission any federally recognized regional planning agency within its region shall provide staff assistance insofar as its resources permit.

27242. All elements of the California Comprehensive Ocean Area Plan, together with all staff and funds appropriated or allocated to it, shall be delivered by the Governor and shall be attached and allocated to the commission at its first meeting.

27243. The commission and each regional commission shall each elect a chairman and appoint an executive director, who shall be exempt from civil service.

CHAPTER 4. CALIFORNIA COASTAL ZONE CONSERVATION PLAN

Article 1. Generally

27300. The commission shall prepare, adopt, and submit to the Legislature for implementation the California Coastal Zone Conservation Plan.

27301. The coastal zone plan shall be based upon detailed studies of all the factors that significantly affect the coastal zone.

27302. The coastal zone plan shall be consistent with all of the following objectives:

(a) The maintenance, restoration, and enhancement of the overall quality of the coastal zone environment, including, but not limited to, its amenities and aesthetic values.

(b) The continued existence of optimum populations of all species of living organisms.

(c) The orderly, balanced utilization and preservation, consistent with sound conservation principles, of all living and nonliving coastal zone resources.

(d) Avoidance of irreversible and irretrievable commitments of coastal zone resources.

27303. The coastal zone plan shall consist of such maps, text and statements of policies and objectives as the commission determines are necessary.

27304. The plan shall contain at least the following specific components:

(a) A precise, comprehensive definition of the public interest in the coastal zone.

(b) Ecological planning principles and assumptions to be used in determining the suitability and extent of allowable development.

(c) A component which includes the following elements:

(1) A land-use element.
(2) A transportation element.
(3) A conservation element for the preservation and management of the scenic and other natural resources of the coastal zone.
(4) A public access element for maximum visual and physical use and enjoyment of the coastal zone by the public.
(5) A recreation element.

(6) A public services and facilities element for the general location, scale, and provision in the least environmentally destructive manner of public services and facilities in the coastal zone. This element shall include a power plant siting study.
(7) An ocean mineral and living resources element.
(8) A population element for the establishment of maximum desirable population densities.
(9) An educational or scientific use element.

(d) Reservations of land or water in the coastal zone for certain uses, or the prohibition of certain uses in specific areas.

(e) Recommendations for the governmental policies and powers required to implement the coastal zone plan including the organization and authority of the governmental agency or agencies which should assume permanent responsibility for its implementation.

Article 2. Planning Procedure

27320. (a) The commission shall, within six months after its first meeting, publish objectives, guidelines, and criteria for the collection of data, the conduct of studies, and the preparation of local and regional recommendations for the coastal zone plan.

(b) Each regional commission shall, in cooperation with appropriate local agencies, prepare its definitive conclusions and recommendations, including recommendations for areas that should be reserved for specific uses or within which specific uses should be prohibited, which it shall, after public hearing in each county within its region, adopt and submit to the commission no later than April 1, 1975.

(c) On or before December 1, 1975, the commission shall adopt the coastal zone plan and submit it to the Legislature for its adoption and implementation.

CHAPTER 5. INTERIM PERMIT CONTROL

Article 1. General Provisions

27400. On or after February 1, 1973, any person wishing to perform any development within the permit area shall obtain a permit authorizing such development from the regional commission and, if required by law, from any city, county, state, regional or local agency.

Except as provided in Sections 27401 and 27422, no permit shall be issued without the affirmative vote of a majority of the total authorized membership of the regional commission, or of the commission on appeal.

27401. No permit shall be issued for any of the following without the affirmative vote of two-thirds of the total authorized membership of the regional commission, or of the commission on appeal:

(a) Dredging, filling, or otherwise altering any bay, estuary, salt marsh, river mouth, slough, or lagoon.

(b) Any development which would reduce the size of any beach or other area usable for public recreation.

(c) Any development which would reduce or impose restrictions upon public access to tidal and submerged lands, beaches and the mean high tideline where there is no beach.

(d) Any development which would substantially interfere with or detract from the line of sight toward the sea from the state highway nearest the coast.

(e) Any development which would adversely affect water quality, existing areas of open water free of visible structures, existing and potential commercial and sport fisheries, or agricultural uses of land which are existing on the effective date of this division.

27402. No permit shall be issued unless the regional commission has first found, both of the following:

(a) That the development will not have any substantial adverse environmental or ecological effect.

(b) That the development is consistent with, the findings and declarations set forth in Sections 27001 and with the objectives set forth in Section 27302.

The applicant shall have the burden of proof on all issues.

27403. All permits shall be subject to reasonable terms and conditions in order to ensure:

(a) Access to publicly owned or used beaches, recreation areas, and natural reserves is increased to the maximum extent possible by appropriate dedication.

(b) Adequate and properly located public recreation areas and wildlife preserves are reserved.

(c) Provisions are made for solid and liquid waste treatment, disposition, and management which will minimize adverse effects upon coastal zone resources.

(d) Alterations to existing land forms and vegetation, and construction of structures shall cause minimum adverse effect to scenic resources and minimum danger of floods, landslides, erosion, siltation, or failure in the event of earthquake.

27404. If, prior to **November 8, 1972**, any city or county has issued a building permit, no person who has obtained a vested right thereunder shall be required to secure a permit from the regional commission; providing that no substantial changes may be made in any such development, except in accordance with the provisions of this division. Any such person shall be deemed to have such vested rights if, prior to **November 8, 1972**, he has in good faith and in reliance upon the building permit diligently commenced construction and performed substantial work on the development and incurred substantial liabilities for work and materials necessary therefor. Expenses incurred in obtaining the enactment of an ordinance in relation to the particular development or the issuance of a permit shall not be deemed liabilities for work or material.

27405. Notwithstanding any provision in this chapter to the contrary, no permit shall be required for the following types of development:

(a) Repairs and improvements not in excess of seven thousand five hundred dollars ($7,500) to existing single-family residences; provided, that the commission shall specify by regulation those classes of development which involve a risk of adverse environmental effect and may require that a permit be obtained.

(b) Maintenance dredging of existing navigation channels or moving dredged material from such channels to a disposal area outside the permit area, pursuant to a permit from the United States Army Corps of Engineers.

(c) **Repair or maintenance activities of any sort; provided, that such activities do not result in an addition to, or enlargement or expansion of, the object of such repair or maintenance activities.**

Article 2. Permit Procedure

27420. (a) The commission shall prescribe the procedures for permit applications and their appeal and may require a reasonable filing fee and the reimbursement of expenses. **All such fees and reimbursements collected heretofore or hereafter shall be credited to, and shall be in augmentation of, the appropriation made in Section 4 of Proposition 20 as approved by the electorate at the general election on November 7, 1972, and are hereby appropriated to the commission for the same period and for the same purposes as set forth therein.**

(b) The regional commission shall give written public notice of the nature of the proposed development and of the time and place of the public hearing. Such hearing shall be set no less than 21 nor more than 90 days after the date on which the application is filed.

(c) The regional commission shall act upon an application for permit within 60 days after the conclusion of the hearing and such action shall become final after the tenth working day unless an appeal is filed within that time.

27421. Each unit of local government within the permit area shall send a duplicate of each application for a development within the permit area to the regional commission at the time such application for a local permit is filed, and shall advise the regional commission of the granting of any such permit.

27422. The commission shall provide, by regulation, for the issuance of permits by the executive directors without compliance with the procedure specified in this chapter in cases of emergency or for repairs or improvements to existing structures not in excess of twenty-five thousand dollars ($25,000) and other developments not in excess of ten thousand dollars ($10,000). Nonemergency permits shall not be effective until after reasonable public notice and adequate time for the review of such issuance has been provided. If any two members of the regional commission so request at the first meeting following the issuance of such permit, such issuance shall not be effective and instead the application shall be set for a public hearing pursuant to the provisions of Section 27420.

27423. (a) An applicant, or any person aggrieved by approval of a permit by the regional commission, may appeal to the commission.

(b) The commission may affirm, reverse, or modify the decision of the regional commission. If the commission fails to act within 60 days after notice of appeal has been filed, the regional commission's decision shall become final.

(c) The commission may decline to hear appeals that it determines raise no substantial issues. Appeals it hears shall be scheduled for a de novo public hearing and shall be decided in the same manner and by the same vote as provided for decisions by the regional commissions.

27424. Any person, including an applicant for a permit, aggrieved by the decision or action of the commission or regional commission shall have a right to judicial review of such decision or action by filing a petition for a writ of mandate **in accordance with the provisions of Chapter 2, (commencing with Section 1084) of Title 1 of Part 3 of the Code of Civil Procedure**, within 60 days after such decision or action has become final.

27425. Any person may maintain an action for declaratory and equitable relief to restrain violation of this division. No bond shall be required for an action under this section.

27426. Any person may maintain an action for the recovery of civil penalties provided in Sections 27500 and 27501.

27427. The provisions of this article shall be in addition to any other remedies available at law.

27428. Any person who prevails in a civil action brought to enjoin a violation of this division or to recover civil penalties shall be awarded his costs, including reasonable attorneys fees.

CHAPTER 6. PENALTIES

27500. Any person who violates any provision of this division shall be subject to a civil fine not to exceed ten thousand dollars ($10,000).

27501. In addition to any other penalties, any person who performs any development in violation of this division shall be subject to a civil fine not to exceed five hundred dollars ($500) per day for each day in which such violation persists.

CHAPTER 7. REPORTS

27600. (a) The commission shall file annual progress reports with the Governor and the Legislature not later than the fifth calendar day of the 1974 and 1975 Regular Session of the Legislature, and shall file its final report containing the coastal zone plan with the Governor and the Legislature not later than the fifth calendar day of the 1976 Regular Session of the Legislature.

CHAPTER 8. TERMINATION

27650. This division shall remain in effect until **January 1, 1977**, and as of that date is repealed.

SECTION 2. Section 11528.2 is added to the Business and Professions Code, to read:

11528.2 The clerk of the governing body or the advisory agency of each city or county or city and county having jurisdiction over any part of the coastal zone as defined in Section 27100 of the Public Resources Code, shall transmit to the office of the California Coastal Zone Conservation Commission within three days after the receipt thereof, one copy of each tentative map of any subdivision located, wholly or partly, within the coastal zone and such Commission may, within 15 days thereafter, make recommendations to the appropriate local agency regarding the effect of the proposed subdivision upon the California Coastal Zone Conservation Plan. This section does not exempt any such subdivision from the permit requirements of Chapter 5 (commencing with Section 27400) of Division 18 of the Public Resources Code.

This section shall remain in effect only until the 91st day after the final adjournment of the 1976 Regular Session of the Legislature, and as of that day is repealed.

SECTION 3. If any provision of this act or the application thereof to any person or circumstances is held invalid, such invalidity shall not affect other provisions or applications of the act which can be given effect without the invaid provision or application, and to this end the provisions of this act are severable.

SECTION 4. There is hereby appropriated from the Bagley Conservation Fund to the California Coastal Zone Conservation Commission the sum of five million dollars ($5,000,000) to the extent that any moneys are available in such fund and if all or any portions thereof are not available then from the General Fund for expenditure to support the operations of the commission and regional coastal zone conservation commissions during the fiscal years of 1973 to 1976, inclusive, pursuant to the provisions of Division 18 (commencing with Section 27000) of the Public Resources Code.

SECTION 5. The Legislature may, by two-thirds of the membership concurring, amend this act in order to better achieve the objectives set forth in Sections 27001 and 27302 of the Public Resources Code.

INDEX